T0219086

Measurement Theory and Practice

To Shelley

Measurement theory and practice

The world through quantification

David J. Hand

WILEY
wiley.com

First published in Great Britain in 2004 by
Arnold, a member of the Hodder Headline Group,
338 Euston Road, London NW1 3BH

John Wiley & Sons Ltd, The Atrium, Southern Gate, Chichester,
West Sussex, PO19 8SQ, United Kingdom

For details of our global editorial offices, for customer services and
for information about how to apply for permission to reuse the copyright
material in this book please see our website at www.wiley.com.

British Library Cataloguing in Publication Data
A catalogue record for this book is available from the British Library

Library of Congress Cataloging-in-Publication Data
A catalog for this book is available from the Library of Congress

ISBN 978-0-470-68567-9

1 2 3 4 5 6 7 8 9 10

Typeset in 10/12 pts Times by Charon Tec Pvt. Ltd, Chennai, India

Contents

Preface

I fear that most readers will find something to dislike about this book. The problem is that measurement is such a vast subject, so ubiquitous, so fundamental to modern thought, that anyone who knows anything about any particular area of measurement will object, rightly, that I have only scratched the surface. Unfortunately, in the interests of actually finishing the book, this seemed inevitable. Entire books could easily be written, indeed many have been written, about most of the topics which I have had to cover in just a few pages or even less. But I sought to present a broad view, lightly touching on as many topics as possible so that I could (at least attempt to) present a unified overview of what measurement is, how it is done, how it has come to be, and what it may be used for.

I originally took a degree in mathematics, specializing in mathematical physics, particularly quantum mechanics and relativity. But then, noting certain truths about the job market, I trained as a statistician. This led to me to work in psychology and medicine where I gradually found myself concerned with issues of measurement in areas such as psychiatric illness, pain, depression, and quality of life. It was clear that both the physical sciences on the one hand, and the behavioural and life sciences on the other, relied fundamentally on notions of measurement, but it was equally clear that what they meant by measurement, and how they handled measurements, differed very significantly between the various disciplines. Moreover, although the physical sciences were generally regarded as more advanced in measurement terms, their literature was almost devoid of serious considerations of what measurement was and how it should be used. Dimensional analysis and the peculiarities of quantum measurement seemed to be the only areas in which these issues had been given serious thought. In contrast, the psychological and sociological literature included entire subdomains devoted to considering such issues. Not only did these disciplines contain long-running debates about what measurement was, and how measurements should be used, but there were subdisciplines, such as psychometrics, whose sole concern was the construction and validation of measurement procedures. However, whatever their differences, all of these various disciplines used the same word: *measurement*. Since, as a statistician, I had to analyze measurements of all kinds, and to be at ease in doing so, I felt I had to be clear about the foundations: I had to be clear about the numerical objects, the results of measurement procedures, the data subjected to my analysis. I felt I needed to understand what measurement was, so that I could decide to what extent different kinds of analysis were legitimate.

The literature of what measurement is and how measurements may be used is widely spread, with contributions from mathematicians and philosophers as well as researchers in the physical, social, behavioural, life and other sciences. In this book I have attempted to present a unified view, and this has required me to formulate my own perspective on what measurement is. I believe I have managed to formulate a definition which does serve to integrate the various disciplines.

I imagine the readership will include psychologists, stung by the occasional criticism that their discipline includes few genuine measurement processes, so that much of psychology is built on shaky foundations. I hope it will also include others from the opposite end of the scientific spectrum, who wish to learn how their ideas on measurement fit into a broader picture – and, indeed, how that broader picture might influence the theories they can construct and the conclusions that may be drawn from them.

Although the book contains mathematics, it is not a mathematics book. Thus, for example, the discussion of axiom systems, which is necessary when we discuss the nature of measurement, is informal, and focuses on the meaning of the axioms, rather than providing a rigorous development. Mathematically inclined readers who feel that the axiomatic development is the right way to the secrets of measurement may follow up the references for the details. Non-mathematically inclined readers may skim or even skip these sections. I hope that the other sections are informative, and even entertaining in places.

This book has been a long time in gestation, and I would like to express my appreciation to my many friends and colleagues who have asked after its progress and who have discussed issues of measurement with me. These include Niall Adams, the late George Barnard, David Bartholomew, Shelley Channon, Martin Crowder, Fergus Daly, Peter Fayers, Harvey Goldstein, John Gower, Michael Healy, Agnes Herzberg, Malcolm Lader, Jim Lindsey, Ivo Molenaar, Sally Morton, John Nelder, Fred Smith, the late John Tukey, and, of course, others too numerous to mention. To all of these I am most grateful. Even if you do not agree with the results of my deliberations as presented in this book, I hope you find them interesting.

In conclusion, although I fear most readers will find something to dislike about this book, I hope that most readers also find something to like about it. I certainly had great fun writing it and I hope the reader has as much fun reading it.

July 2004

David J. Hand
Imperial College London

Acknowledgements

I am grateful to the following for permission to reproduce the letters to *The Times* in Section 4.1: Mr Tim Bloomfield, the Reverend Brian Phillips, Mr Bill Cairns, Mrs Janis Mason, Mr Godfrey Curtis, Mr Derek Cannell, and Mr Andrew Blackmore. I am grateful to Mr Andrew Bernard for permission to reproduce the letter to *The Times* in Section 7.2.15.

1

Introduction

Measurement is the contact of reason with nature.

Henry Margenau (1959)

1.1 The need

Measurement is ubiquitous. Cooking recipes specify the *volume* or *weight* of the various ingredients to use, the *time* for which they must be cooked, and at what *temperature*. Students are assessed by their *scores* on tests and in examinations. We drive safely by keeping below the *speed* limit. We decide *how many staff* we need to run our companies, what their *salaries* should be, and *how much output* we can produce. And we take carefully measured *doses* of medicines when we are ill.

The ubiquity of measurement, and the facility with which we use measurements and units of measurement, means that concepts of measurement are often invisible. They are simply part of the conceptual universe in which we function. However, as we show in later chapters, it has not always been thus. Indeed, even now, rips in this usually unnoticed background fabric occasionally make themselves apparent, casting, for a moment, doubts on one's view of the way the world works. Sometimes, as we will illustrate in the pages that follow, these tears can have serious, even fatal consequences. At a simple level, different units are sometimes used in parallel, requiring intellectual effort to convert from one to the other – between imperial and metric units in cooking or dieting, for example, or between Fahrenheit and Celsius in weather forecasting. And this can sometimes have dramatic consequences: 'The official report on the loss of the earlier Mars probe, the $125 m (£78 m) Climate Orbiter concluded it had burnt up in the Mars atmosphere because of an elementary miscalculation: metric and imperial units had been confused' (the *Independent*, 23 March 2000).

More fundamentally, sometimes it is not clear quite what a unit means: we are familiar with units of weight and length, such as the pound or metre, but just how much is a 'teaspoonful'? Worse than this, even the units with which we think we are familiar can wriggle and twist in our grasp. It is not merely the strength of British beer which misleads American visitors to the UK, but also the fact that the British pint has a volume just over 20% greater than the US pint. Even such familiar concepts as length may not be all that we think: the American foot is 0.0003% longer than the British foot – to paraphrase Bernard Shaw: 'two cultures separated by common feet'.

Even when we have clearly defined and generally accepted measurement procedures and units, there are sometimes puzzling questions about how we may use this information. Again, at the superficial level of everyday life such questions seldom arise, or if they do they are glossed over, but at a deeper level they have serious implications for our understanding and the actions we may take. Examples of this occur throughout the book. They are related to deep issues of exactly what measurement is and what can be measured. Weight and length are all very well, but what about depression, opinion, attitude, happiness and love? And, while I may talk about the total weight of students in my class, can I talk about their total happiness? Ideas of what can be measured have changed dramatically over the course of time. For example, early eighteenth-century France saw a lengthy discussion about changes in the size of the French population, without actually trying to measure it.

As we probe deeper, so we will see that the usually unnoticed fabric not only has rips, but is also folded and stretched. It is not the simple background infrastructure we generally assume: there are issues which need to be spelt out explicitly. The physicist Eugene Wigner eloquently drew attention to this when he commented on 'the unreasonable effectiveness of mathematics in the natural sciences' (Wigner, 1960). Just why is mathematics so effective at describing the way the physical universe functions? Lest the reader suggest that I am making careless assumptions, the fact that mathematics *is* so effective is demonstrated by the technological world around us: our machines generally work, our hugely complex industrial and governmental bodies generally function, maybe not perfectly, but at a level which requires organizational complexity on an awe-inspiring scale. And those machines and organizations are built on a foundation of counting and measuring. The proof of the measurement pudding lies in the modern world: something about measurement is 'right' or progress over the last few centuries would not have been so dramatic.

One answer to Wigner's question is that human mathematics has been developed by human brains, which themselves are evolutionary products of the natural world. No wonder, then, that our mathematics matches the world about us. This may be true, but it does seem to imply a certain limitation. The Whorf–Sapir hypothesis of linguistic relativity says that our languages influence and constrain what we can think about. The same must be true of measurement. The reader may be familiar with the comment by William Thomson, later Lord Kelvin, who wrote: 'I often say that when you can measure what you are speaking about and express it in numbers you know something about it; but when you cannot measure it, when you cannot express it in numbers, your knowledge of it is of a meagre and unsatisfactory kind; it may be the beginning of knowledge, but you have scarcely, in your thoughts advanced to the stage of science, whatever the matter may be' (Thomson, 1891 pp. 80–81; see also Merton *et al.*, 1984, for something of the history of this quotation).

Wigner's question about why measurement works is but one of the deep questions we need to examine. Others relate to questions of what exactly measurement is. Classical physics has a fairly narrow view of the subject, as is illustrated in Section 2.5 and Chapter 7. The modern perspective on measurement is much broader, and covers areas such as measuring opinion, attitude, software complexity, quality of life, organizational efficiency, and such. These are areas which at first might seem so different from the physics of weight and time that one might be unsure if the same word 'measurement' should be used. We explore these issues further, especially in Chapters 2 and 3.

We measure so that we can understand and manipulate the universe – where we are using the term 'universe' in a general sense. It could be a physical universe, social universe, psychological universe, and so on: the universe of discourse. Manipulation, in itself, can cause problems, as it can interact with measurement. Feedback loops may be established. This can be good – the system may settle towards an ideal level of the measurement. But it can also have unfortunate consequences: it is crucial that the right thing is measured, or the system will go astray. This is probably particularly important in human behavioural contexts, such as social, psychological and economic measurement situations, where, as we will see in later chapters, defining the concepts to be measured is often far from straightforward. Beyond this, however, changes induced by modifying a system so that its measured properties have certain values may lead to those very measurement tools becoming inappropriate. McDowell and Newell (1996, p. 11) illustrate this in the context of social health measurements, remarking that such measures often

> focus attention on the problem they record, and interventions are developed to resolve the issue. As an example, the infant mortality rate (IMR) is often used as an indicator of health levels in preindustrial societies, where high rates are an important concern, and where reductions can be relatively easily achieved. As infant mortality declines, however, a law of diminishing returns begins to apply, and further reductions require increasingly large expenditures of resources. As the numerator becomes smaller, it also becomes less representative as an indicator of the health of the broader population. In effect, the sensitivity of the IMR to the success of interventions explains both its value as an indicator of progress, and also its eventual demise. The resolution of one type of health problem reveals a new layer of concerns, a process that Morris has called the onion principle.

> (Morris, 1975)

The Research Assessment Exercise, to which UK university departments are subjected every few years, shows signs of a similar phenomenon.

We might make an initial attempt at defining measurement as 'quantification: the assignment of numbers to represent the magnitude of attributes of a system we are studying or which we wish to describe'. If we cannot measure, we cannot discuss the quantitative, and we are forced merely to discuss the qualitative. The qualitative is all very well, but it is very susceptible to misunderstanding and distortion, and its meaning may change with context. The same man may be tall or short, according to whether he is a jockey or a basketball player; but if he is 6 feet tall, he is 6 feet tall regardless of his profession. A job applicant may describe himself as having a good knowledge of French, but a quantitative test may reveal otherwise. The quantitative lifts our discussion from the untidiness of the real world, to a plane which we can manipulate and control, and use to predict and understand. That is, in translating our observations about the world into numerical form, we are mapping from the real world to an artificial one and, in particular, one in which we can apply mathematical tools. This book is about such mappings, describing how they are carried out, the properties they must have, and how manipulating the results of such mappings tells us about the domain they represent. We shall see that there are many different kinds of measurement mappings, with different properties, and that often not all of the properties of the representing numbers are used. As is discussed in detail in Section 2.5, this has caused major controversy in the past, in particular in the form of a sometimes heated debate about the relationship between measurement and statistics.

Measurement is not only a translation from the real world into an ideal one. Measurements are also conventions: they are widely recognized, so that people throughout the world can talk to each other using the same language. Measurement provides this *lingua franca*: measurements travel. But it was not always thus. Throughout history people have resisted the move towards universal standardization and the process is still far from complete at the beginning of the twenty-first century: 'Colin Hunt, a greengrocer at Hackney market, north east London, was given a 12 month conditional discharge in June last year for pricing pumpkins, sweet potato, cassava and plantain by the pound' (*Daily Telegraph*, 19 February 2002). Chapter 7 hints at the complex evolution of familiar modern units such as those used for length and weight.

There are also other deep issues discussed in this book. Often measurement does not merely *represent* an underlying reality, it also *imposes* constraints on the resulting numbers which do not exist in the underlying reality – another aspect of the fact that measurements are also conventions. This distinction, and the way the two concepts of representation and imposition come together in measurement systems, is introduced in Section 1.3 below.

Beyond all of the above, if our quantification process is faulty, then our conclusions, though they may be true about the imaginary world we have created, may well be false about the real world we sought to understand. This must lie at the root of many of the attacks on measurement as a philosophical basis from which to view the universe – as discussed in Section 1.5. To be effective, measurement must be carried out with great care and attention to detail. This has various aspects. Our measurements must be *valid*: they must measure what we want to measure. It may be no good using wealth as an indicator of happiness, because, we are told, the one does not lead to the other. Notice, however, that the very fact that we can say that they are not related implies that one has some kind of quantification of happiness. It may be a weak kind, and one of the aims of this book is to clarify what we mean by strong and weak kinds of measurements. Validity tells us that, although our measurement procedure may be subject to chance variation, in general ('on average', perhaps) it gets it right.

Our measurements must also have *precision*. Precision tells us about the chance variation in our measurement. If a measurement is precise, measuring the same attribute of the same object under effectively identical conditions should lead to similar results. In an ideal world, of course, there would be no such chance variation, but the world we are discussing is the real one, not an ideal one: there are always sources of uncontrolled variation. In different contexts, other terms are used for these various aspects of measurement quality, and these are discussed in Chapter 4.

Points such as the above have been made throughout the ages. They also illustrate how ideas about measurement have evolved over time, developing into the current views. Francis Bacon (1561–1626), regarded by many as the father of the experimental approach to science, in Aphorism XLVIII of his *Novum Organum* (1620) described the role of quantity:

> Among Prerogative Instances I will put in the twenty-third place *Instances of Quantity*, which (borrowing a term from medicine) I also call *Doses of Nature*. These are they which measure virtues according to the *quantity* of the bodies in which they subsist and show how far the *mode* of the virtue depends upon the *quantity* of the body.

The phrase 'doses of nature' has an attractive ring to it! Earlier, in Aphorism VIII, he remarks, when discussing notions of continuity:

> Nor again is there any reason to be alarmed at the subtlety of the investigation, as if it could not be disentangled. On the contrary, the nearer it approaches to simple natures, the easier and plainer will everything become, the business being transferred from the complicated to the simple; from the incommensurable to the commensurable; from surds to rational quantities; from the infinite and vague to the finite and certain.

Galileo (1564–1642) drew attention to the fact that to understand the natural universe it was necessary to translate it into an ideal one:

> Philosophy is written in this grand book, the universe, which stands continually open to our gaze, but it cannot be understood unless one first learns to comprehend the language and interpret the characters in which it is composed. It is written in the language of mathematics, and its characters are triangles, circles, and other geometric figures without which it is humanly impossible to understand a single word of it; without these, one wanders about in a dark labyrinth.
>
> (Drake, 1957)

John Arbuthnot (1667–1735) had earlier made a remark similar to that of Lord Kelvin, quoted above: 'There are very few things which we know, which are not capable of being reduc'd to a Mathematical Reasoning, and when they cannot, it's a sign our Knowledge of them is very small and confus'd' (quoted in Bennett, 1998, p. 88). And Lord Kelvin himself, in his Presidential address to the British Association for the Advancement of Science in 1871, also said: 'nearly all the grandest discoveries of science have been but the rewards of accurate measurement and patient long-continued labour in the minute sifting of numerical results' (quoted in Skinner, 1991).

Kepler (1571–1630) was even more robust in his adoption of the numerical perspective. Burtt (1954, pp. 67–68) writes:

> For Kepler, of course, the real qualities are those caught up in this mathematical harmony underlying the world of the senses, and which, therefore, have a causal relation to the latter. *The real world is a world of quantitative characteristics only; its differences are differences of number alone.* ... 'Just as the eye was made to see colours, and the ear to hear sounds, so the human mind was made to understand, not whatever you please, but quantity.'

More recent thinkers have also emphasized the fundamental role of modern measurement concepts. The philosopher E.G. Boring wrote 'we hardly recognise a subject as scientific if measurement is not one of its tools', and Edward Wilson (1998, p. 48) brought this home by showing the consequences of measurement ideas:

> Today the greatest divide within humanity is not between races, or religions, or even, as widely believed, between the literate and illiterate. It is the chasm that separates scientific from prescientific cultures. Without the instruments and accumulated knowledge of the natural sciences – physics, chemistry, and biology – humans are trapped in a cognitive prison. They are like intelligent fish born in a deep, shadowed pool. Wondering and restless, longing to reach out, they think about the world outside.

They invent ingenious speculations and myths about the origin of the confining waters, of the sun and the sky and the stars above, and the meaning of their own existence. But they are wrong, always wrong, because the world is too remote from ordinary experience to be merely imagined.

The psychologist Donald Laming, in a review of Michell (1999), wrote: 'If the sums do not add up, the science is wrong. If there are no sums to be added up, no one can tell whether the science is right or wrong. If there are no ratio-scale measurements, there can be no sums to add up' (Laming, 2002).

Measurement technology in the natural sciences developed before that in the social sciences. Interestingly, however, the incentive for this development did not lie so much in the needs of physics and chemistry for accurate measurements, but in social and economic needs, in the need for proper accounting, measuring and weighing goods and products, and trade – in measurement for management, one might say. In this book we will describe and illustrate measurement in both the hard and the soft sciences, as well as in many other areas.

1.2 The origins

1.2.1 The origins of numbers

The concept of number is necessary for, and in that sense precedes, the concept of measurement.

Normal human perception can directly distinguish only between integers up to about 4. Beyond that, a symbolic representation – a tally – is necessary. Note, however, that a tally is not a concept of number. It is just a record, a one-to-one match between the objects and a symbolic representation. Prehistoric humans developed many ways to do this: notches on sticks (the earliest such sticks were found in western Europe and may be 35,000 years old) or antlers, knots on cords, scratches on stones, and so on. Such systems continue to be used in various walks of life nowadays. This idea of a tally is the origin of the system of Roman numerals (i, ii, iii, iv, v, vi, etc.) a system which is still in use, although merely as a system for enumeration, and not for computation. Perhaps we should emphasize the motivation and stimulus underlying these systems: it was not 'science', an attempt to understand the way things worked. Rather it was accounting, an attempt to keep a record of things. This point will recur in what follows.

Short-term tally systems were also invented based on existing natural sets of objects. The 10 fingers on a pair of hands provide the most important example, but other natural groups were also used, such as the four legs on animals. They are 'short-term' because they can be used to hold a number for a few moments, but are inconvenient for a longer period. Tally systems can also be oral. If we have a sequence of distinct sounds, then we can match each sound against a different member of the set of objects.

The use of groupings of objects into fixed size sets is the starting point for the giant intellectual leap of grouping numbers: if each stick can contain five notches, I can count the sticks themselves and hence count large sets of objects. Grouping of objects into fixed size sets essentially provides the idea of a number base. The use of base 10 is thought to derive from the ten fingers on the human hand (though the Yuki of California use base 8, thought to be based on the eight gaps between the digits on the hands). Things are then simplified yet further if, when writing down the numbers, the *position system* is used. This system, in which the position of a number token indicates the level of group it represents (the '1' in 1, 10, 100, represents different things by virtue of the differences in position, for example), appears to have been discovered (or should I say 'adopted'?) by only a few civilizations. In the introduction to his remarkable book on the history of numbers, Georges Ifrah (1998, p. xxiv) says four civilizations: the Babylonians in the second millennium BC, the Chinese around the dawn of the Christian era, Mayan astronomers around AD 300–500, and in India around AD 500. Some concept of zero is a prerequisite for such a system. The position system means one can count an arbitrarily large set of objects using only a very small set of distinct symbols: just two symbols at the extreme of binary arithmetic.

Short-term tally systems are the foundation for calculating: I can match the five sheep in my flock to the five fingers on my hand, and fold two fingers down to represent the two we ate at our feast, to reveal that I have three left. This sort of calculation is not so easy with notches on sticks: while it is easy to add another notch, it is difficult to subtract one. If we eat two of the five sheep represented by notches on our stick, we'll have to carve another stick to represent the number remaining in the flock. But if we represent the sheep by pebbles or shells on a board, or beads on a wire, we can easily remove them or transfer them to the other end of the wire. The step beyond this is to use differently shaped pebbles, or clay tokens, to represent groups of objects of different sizes: perhaps a small token to represent a single object and a large one to represent 10 such objects. People in what is now Iran used such a system before 3000 BC, with a stick shape representing a single object, a pellet representing 10 sticks, and a ball representing 10 pellets, but other civilizations also adopted this system. This appears to have been one of the origins of writing numbers down. Transactions could be recorded by putting the tokens in a closed clay shell. This could be transferred from place to place, and broken open to reveal the number. Extending this yet further, a representation of the contents could be scratched on the outside of the shell, and this is the origin of the Sumerian numerals, which appear to be the oldest written form of numbers.

By this stage, the numbers have become entities in their own right. The commonality of four legs, four loaves, and four houses is apparent. Once we have reached this stage, we can juggle with the numbers themselves: we can do arithmetic and know that our conclusions will apply to any groups of objects of the specified sizes. The world we are studying has been effectively mapped into a representing symbolic system. The relationships between sets of objects in the world have been mapped to relationships between our representing symbols. That is, the rules by which we manipulate our numbers correspond to physical manipulations which could be carried out in the real world.

Once the natural numbers and zero have been accepted, the next step is to accept negative numbers. Initially, people were suspicious of these, since they appeared to correspond to nothing in the real world. Gradually, however, they began to be accepted, not least because a negative number from one person's perspective is a positive number from another's (your negative bank balance represents real money owed to the bank).

Zero and the negative numbers are but two examples of extensions to the number system which were initially regarded with suspicion and resisted. Many other examples exist. One example arose in the sixth century BC, when an Ionian philosopher called Pythagoras established a movement (the Pythagorean Brotherhood) in Sicily. Pythagoras believed that everything could be understood through mathematics. Whole numbers played a key role, with rational numbers being legitimate because they were ratios of whole numbers: the concept of two equal parts of a loaf of bread making up the whole is straightforward.

Unfortunately, towards the end of Pythagoras's life, the Brotherhood discovered that there were some things in particular, the diagonal of a unit square – which could not be described this way. 'Irrational' is a good term to use for such counterintuitive objects. This was a dramatic blow to the Pythagorean Brotherhood, which tried to conceal the discovery. Eventually, mathematicians hit on the idea of expanding the number system to include irrational numbers, and represented numbers by the lengths of line segments in a geometrical representation. In a sense, this is the reverse of the measurement process, which represents the length of line segments by numbers. This geometric representation of quantitative concepts held sway throughout most of history, and it was not until the end of the nineteenth century that it began to be seriously expanded.

The resistance and suspicion stimulated by negative and irrational numbers was repeated yet again when 'numbers' such as $\sqrt{-1}$ proved to have a useful role. Again this suspicion is reflected in their name – 'imaginary' numbers, numbers which have no real existence. The adoption of names reflecting the way these numbers extended the number system is also apparent in 'irrational' numbers, which cannot be expressed as the ratio of two integers, and 'transcendental' numbers, numbers such as e and π which are irrational numbers and are not roots of a polynomial equation with integer coefficients.

Imaginary numbers provide a nice illustration of a construction in which formal manipulations go beyond the realm of the real numbers but then return to give real results which can be readily interpreted. Here is an example. In 1545 Girolamo Cardano (1501–1576) published a solution to the cubic equation $x^3 + mx = n$. His solution, which represented a great breakthrough, was

$$x = \left(\frac{n}{2} + \sqrt{(n/2)^2 + (m/3)^3}\right)^{1/3} - \left(-\frac{n}{2} + \sqrt{(n/2)^2 + (m/3)^3}\right)^{1/3} \qquad (1.1)$$

For many values of m and n this gives straightforward results. But for certain values of m and n we find ourselves faced with the square root of a negative number when carrying out the calculations. Fortunately, if we are prepared to manipulate such entities, regardless of what they may 'mean', these awkward square roots cancel out and we obtain a real result. Nowadays this is made explicit by the introduction of the symbol i (or j in some disciplines) to represent $\sqrt{-1}$.

More generally, we may find that we have readily interpretable measurement concepts, which can be handled in formal way using manipulations and intermediate concepts which go beyond the ready interpretation of the measurements, but which, at the end of the manipulations, yield an interpretable result. Indeed, one might regard the multiplication of two negative numbers as an even simpler example of this. ('Unlike the operation of adding positive and negative numerals, it is difficult to see how Indian mathematicians could have interpreted the multiplication of a negative numeral by a negative numeral as other than a purely formal operation' (Roche, 1998, p. 20).)

The two main philosophical schools of the nature of mathematics are *formalism* and *Platonism*. Reuben Hersh (1998, p. 7) says: 'Platonism is dominant, but it's hard to talk about it in public. Formalism feels more respectable philosophically, but it's almost impossible for a working mathematician to really believe it.' The formalist view is that mathematics is a system in which objects are manipulated according to rules, and that the objects have no 'real' existence. In contrast, the Platonist (or realist) perspective is that mathematical entities exist on a plane 'above' that of the natural universe: they existed before humans started to think about them, and they will continue to do so after humanity has gone. Of course, as with all philosophical schools, each of these has many variants, and there exist other less important schools (for example, *intuitionists* and *empiricists*). Hersh (1998) argues that mathematics is necessarily a product of human culture and ways of thinking – we are reminded of the answer to Wigner's puzzle mentioned above.

1.2.2 The origins of measurement

I mentioned above that the measurement concepts in the backdrop of modern lives are so ubiquitous and familiar that they are seldom explicitly noticed. This book aims to look deeply at these concepts, but both measurement and counting were used long before philosophers began to contemplate their meaning. This inevitably prompts one to ask whether it is necessary to contemplate their meaning. If they can be used perfectly adequately, why make life difficult? The answer, of course, is that by understanding these ideas we can extend them, identify their limitations, and generally make sure we use them properly. The trouble is that the simplistic concepts used in everyday life can conceal ambiguities and uncertainties, and these, of course, can have real practical implications: corporations do go bankrupt, bridges can collapse, space missions may fail, and sometimes public services do not serve the public.

As a second-order effect, by understanding the notions of measurement properly, we can teach the ideas and methods effectively. Without such deep understanding, learning relies on apprenticeship systems, whereby a student learns from watching and copying a master. This is all very well, but it is labour-intensive and does not equip the student (or the master, for that matter) with the facility to handle situations out of the ordinary, or apply the ideas reliably to different situations.

The existence of measurement concepts in the infrastructure of human civilization is illustrated by the ubiquity of metaphors which use measurement ideas. Klein (1974, pp. 92–93) draws attention to a passage in the Bible:

The most remarkable metrological story in the Bible is in Daniel. The time is 539 BCE; the place is the great feast given by Belshazzar, ruler of Babylon, attended by thousands of his lords, captains, their wives, concubines, serving women, and men-at-arms. The monarch orders his guests to drink from golden vessels that had been taken from the Temple of Solomon in Jerusalem when the Babylonians (Chaldeans) destroyed it in 587 BCE. They drink wine from the sacred vessels, then praise their own 'gods of gold, and of silver, of brass, or iron, of wood, and of stone.'

As they do so, the fingers of a man's hand appear and write on the palace wall. Seeing 'the writing on the wall,' Belshazzar is terrified. 'The joints of his loins were loosed, and his knees smote one against the other.' He calls his astrologers and soothsayers, but they cannot interpret what has been written. At last the captive Jew, Daniel, is brought. He dares to decode for the king the message on the wall:

MENE MENE TEKEL UPHARSIN

In translation, a metrological metaphor appears. In somewhat modernized language: MENE: God has measured your kingdom and reduced or terminated it. TEKEL: You yourself have been weighed in the balance and found wanting. UPHARSIN: (or PERES): Your kingdom [shall be] divided and given to the Medes and the Persians.

There are plenty of other references to measurement in the Bible. Deuteronomy 25: 13–14 previsions a whole human history of confusing (sometimes deliberately so) weights: 'You shall not have in your bag two kinds of weights, a large and a small. You shall not have in your house two kinds of measures, a large and a small.' Leviticus 19: 35–36: 'You shall do no wrong in judgement, in measures of length or weight or quantity. You shall have just balances, just weights, a just ephah, and a just hin.' Chapter 7 has more to say about attempts to unify length and weight measurement.

Likewise, the Koran includes references to measurement (McGreevy, 1995, p. 27):

Woe to the unjust who, when others measure for them, exact in full, but when they measure or weigh for others, defraud them!

Give a just weight and measure and do not defraud others of their possessions.

These scriptural quotations illustrate the fact that the origins of measurement lie not in scientific or even prescientific attempts to understand the universe about us, but in more immediate matters of trade and commerce, not the least in fraud prevention. The fact that hard measurements impose a consistency (that measurements travel, as we put it above), allows them also to impose a uniformity of action, and commerce and trade were consequently a prime mover in the drive towards quantification. How many bales of wool? How many yards of cloth? How long to deliver them? How much would they cost? Of course, money is a key concept here, because measuring or counting money drives many economic activities. Crosby (1997, p. 71) remarks that 'price quantified *everything*', quoting Walter Burley of Merton College as saying in the fourteenth century that 'every saleable item is at the same time a measured item', and going on to note that through service and interest on debts, time also had a price, so that 'if time had a price, if time were a thing that could have a numerical value, then what about other unsegmented imponderables, like heat or velocity or love?'

Keeping track of goods ordered, dispatched, received, delivered, owed and paid for required an accurate and reliable system of bookkeeping. Double-entry bookkeeping began to be used in the thirteenth and fourteenth centuries. Luca Pacioli (1445?–1517?) is often called the father of double-entry bookkeeping, and appears to be the first to have published a printed description of it. Here is Pacioli's example of how a memorandum should be recorded (*Pacioli on Accounting, 40*, from Crosby, 1997, p. 221):

On this day, we have (or I have) bought from Filippo de Ruffoni of Brescia, twenty pieces of white Bresciani cloth. They are stored in Stefano Tagliapietra's vault and are of so many arm lengths apiece as agreed upon. They cost twelve ducats each and are marked with a certain number. Mention if the cloth is made of triple warp-cord, four to five arm lengths square, wide or narrow, fine or medium, whether Bergamene, Vicenzan, Veronese, Paduan, Florentine, or Mantuan. State whether the transaction was made entirely for cash, or part only for cash and part on time. State when the balance is due or whether payment was partly for cash and the remainder in goods.

Note how many counts and measurements appear in this example. And one might easily add more: how long would it take the goods to arrive? how would ships carrying goods know their location?

If there is an infinite number of attributes which may be measured, then there is also an infinite number of units by which they may be measured. We are familiar with units such as the inch and the centimetre for measuring length, but over the course of history an astronomical number of different units has been defined and adopted. This will be especially apparent in the discussion of the physical sciences in Chapter 7 (Alder, 1995, refers to 700–800 different metrical names in eighteenth-century France), but it is also true of the social and behavioural sciences. On the other hand, as we discuss especially in Chapter 3, the social and behavioural sciences are complicated by the fact that slightly different measurement scales will generally be measuring slightly different things. Having said that, it is not a priori obvious that it is appropriate to use the same weighing scales for measuring out grain and apples, and typically different units were used for measuring (what we would now regard as) the same attribute of different substances. The difficulty, as much with the physical sciences as with the social and behavioural sciences, is in deciding exactly what the attribute is: is the attribute 'area' when applied to poor agricultural land the same as the attribute 'area' when applied to rich agricultural land? And is it thus appropriate to use the same unit of measurement for these two attributes? Indeed, in modern manufacturing, we might pay by piece-work, that is, in terms of the number of items produced, rather than by the conventional hour: our way of measuring work differs according to our purpose.

The creation of a huge number of different physical measurement units does not reflect an inadequacy or laziness on the part of those using them. The units generally had local meaning, and were typically tied to everyday human life: the yard, the metre, the mile, the pound, the kilogram, and so on all were of *human* size, in the sense that they had relevance. Until the advent of mass long-distance communication and travel, the units also only had *local* relevance, so that there was no reason for people in different communities not to use different units.

Of course, trade and commerce were not the only drivers towards quantification. Although carpenters and architects can construct tables and churches without using measurement concepts, the advantages of adopting them are readily apparent.

If science was not the prime driver towards the development and adoption of quantitative methods, it is certainly one of the activities which has taken it most closely to heart. Again the consistency permitted by rigorous measurement means that an objective science, one which people everywhere can use in the same way, can be developed. Porter (1995, p. ix) says: 'Quantification is preeminent among the means by which science has been constructed as a global network rather than merely a collection of local research communities.' Beyond consistency in description and communication, measurement also permits the description of experimentation as scientific method by such thinkers as Robert Grosseteste (1169–1253) and Francis Bacon (1561–1626). Only when sufficiently precise measurements are made can one make effective comparison of theory with experiment. Without precision, one can always argue that a theory is correct – and this does little to help either understanding or prediction.

The history of measurement is far too large a subject to treat here. The early history is clouded in obscurity. Glances at the more recent history will appear in later chapters of this book. It is a history of the human race, of human intelligence, and of human civilization. It is a story in which sparks flickered to life, burst into flame, and then died out and disappeared – only to reappear elsewhere. It is the story of a struggle between the Aristotelian qualitative and the Platonic-Pythagorean quantitative views of the universe. And it is a story in which, over the past 1000 years, the quantitative view has gradually become accepted. The sparks, the small tongues of flame, have now merged and become a roaring inferno, so powerful that no one doubts their power as a way to see the universe around us.

1.3 But is it measurement?

Whereas it might seem obvious that *counting* is useful in life, it is perhaps not so evident, a priori, that *measurement* can be applied to the real world or is useful. To start with, by definition the measurement process extracts just one aspect of the object it is measuring and assigns a number to that aspect. But things in the real world are not described by a single aspect. Different people have different height, weight, colour, amount and thickness of hair, size of nose, accent, sense of humour, and so on through an infinite list of attributes. Certainly we can categorize, describe and then even measure individual attributes of people, but people are not characterized by individual attributes.

We can count stones, sheep and people, but we cannot *count* length, weight, or time. In order to apply arithmetic to continua such as these we have to discretize them: divide them into chunks which can then be counted. Attributes such as length, weight, and time are *quantifiable* simply because they can be divided into *quanta*, which can then be counted (and upon which counts we can then erect the superstructure of mathematics and mathematical concepts). To measure the length of a stick, we have a (notional) collection of short rods of identical length, and we count how many of these span the same distance as our stick. Children beginning to learn about the quantification infrastructure upon which our civilization is built do this explicitly, but adults, who have absorbed the practice (and sometimes the principles) into their subconscious do it automatically.

There are certain operations implicit in this process of discretizing continua so that counting can be applied. We need to be able to compare objects, to see that more than one short rod will be necessary to match the extent of our stick. And we also need to be able to concatenate rods properly, so that they do divide the extent of the stick up – after all, if I lay the rods in parallel, then, no matter how many I have, their combined extent will never match that of the stick. These sorts of operations, as well as others which we discuss in this book, are what makes a process a *measurement* process.

Some researchers have taken the position that the counting operation is fundamental to measurement (e.g. Michell, 1986, 1999), and that this hinges on the ability to define a unit of measurement and to aggregate such units. Others (in recent years, most others) have taken a more relaxed perspective, in which a broader range of empirical relationships, mapped to homomorphic mathematical relationships, are said to constitute measurement. These issues are discussed in detail in Chapter 2. In this book, we have deliberately set out to construct a model of measurement *practice*, not of what measurement, in an imaginary ideal world, should be. That is, we seek to be descriptive about measurement practice, rather than proscriptive. In particular, we take the view that that measurements are mappings from objects in the universe being studied to a numerical representation called a *variable*, and that measurement procedures have two fundamentally different aspects, which we will term *representational* and *pragmatic*. The representational aspect maps observed relationships between objects, in terms of a particular attribute of interest, to relationships between numbers. For example, observation that one standing person can look down on another means that their height attribute can be mapped to a numerical representation, the variable *height*. The height attribute of the former person will be mapped to a larger number of *height* than the height attribute of the latter person. The representational aspect thus places constraints on the numbers we can use, based on relationships within the empirical system we are studying. The pragmatic aspect also places constraints on the numbers we can use, but these constraints are chosen for reasons of convenience or practicality. For example, we might construct a measure of attractiveness to the opposite sex as a weighted sum of various attributes, including physical appearance, personality, financial situation, and so on, including the various components because we believe they are an important aspect of what we want to capture, and weighting them according to what we see as their importance. Pragmatic constraints span the gap between the range of possible numerical assignments which arise from the representational approach, and the unique assignment used in a particular study. They *crystallize* the numerical assignment, so that we know exactly what we are talking about. A very simple example is the choice of unit in the physical sciences: we cannot report length measurements without also reporting a unit of length.

The extent to which a variable will be representational or pragmatic will depend on what we are trying to do. In any case, and always, the variable needs to be made precise before it can be regarded as measuring any attribute. To do otherwise would be unscientific, and would potentially invalidate any conclusions. Some variables correspond to well-defined empirical attributes. At an extreme, simple counts ('counting' is sometimes called an 'absolute' measurement scale) of well-defined empirical objects (e.g. the number of sheep in a flock) require no pragmatic constraints. Less extreme examples are given by weight and length, which simply require one to choose the units. Other variables require more elaborate models in order to measure them, and sometimes these models require extensive choices to be made which are not based on representational considerations, but rather on pragmatic ones. For example, in measuring short-term memory, there is a collection of tests which could be used. One might choose just one of these (e.g. recalled digit span) or one might attempt to integrate scores from several, via a model such as factor analysis. Choosing just one aspect is equivalent to an assertion that this one aspect of short-term memory is of interest to you (even 'this is what I mean by short-term memory for present purposes'), and other people are, of course, at liberty to disagree or choose otherwise. It is thus a pragmatic choice. In the case of health-related quality of life (HRQoL), one asserts at the start that a specified combination of given variables is what is of interest. One might say that numerical ratings of extent of constant pain, feelings of worthlessness, and feelings of no possibility of improvement, combined in a given weighted sum, are what one means by HRQoL. We are saying here 'I choose to include this variable, with this weight, because that is the extent to which I want it to contribute to my measure'. Again, others may prefer other variables, combined differently. The choice is a pragmatic one.

Of course, one may attempt to restrict one's subsequent analysis to statements which are invariant to changes in the pragmatic constraints, so that the statements only reflect the empirical reality being studied. This is necessary if there is disagreement or lack of convention about choice of pragmatic constraints. In the case of measurement in the physical sciences, for example, people use a wide variety of units, and one would not want one's conclusions to vary from unit to unit. Section 7.1.3.2.2 gives an example of a model built without taking the wish for this invariance into account. It is a perfectly adequate model for the particular units recorded, but cannot be a model of the real world because of its lack of invariance – and this matters in this case because there is no standard convention about what units (of time duration, in this example) should be used. Issues such as this are discussed in detail in Chapter 2.

Norm groups in psychological measurement provide another example of pragmatic constraints – for example, the population used to produce the mean of 100 and standard deviation of 15 in IQ tests. This is a pragmatic constraint rather than a representational one since it is a choice made by the user about how the system should be constructed. There is nothing fundamental in the attribute being measured which dictates what population should be used as the norm. So, just as in the HRQoL scale above we said 'I choose to include this variable, with this weight', so here we say 'I want to graduate the scale using this population because I regard this as the key population'. It is a choice made by the constructor of the scale, not a product of empirical relationships between objects in terms of the attribute being measured.

Pragmatic choices are often imposed so that measurements are readily interpretable. For example, Binet (Chapter 5) introduced the system of interpreting children's IQ scores in terms of the average age at which normal children would achieve those scores, so that, for example, we can describe an 11-year-old as having the mental age of an 8-year-old. In medicine an example is the number needed to treat, or NNT (Section 6.7). The relative risk arising from different threats is a notoriously difficult concept to communicate, and one way is to describe it in terms of how many years one would expect to wait before a fatality. Likewise, many people lack facility with manipulating probabilities, but are quite at home with the equivalent odds formulation.

As a final and more subtle example, if I believe that the conditional independence assumption in a factor analysis model is correct, then the choice of manifest variables in that model is a representational constraint. If, however, I do not believe it (for example, I might be finding the best-fitting single-factor model when I really believe that more factors are needed) then my choice of variables is a pragmatic decision.

In considering the difference between the representational and pragmatic aspects of measurement, it is useful to note that, in statistics, the word 'model' is used to describe structures derived in two rather different ways. In the first, the model is based on some kind of theoretical belief about the relationships in nature (i.e. a representation). Such models are called *mechanistic, realistic, substantive, iconic* or *phenomenological*. In the second, the model is derived purely as a convenient statistical summary of a data set. These models are called *empirical, descriptive,* or *operational*. Discussions of the distinction appear in Neyman (1939), Box and Hunter (1965), Lehmann (1990), Cox (1990) and Hand (1994, 1995). Mechanistic models are useful for elucidating the mechanism underlying a process – for *understanding* what is going on. Empirical models are useful for prediction or decision-making (mechanistic models may also be useful for this), but not for understanding the underlying process. However, since empirical models do model observed relationships in the data, albeit not relationships based on any underlying theories, they are also typically properly regarded as representational. To be pragmatic, the model would be derived independently of the data, and imposed on them. Only if we are forcing the model on the data (regardless, if you like, of whether it reflects the important structures within the data) is it pragmatic.

In general, variables are constructed as a mixture of representational and pragmatic constraints: one creates a model for what one believes is the empirical situation, or as near as one can get to it (using representational models), and then adopts certain assumptions or conventions for the remainder, for the gap between the model and the reality (using pragmatic constraints). The gap-bridging pragmatism is needed for almost all measurements, though it plays a larger role in some. For example, in length measurement one's model is that the objects obey the order and concatenation relationships, but, of course, there are too many objects to test them all, and nor can one test objects which are astronomically long or microscopically short. One therefore adopts, as a pragmatic assumption, the assumption that the properties also apply to such objects.

The representational aspect of measurement has also been called 'theoretical' and, in a particular domain, 'psychometric'. The pragmatic approach (or ideas closely related to it) has also been called 'operational', 'clinimetric', or simply 'non-representational'. As we have seen, a representational approach seeks to construct a numerical system in which the relationships between the numbers match the relationships between the objects arising from some attribute (or, more generally, attributes). It is thus an idealized mathematical construct, a representation, which provides a reasonable approximation to the behaviour of natural objects. It is idealized just like the points, lines, and circles of geometry are idealized, but its ideal nature does not mean it is not valuable, any more than geometry is not valuable because of its failure to represent the imperfections of physical reality.

In contrast, the pragmatic aspect of measurement defines the variable being measured. Instead of beginning with the empirical system, and noting relationships between objects which must be preserved by the relationships between the representing numbers, pragmatic measurement identifies which aspects of the empirical system one wishes to contribute to a high measured value and combines them so that they do. The essence of a pragmatic model is that the definition is chosen to have some useful practical purpose.

We have already noted the representational and pragmatic aspects of HRQoL measurement. In this context Fayers and Hand (2002) say:

> The objective of the psychometric approaches outlined above might be characterised as *attempting to measure a single attribute using multiple items.* In contrast, clinimetric methods *attempt to summarise multiple attributes with a single index.*

They then go on to say:

> Now, of course, [the second of these] cannot be done – a single index loses the intrinsic differences between the attributes and it also sacrifices any possibility of allowing for an interaction (in the presence of attribute *A*, attribute *B* leads to poor QoL, but in the presence of *C* it does not). Instead the researchers aim their strategies at choosing and suitably emphasising the most important attributes to be included in the index (Wright and Feinstein, 1992). In fact, of course, one is choosing how to *define* the concept being measured by one's choice of variables and the way of combining them.

Thus representational measurement *represents* an attribute in the empirical structure by a variable in the mathematical structure; pragmatic measurement *defines* a variable.

The distinction between the representational and pragmatic aspects of measurement is an important one. One consequence of the distinction is that, at the extremes, entirely different techniques are needed to construct scales of the two types. This fact, which has not always been recognized, has often led to confusion. Construction of purely representational measurement models *assumes* the existence of a mapping which relates observable or manifest variables to the concept one is seeking to measure, and then seeks to find that mapping (i.e. find numbers which match the empirical relationships). This requires an empirical investigation of the subject matter to explore whether or not the objects do have the relationships required for the postulated mathematical representation. If we cannot order the objects according to some attribute, then we cannot assign numbers to them to represent their order. In contrast, pragmatic scale construction requires one to state clear and explicit restrictions on the numerical assignments. These restrictions can come from a variety of sources, including practical convenience and mathematical convenience (e.g. the adoption of axioms for price indices, discussed in Chapter 8).

Other high-level distinctions between types of measurement models are also possible. As with the distinction between representational and pragmatic aspects, these classifications are not perfect – the classes in them overlap at the edges – but this lack of perfection does not detract from their usefulness. Categorizations of the real world are seldom perfect, and yet they are fundamental to human thought and progress. At the very least, the distinctions allow us to structure this book.

One such further distinction is that between *direct* and *indirect* measurement. Direct measurement involves the assignment of numbers to objects so that the relationships between the numbers represent the relationships between the objects in terms of a *single* attribute (the representational case – e.g. using weighing scales to assign numerical measures of weight) or so that the numbers define the scale on which the attribute is to be measured (the pragmatic case – e.g. the choice of minerals in the Mohs scale of hardness). Indirect measurement, in contrast, determines which numbers to assign on the basis of the postulated relationships between the variable to be measured and others which can be observed directly. Intelligence, for example, cannot be observed directly, but it is thought to be related to other measures – such as the ability to answer certain kinds of questions. If one has a model which relates intelligence to these other measures then a score for intelligence can be derived from the scores on these other measures. In indirect models the variables which are measured, and which are combined in some way to yield the measure of the target concept, are called *observable* or *manifest* variables. The target concept is sometimes called a *latent* variable – although this is not always appropriate. For example, in pragmatic situations there is little notion of a pre-existing (i.e. latent) variable, just awaiting a sufficiently clever analysis to reveal its value. Instead it is regarded as an explicit construct, deliberately created by combining the manifest variables in a certain way. The terms 'fundamental' and 'derived' measurements are used by Campbell and others in what are broadly our meanings of direct and indirect. Adams (1966, p. 145) uses the terms 'intrinsic' and 'extrinsic' to mean roughly what we call direct and indirect, respectively. ('By an "intrinsic" measurement procedure (or better, a procedure of primarily intrinsic significance), I mean, roughly, one which involves only observations of the same kind of phenomenon which the measurement is used primarily to give information about.')

Broadly speaking, direct measurement corresponds to simple models and indirect to more complicated models. The fact that one may disagree on the boundary between simple and complicated accounts for the ambiguity between these two types of measurement. Note also that the distinction is blurred through the use of sophisticated modern measuring instruments – an electronic device might measure several variables directly and combine them, without any human intervention, to yield a single outcome variable. Subtleties also begin to enter – and the distinction blurs yet further – when we start to think about modern measurement techniques which use electronic scales which output direct numerical weight values, rather than (say) positions on a dial. In any case, one might note that the observations in even simple direct models are made with error, so that what one is really interested in here is the 'latent' 'true' variable. The steps from the empirical relationship to the numerical output are no longer simple and obvious in terms of relationships between objects in terms of a single attribute. Despite such complications, the distinction is often useful.

Since indirect models define the variable of interest in terms of other variables, it is natural to describe the former verbally in terms of the contributions of the latter. If a test of intelligence is composed almost entirely of questions about arithmetic, we might call it a measure of 'arithmetical intelligence'. This process, known as *reification*, is a convenient shorthand way of describing the content of the measured variables, but it can be the subject of excessive enthusiasm, with rather contrived meanings sometimes being attributed to the resulting derived variable. In the past, this has been one of the critical sticks used to beat factor analysis.

We should stress the fact that the indirectness of indirect variables means that they cannot be measured directly. In the case of representational measures (when one may talk of 'latent' variables), they are part of the theoretical construct relating them to the manifest variables. This means that any 'real existence' they may have is predicated on the truth of the theory surrounding them. The current fashion is to regard theories not as being 'true' or 'false' descriptions of reality, but merely 'more adequate' or 'less adequate' models of it. Taking this to an extreme, one might argue that what is important is not the reality or otherwise of the latent variables, but their effectiveness in permitting prediction of empirical phenomena. If one can accurately explain what is happening in terms of some latent variable then that variable is valuable. Classic examples of this from physics abound: the very notion of a field of force is an artificial mathematical construct which is only measurable by virtue of its effect on other bodies. And yet this latent construct pervades physics and is extraordinarily effective in permitting prediction of empirical phenomena. The same idea occurs elsewhere. In psychology, for example, Michell (1999, pp. 159–160) remarks:

> As Cronbach and Gleser (1957) later came to acknowledge, the usefulness of tests in predictive or decision-making contexts is an issue independent of whether or not tests measure psychological attributes. Cronbach and Gleser advocated replacing the 'measurement model' by the 'decision theory model' and, while this recommendation may have value in specific, practical contexts, the issue of whether or not such tests measure anything remains a genuine scientific issue.

The attitude which disregards the question of the reality of the measured attribute and focuses solely on its observed interactions and relationships with other variables is very much a pragmatic perspective.

It is important to note, in the context of taxonomies such as those above, that the distinctions we are making relate to the types of scales, and not to the underlying concepts. Thus, for example, one could formulate a representational measurement scale for anxiety, based on observed empirical relationships, or one could formulate a pragmatic measure, based on an explicit definition of what one wants to mean by a measure of anxiety. Indeed, one could formulate several measurement scales of each type. They would each measure different 'kinds' of anxiety but I (here, in this book) would not describe them as 'representational anxiety' or 'pragmatic anxiety', but merely say that they were 'representational' or 'pragmatic' measures of anxiety, leaving the question of what exactly anxiety is to those concerned with such questions. Typically, in any case, any given anxiety scale would have both representational and pragmatic aspects.

Likewise, the notions of direct and indirect measurement relate to the measurement scales and not to the concepts: pressure can be measured directly, or it can be measured indirectly by combining measures of weight and area; length can be measured directly (via a count of end-to-end concatenated rods each of identical length) or indirectly (by combining the speed of transmission of a signal and the time it takes to traverse the length in question); preference can be measured directly (by asking subjects to mark their preference for one object compared to another on a visual analogue scale) or it can be measured indirectly (via a complex statistical analysis of a set of interrelated questions).

A taxonomic distinction which it is tempting to make, but which does not serve us well as we dig deeper into the subject, is that between measurement in the natural sciences and measurement in the social and behavioural sciences. The canonical examples of direct measurement, where the relationships between objects in terms of a single attribute determine the values, are length and weight – as discussed in detail in Chapters 2 and 7. These seem simple, at least compared with the models required to extract reliable measurements in the behavioural and social sciences (Chapters 5 and 8). Moreover, measurement in the social and behavioural sciences seems to contain ambiguity and even arbitrariness which measurement in the physical sciences does not. For example, the very choice of manifest variables which are fed into a factor-analytic model is a choice to be made by the investigator and is not some natural function of the universe. However, such comparisons are unfair. One might instead compare the measurement of the time a particle takes to transit given distances with the measurement of enjoyment of musical performances where the latter is on the basis of a single question asking how much the subject enjoyed the performances (to be scored on, say, a numerical scale from 0 to 1). In both cases a single number is produced, which reflects the relationships between the empirical objects (distances in the first case and performances in the second). Admittedly the physical measurement probably has a stronger representational component (mapping to a ratio scale instead of an ordinal one, perhaps). Also, the psychological measurement is likely to be subject to greater distortion from bias and noise, not to mention the fact that it will doubtless evolve over time so that replicating the measurements is difficult, but this is merely a reflection of the fact that the substantive domain of the second problem is intrinsically more complicated than that of the first. Presumably the subjectivity in the second situation is part of the object being measured – why else ask how much an individual enjoyed a performance?

This question of subjectivity raises several important issues. If the aim of the measurement is to see how someone feels, then it is pretty clear that the subjective component should be an integral part of the result. For example, in studying obesity one might want a measure of the subjective perceptions of how fat the subjects were. This is different from an objective measure of body fat content or skin fold thickness. Either approach could be right, depending on the aims of the study. Indeed, more sophisticated extensions of this are possible: we might ask someone else to give us their subjective assessment of how fat the subject is. This is subjective from the point of view of the person giving the rating, but objective from the subject's viewpoint. It is worth noting parenthetically here that weight, the most common measure used in such circumstances, is not really relevant. It is merely a proxy variable for the attribute of interest, be that a subjective perception of how fat someone is or an objective measurement of their body fat content.

It is clear from this example, that the notion of the 'true' value for a variable should take account of the aims of the study. But notions of 'true' values are more subtle than that. Early work, for example that in psychological test theory (see Chapters 3 and 5), was predicated on the existence of a true value, with the observed measurements regarded as being perturbed from that because of random measurement error or bias from various sources. The trick was to remove this perturbation to get at the true value. Measurement in the physical sciences is typically regarded as resulting in observed values composed of the true value plus an additional random component (and, in the schoolroom view of classical physics at least, this random component is small so that it can often be ignored). However, even leaving aside the fact that things evolve over time, so that the true value now may not be the true value when the object is measured again (no matter how soon), it is not clear that the notion of true value is always useful. In many situations the more fundamental concept which results from measurement is a *distribution* over values. One can, of course, take some location measure from this distribution as the 'true' value, but this is a rather arbitrary procedure.

1.4 Measurement and statistics

Measurement is the activity which produces the raw material which statistical methods analyze. Thus one might expect to see much discussion of the interplay between measurement and data analysis in the statistical literature. It is therefore perhaps strange that the statistical literature contains very little of such discussion. Such discussion does exist, but it is mostly in the social and behavioural science literature, and it has sometimes been acrimonious. ('The most distinguished physicists, when they attempt logical analysis, are apt to gibber, and probably more nonsense is talked about measurement than about any other part of physics' (Campbell, 1938, who was himself a physicist). 'This statement of Siegel's is completely incorrect. This particular question admits of no doubt whatsoever' (Anderson, 1961). 'The height of absurdity' (Gaito, 1980).) Highlights (if that is the correct word) of the debate are summarized in Section 2.5.

The debate hinged around the appropriateness of applying different kinds of statistical methods to data obtained by different kinds of measurement procedures. It is complicated by issues of transforming variables and the precise interpretation of research questions. For example, the validity of parametric statistical techniques depends on assumptions about the underlying distributional forms for the random variables. These techniques are concerned merely with making statements about the numbers which are subjected to the statistical analysis, and the distributions from which they were drawn. In particular, such tools require no conditions on the measurement scales – on 'where the numbers come from' – all they need are the numbers. However, generally one will be interested not merely in the properties of the numbers *per se*, or the populations of numbers from which they were drawn, but in the attributes that they purport to measure. The link between the numbers one is analyzing and the attribute they represent is crucial. If the pragmatic constraints are not so strong that a single scale is universally used for the attribute, and if one wants one's conclusions to be valid whatever scale is used for that attribute, then one must constrain the numerical operations involved in the statistical analyzes, so that the statements one makes are consistent between alternative legitimate numerical representations. For example, people's height can be measured in inches or in centimetres. The choice of unit is external to the attribute being measured, and is not determined by that attribute. If one has not specified (by pragmatic constraint) that a particular unit should be used, then one might legitimately require that the conclusions of any analysis should be invariant to switching from inches to centimetres. It would hardly be acceptable to have one's analysis conclude that 'on average' one group of people was taller than another group when they were measured in inches, but shorter when they were measured in centimetres. Problems seldom arises with length measurement, but often arise with measurement procedures which merely order values of an attribute. One might, for example ask people to place foods in order of preference. One could then assign numbers to the foods such that the order of the numbers represented the order of preferences. However, unless rather more elaborate procedures restricting the values of the numbers are adopted, or unless additional, pragmatic, constraints are adopted, then any set of numbers will do, provided they preserve the order relationship. This is not a problem if one restricts one's statistical analysis to statements which are invariant to arbitrary ordinal transformations, but a vast amount of modern statistical analysis makes statements which are not invariant in this way. For example, as we later illustrate in Section 2.4.1, one might conclude that one group had a higher arithmetic mean than another group when using one ordered set of numbers, but a lower mean when using another, equally legitimate ordered set. 'Equally legitimate' here means that both sets of numbers properly represent the preference order. This state of affairs is rather unfortunate: it means that the conclusions of the statistical analysis depend on an arbitrary choice of those taking the measurements.

It is because of the acuteness of the problem with ordinal measurement that the issue has attracted so much attention in the behavioural and social sciences: much of the data in these disciplines lies on ordinal scales. In contrast, in the physical sciences most attributes (there are a few exceptions) are measured up to changes of unit, with the choice of this unit being the only arbitrary nature of the measurement process.

For ordinal data, a possible resolution is to use nonparametric or distribution-free statistical methods. These generally use only the rank order information in the data, sacrificing any information about the sizes of the differences between consecutive values, so that their conclusions are indeed invariant to alternative choices of order-preserving scales. As far as statistical testing is concerned, the sacrifice of information is generally not large, with the tests typically having quite a high proportion of the power of parametric alternatives. However, the interpretation of the results of such tests is generally not as straightforward as the interpretation of parametric tests (e.g. concluding that one distribution 'stochastically dominates' another, rather than that one distribution has a larger mean than the other). Implicit in this is the fact that nonparametric and parametric tests will typically not have the same meaning. This is something to be wary of when conducting analyses.

Implicit in the above discussion is that there are different *levels* of measurement, corresponding to different degrees of stringency in the type of transformation which may be made while still preserving the empirical relationship between observed values of the attribute. Thus, for example, in the preference case, arbitrary order-preserving transformations were acceptable, while in the case of length, only rescaling (multiplication by an arbitrary positive constant) was acceptable. Such issues are discussed in depth in Chapter 2. An important early attempt to produce a taxonomy of levels of measurement was made by the psychophysicist S.S. Stevens, who defined *nominal, ordinal, interval,* and *ratio* scales (Section 2.2.5). Nominal scales permit any one-to-one transformation, ordinal scales permit any order-preserving transformation, interval scales permit any linear transformation, and ratio scales permit any rescaling transformation. These scales types have now been embedded in a powerful representational measurement theory.

However, Stevens's scales are not the only way to classify variables. They were devised in an attempt to resolve the issue of the legitimacy of statistical analyses, but other categorizations have alternative, typically more pragmatic origins. Van den Berg (1991) described many different classifications of measurements: counts versus measurements; nominal, ordinal and numerical; dichotomous as a separate category; qualitative versus quantitative; qualitative levels sometimes being called categorical or non-numerical; quantitative levels sometimes being called metric, numerical, or simply measurement. Nelder (1990) described various 'modes' of data, distinguishing continuous counts, continuous ratios, count ratios, and categorical, with the last being divided into three subtypes (nominal, ordered on the basis of an underlying scale, and ordered without an underlying scale). Bartholomew (1987) distinguished metrical from categorical. Mosteller and Tukey (1977) identified grades, ranks, counted fractions, counts, amounts, and balances. And other classifications have also been suggested. Many, if not most, of these classifications are based on pragmatic data-analytic grounds: the statistical techniques used to analyze a mere classification are different from those used to analyze a 'continuous numeric' variable such as length.

1.5 Against measurement

There appears to be an increasing tendency, in the modern age, to disparage things as being 'run by accountants', with the implication that quantification leads to the loss of soul, a sacrifice to 'cold' numbers, as it were. Likewise, it is not a compliment to describe someone as 'calculating'. In the context of medical measurement, Feinstein (1987b, p. 1) says: 'The distress probably arises from the fear that formal clinimetric attention to these acts of measurement may threaten the few remaining features of humanistic clinical art that have managed to survive the technologic transformations of modern medical science.'

This tendency is not new: 'There are certain ideas of uniformity which sometimes seize great minds (as they did Charlemagne's), but which invariably strike the petty. They find in them a kind of perfection which they recognise because it is impossible not to discover it; the same weights and measures in commerce, the same laws in the state, the same religion in all parts. But is uniformity always appropriate without exception … ?' (Montesquieu, 1875).

Boyle (2000, p. xix) in his uncompromisingly titled book *The Tyranny of Numbers: Why Counting Can't Make Us Happy*, says: 'Every time a new set of statistics comes out, I can't help feeling that some of the richness and mystery of life gets extinguished.' Well, he is right that some of the mystery goes. But my *Concise Oxford Dictionary* defines mystery as 'A secret, hidden, or inexplicable matter'. Surely it is all to the good to extinguish such things. As to whether the richness goes, on the contrary, the statistics (assuming, of course, that they are properly collected and interpreted) add to the richness, depth and understanding of life, deepening our appreciation of it, with the potential for making it better. The fact is that improved measurement procedures lead to enhanced understanding and improved prediction. This is illustrated in Section 5.8, where we review some areas in which formal measurement procedures have taken over from the subjective assessment procedures which were formerly used (not without a strong rearguard action from those wedded to subjective approaches, it has to be said).

The criticism directed at measurement is often misdirected criticism. It would often be better directed at the inadequacy of the measurement procedure – the lack of accuracy, precision, or care with which the measurements are taken, or the use of inappropriate measurements. Perhaps the latter is particularly problematic in social contexts, where it is notoriously difficult to get an empirical handle on what one really wants to measure. Boyle (2000, Chapter 3) provides several illustrations of this. However, the truth is that measurements are the only way to get a grip on what is happening, on what matters. The fact that sometimes measurements are inadequate, or may be misused, is not a reason for criticizing or abandoning the entire enterprise. There simply is no alternative. Of course, one must not go too far. One must be wary of elevating the process of measurement to unrealistic heights. Just as the Caesars in ancient Rome employed a slave to whisper into their ears 'remember, you are only human', so we must remember that no measurements are entirely accurate. We must avoid the mistake illustrated in the poem from Chambers (1997):

Economists have come to feel

What can't be measured isn't real

The truth is always an amount

Count numbers, only numbers count.

What we should do is use the awareness of the inadequacy of our measurement technologies as an incentive to strive to improve them. Having recognized that they are not perfect, identify in what way they fail to measure what we want to measure (we might say, identify in what way they fail to 'measure up'), and adapt them. If we abandoned something immediately it failed to be perfect, human civilization would not progress very fast.

Of course, there are additional difficulties in measuring social phenomena, which do not arise when one is not measuring thinking beings. In particular, people, either individually or *en masse*, can adjust themselves to match the measuring instrument, and this inevitably attracts much criticism. League tables (discussed in Chapter 8), are an easy target for criticisms of this kind. They are not perfect, and they never can be. But they are certainly better than nothing, and they can be improved. One of the problems is that humans and human institutions being measured can adapt to optimize their measurements: surgeons can refuse to operate on the difficult cases, schools can refuse to enter those pupils likely to do poorly in examinations, health authorities can defer making appointments for some patients, so that the waiting lists look smaller, and so on. No wonder we see 'grade inflation' (Chapter 5) in various situations, as people optimize their performance to the thing being measured. Boyle (2000, p. 49) quotes the business psychologist John Seddon: 'People do what you count, not necessarily what counts.' The criticisms are justified, but, again, they should not be transferred to the measurement enterprise. Rather, it simply means that more sophisticated measurement tools need to be devised.

Another difficulty perhaps especially critical in social and behavioural contexts is that the phenomenon under investigation may be complex. It will typically not be obvious what should be measured – and if it is not obvious what should be measured then it cannot be obvious how to measure it. But, again, the mere fact that the enterprise is difficult is not a reason for abandoning it. Refining the measurement process goes hand in hand with refined understanding of the thing being measured. McDowell and Newell (1996, p. 37) illustrate: 'There have been clear links between conceptual work on the definition of pain and the development of measurement scales, so the results obtained using these methods have led to refinements in the conceptual definitions of pain.'

Perhaps the last word on this topic should come from Stevens himself:

> The business of pinning numbers on things – which is what we mean by measurement – has become a pandemic activity in modern science and human affairs. The attitude seems to be: if it exists, measure it. Impelled by this spirit, we have taken the measure of many things formerly considered to lie beyond the bounds of quantification. In the process we have scandalised the conservatives, created occasional chaos, and stirred a ferment that holds rich promise for the better ordering of knowledge.
>
> (Stevens, 1959)

The last clause here tells of the power that effective measurement can bring: the attempt to construct a measurement model implies careful investigation of the phenomenon being studied. Successful construction of a measurement procedure does not mean that the phenomenon has been 'reduced' in any sense. Rather it means that that phenomenon has been understood.

1.6 And more

Of course, as with anything else, it is possible to take the ideas of measurement too far; however, this overextension is generally not in the process of defining and taking measurements, but in the interpretation put on them. In particular, it is all too easy to forget the pragmatic component of measurement and regard it as purely representational. As John Stuart Mill (quoted in Gould, 1996, p. 350) wrote:

> The tendency has always been strong to believe that whatever received a name must be an entity or being, having an independent existence of its own. And if no real entity answering to the name could be found, men did not for that reason suppose that none existed, but imagined that it was something peculiarly abstruse and mysterious.

Measuring things – the data collection process – does not come free. In fact, there is a growing awareness, again in social and behavioural contexts, of what we might term the 'measurement burden'. For example, at the time of writing it is possible for a UK citizen to enter the United States without a visa. To do this, one completes a 'visa waiver form', on the back of which there is a short passage which says: 'Public reporting burden – The burden for this collection is computed as follows: (1) Learning about the form 2 minutes; (2) completing the form 4 minutes; for an estimated average of 6 minutes per response.' It then goes on to say that 'If you have comments regarding the accuracy of this estimate, or suggestions for making this form simpler, you can write to …'. The point is that measurement costs. And if the aim of measurement is to lead to an improvement of some process, one must be certain that the cost of taking that measurement does not outweigh the gains to be made through knowing it. This point is particularly crucial in managerial and governmental contexts: if the workers spend all their time monitoring what they are doing, then nothing useful gets done. I am reminded of an apocryphal (one hopes!) story about the manager who declared: 'This meeting is going to continue until I find out why so little is getting done around here!' At a more mundane level, social researchers are acutely aware that asking respondents to complete a questionnaire of 5000 items is likely to be counter-productive. It is not always so obvious that governmental bodies are as aware of these issues. Staff at the university at which I used to work had imposed upon them, over the course of a year or two, appraisal schemes, equal opportunities monitoring processes, the Research Assessment Exercise, course production audits, research student monitoring systems, and, of course, a teaching quality assessment. All of these things are fine and good in isolation, with commendable and laudable aims, but taken together they represent a massive measurement burden which might well adversely impact staff productivity. (Of course, I'm sure the cumulative burden of these things is carefully thought through nowadays at that university.)

1.7 This book

The measurement enterprise is a vast one. It underlies our science, our civilization, our commerce, and our very lives. It forms the infrastructure on which modern humans exist, in the sense that our mental picture of the world is very much based on measurement (hence the subtitle of the book). Clearly, no single book could possibly deal adequately with such an entity. What I have tried to do in this book is demonstrate the universality and fundamental nature of measurement, illustrating how measurements are constructed and how they are used. The first half of the book investigates measurement in general, while the second examines measurement in some specific areas. In Chapter 2, I explore what measurement is, showing how it combines two complementary ideas, one being the need to represent empirical realities of the world, and the other the need to define the attribute we wish to measure. In Chapter 3, I explore how measurement is done, how numbers are assigned to objects to represent the magnitudes of their attributes. Of course, no such process is perfect. This book is about the way humanity understands the empirical world, and no such understanding is without error. Chapter 4 explores inadequacies and inaccuracies in measurement procedures. Chapters 5–9 then examine the particular problems of measurement in psychology, medicine, the physical sciences, economics and the social sciences, and other areas. Of course, these are merely examples. It goes without saying, that the number of areas which I have not examined is infinite.

Our view of life is gained through the spectacles of measurement. The aim of this book is to sharpen that view. Theodore Porter (1995, p. 32) put it nicely: 'Karl Pearson was neither the first nor the last to worship quantification, which he regarded as integral to scientific method. Its appeal has been the appeal of impersonality, discipline, and rules. Out of such materials, science has fashioned a world'.

2

The nature of measurement

What has happened in measurement theory is convincing evidence that powerful advances in scientific metatheory are possible when rigorous thinkers are willing to put some intellectual muscle into the enterprise.

Rozeboom (1966)

2.1 Introduction

As outlined in Chapter 1, there are two fundamentally different aspects to measurement, and the relative emphasis placed on each aspect will depend on one's aims and on the system one is measuring. I call these the *representational* and *pragmatic* aspects. The representational aspect refers to the extent to which the numbers are chosen so that the relationships between them reflect the relationships between the objects being studied, in terms of the attribute in question. This may be a simple representation (*direct measurement*): we may assign numbers to people so that taller people are assigned larger numbers. Or it may be a complicated representation (*indirect measurement*): we may assign numbers to people to represent some overall summary of their scores on a series of questions in a simple test, so that people who answer more questions correctly are assigned larger numbers. The pragmatic aspect refers to choices which must be made by the person making the measurement (or developing the measuring system or instrument) which are not dictated by the empirical system being represented. Once again, these may be simple (*direct measurement*), in that they are related to single attributes: in constructing a hardness scale we might decide to position ten different minerals at the positions 1, 2, 3, ..., 10 according to which of the others they can scratch or are scratched by. Or they may be complicated (*indirect measurement*), in that they relate multiple aspects of the objects together to define the attribute in question: in quality of life measurement, we may decide that factor A is twice as important as factor B, and define our measure as an appropriately weighted combination of measures of A and B; in defining 'wind-chill factor' we must decide how to combine temperature and wind speed. Typically, measurement procedures involve a mix of representational and pragmatic aspects, although this may not always be immediately apparent.

This chapter explores the nature of representational and pragmatic measurement in detail. The actual process of assigning numbers to objects to represent the values of their attributes is explored in Chapter 3.

The representational aspect of measurement is based on the principle of constructing a numerical model of the empirical system under study. The approach began its formal development around the end of the nineteenth century and the beginning of the twentieth century with the work of Helmholtz (1887), Hölder (1901), and Russell (1903), although the ideas can be traced back to Euclid and beyond. Helmholtz gave conditions of order and combination which objects must satisfy for these relationships to be represented by order and addition of positive real numbers, and also showed that such representations must be related by multiplication by a positive factor, and Hölder developed the ideas further.

The physicist Norman Campbell (1920) described this approach in an influential book. In the summary of his Chapter 10, he wrote:

> measurement is the assignment of numerals to represent properties. Why is the process important and why is it applicable to some properties (e.g. weight) and not to others (e.g. colour)?
>
> The answer must lie in some relation between numerals and measurable properties which does not apply to non-measurable properties. This relation is found in the common possession of order. The conception of order is analysed, as well as the relation between numerals and numbers. All measurable properties are capable of being placed in a natural order by means of definite physical laws which are true of them.
>
> But the possession of order alone will not enable a property to be measured, except possibly by the use of previously established systems of measurement for other properties. In order that a property should be measured as a fundamental magnitude, involving the measurement of no other property, it is necessary that a physical process of addition should be found for it. By a physical process of addition is meant an operation which is similar in a certain manner to the mathematical operation of addition.

Thus, to Campbell, 'fundamental' measurement solely related to those situations in which empirical relationships of order and 'physical addition' could be represented by numerical relationships of order and addition. Campbell also remarked that 'the difference between properties that are and those that are not capable of satisfactory addition is roughly that between quantities and qualities', a distinction similar to one made by Aristotle.

Campbell's definition of 'fundamental' measurement requires an isomorphism between a physical way of combining objects (or 'concatenating' them, as we nowadays say) and addition of numbers. Length and weight are classic examples. Fundamental measurements could be combined to yield derived or indirect measurements, with density, defined in terms of mass and volume, being an example. All of this hinges on the ability to model certain empirical relationships by addition. Without that, by Campbell's definition, measurement is not possible. In particular, measurements of sensory events such as 'loudness' are not possible. This was the conclusion reached by a committee of the British Association for the Advancement of Science, set up in 1932 to report on the feasibility of such measurement, which included Campbell amongst its members (Ferguson *et al.*, 1938, 1940). One member wrote 'I submit that any law purporting to express a quantitative relation between sensation intensity and stimulus intensity is not merely false but is in fact meaningless unless and until a meaning can be given to the concept of addition as applied to sensation', and Campbell asked 'Why do not psychologists accept the natural and obvious conclusion that subjective measurements of loudness in numerical terms (like those of length ...) are mutually inconsistent and cannot be the basis of measurement?' Additivity was regarded as fundamental.

Psychologists responded in various ways to this attack on the foundations of their discipline. Some ignored it, as far as the science of psychology went. For example, (Michell, 1999, p. 156): 'Bartlett [President of Section J: Psychology, of the British Association for the Advancement of Science] recommended that quantitative psychologists proceed as usual, describing their procedures as measurement, despite the Campbellian critique, in the confident hope that others outside the discipline would eventually describe them in the same way.' Others, for example Guilford and Comrey (1951), proposed adopting an extreme operational perspective, in which the 'numbers are used to represent the results of certain operations that have been performed' (p. 507) and 'the meanings given numbers assigned in any type of measurement are merely an expression of the operations performed'. Operationalism, which is closely related to the pragmatic perspective, is discussed in more detail below.

Undoubtedly the person to have most impact was S.S. Stevens. At about the same time as the British Association work was in progress, Stevens was involved in discussions at Harvard with a group which included such eminent thinkers as Birkhoff, Carnap, Feigl, Hempel and Bergmann. As with many subsequent contributors to the debate (see Section 2.5 below), Stevens was not afraid to state his opinions in unequivocal terms. Writing in 1951, he noted that Campbell, himself a distinguished physicist, had earlier written (Campbell, 1938): 'The most distinguished physicists, when they attempt logical analysis, are apt to gibber; and probably more nonsense is talked about measurement than about any other part of physics.' Stevens (1951) remarked that Campbell 'seems to contribute his fair share' to this nonsense.

Stevens wrote (1959, p. 23): 'What I gained from these discussions was a conviction that a more general theory of measurement was needed, and that the definition of measurement should not be limited to one restricted class of empirical operations.' That is, as well as or instead of representing the empirical concatenation operation between objects, one might represent some other empirical relationships by suitable mathematical relationships. Following this approach, Stevens proposed classifying measurement scales in terms of the relationships between the numerical assignments which could be adopted to model the empirical relationships; that is, in terms of the operations. Publication of this work was delayed by the War until 1946 (Stevens, 1946). Stevens's strategy was a clear attempt to respond to Campbell's criticisms. In Stevens (1951, p. 19), for example, he struck an analogy between the way in which complex extensions of the real numbers had enriched mathematics and the way in which the relaxation of what was recognized as a legitimate measurement procedure by Campbell's camp might enrich science.

2.2 Representational measurement

The *magnum opus* of representational measurement theory is the three-volume work *Foundations of Measurement* (Krantz *et al.*, 1971; Suppes *et al.*, 1989; Luce *et al.*, 1990; for a critical assessment, see Niederée, 1994). The approach adopted by this work, and by representational measurement theory in general, is made clear in the opening sentences of the preface:

Scattered about the literature of economics, mathematics, philosophy, physics, psychology, and statistics are axiom systems and theorems that are intended to explain why some attributes of objects, substances, and events can reasonably be represented numerically … Although such systems are of some mathematical interest, they warrant our attention primarily as empirical theories – as attempts to formulate properties that are observed to be true about certain qualitative attributes.

In representational measurement theory we begin with a set of *objects*, each of which possesses one or more common *attributes*, each in turn of which can be divided into mutually exclusive and exhaustive equivalence classes. The notion of 'equivalence' is formally defined below; for now, two objects are in the same equivalence class for a given attribute if, in terms of that attribute, they are indistinguishable. We should perhaps note parenthetically here that objects can possess *intrinsic* and *extrinsic* attributes. Intrinsic attributes are part of the definition of the object – the charge of the electron, for example. Extrinsic attributes are aspects of the interaction of the object with its environment – the velocity with which an electron is travelling, for example. Intrinsic attributes cannot be changed without changing the object being discussed (if I change the weight of a rock by cutting off a part, it is a different rock), but extrinsic attributes can (if I move the rock to a different place, it is still the same rock). Measurement scale issues apply to both types of attribute.

This section is concerned with single attributes (hence, *direct* measurement). Later sections discuss extracting attribute values from multiple measurements to which they are related (*indirect* measurement). Thus, in this section, each object can be uniquely allocated to a single equivalence class according to the 'value' of its attribute. For example, we can allocate books to equivalence classes according to how much they project above the top of the desk when laid flat upon it. Here the attribute in question is the books' width, and two books are in the same class if neither projects above the other. Or we can assign them to equivalence classes according to whether a given reader has no preference for them compared to other books. Here the attribute is the subjective preference of the reader.

There are relationships between the equivalence classes. In the above examples we can note that some classes contain books which project further from the desk than other books, and that some books belong to a class which is said to be preferred more than others by the reader in question. The objects and the relationships between them induced by the relationships between the equivalence classes for the attribute constitute an *empirical relational system* (ERS). Of course, since the ERS describes the real world, it is not perfect. There will be situations where we would like to assign object *a* to the same class as object *b*, and object *b* to the same class as object *c*, but not assign *a* to the same class as *c*, simply because our instruments and observations are not infinitely accurate or reliable. Moreover, our ERS will be based on just a sample of objects – one cannot put *all* books on the desk or ask the reader to read *all* the books in the world in order to assign them to classes. This implies that there is an inductive aspect to representational measurement. We shall return to problems such as these later, once we have established the basic framework for representational models.

Based on our (limited) empirical experience of the real world, we seek to formulate a model which adequately describes the relationships between the sort of objects with which we are concerned in terms of the attribute in which we are interested. This will be an idealized model. That is, it will make simplifications and approximations, but we hope that it captures the essence of the relationships we are interested in. If it does not do so then it must be modified, extended, or rejected and replaced by some other, in just the same way that any other descriptive model or scientific theory must. In keeping with other work on representational approaches to measurement, we shall call the idealized model a *numerical relational system* (NRS). This is because we are mostly concerned with numerical representations: we will be seeking a set of numbers such that the relationships between those numbers match the relationships between the objects in terms of the attribute in which we are interested. As we shall see, at times we will be interested in only some of the relationships between the numbers – perhaps their order, perhaps their relative size, and so on.

A parenthetical comment might be appropriate here. The representational aspect of measurement is concerned with a mapping from the ERS (the 'real world') to an NRS (the 'idealized representation'). Implicit in this perspective is that mathematics is a separate domain from the real world. However, not everyone holds that view:

> Numbers, [the mathematician] Hardy believed, constituted the true fabric of the universe. In an address to a group of physicists in 1922, he took the provocative position that it is the mathematician who is in 'much more direct contact with reality. This may seem a paradox, since it is the physicist who deals with the subject-matter usually described as "real".' But a chair or a star is not in the least like what it seems to be; the more we think of it, the fuzzier its outlines become in the haze of sensation which surrounds it; but '2' or '317' has nothing to do with sensation, and its properties stand out more clearly the more closely we scrutinise it ... 317 is a prime, not because we think so, or because our minds are shaped in one way rather than another, but *because it is so*, because mathematical reality is built that way.
>
> (Hoffman, 1998, p. 32)

The representational approach to measurement seeks to establish a mapping from the objects, via the equivalence classes to which they belong, to a number system in such a way that the relationships between objects are matched by relationships between numbers. These numbers form the values of a *variable*. In order to establish such a mapping we need to describe the properties which the empirical system must satisfy. That is, we need to present *axioms* such that, if the ERS satisfies them, then we can set up an NRS which models it. Our ultimate aim is that we can then manipulate the numbers in the NRS, drawing conclusions which will reflect corresponding relationships between objects in the ERS. In particular, of course, we are often concerned with statistical manipulations, and often we aim to make inferential statements about objects which are not included in the limited sample we have examined.

The problem of formulating sets of axioms which permit mapping to different NRSs is termed the *representation problem*. To motivate things, we begin with a simple informal but fundamental example. Consider a set of rigid rods. These have a length attribute, and we can allocate rods to equivalence classes according to whether or not, when laid side by side so that the left-hand ends coincide, the right-hand ends also coincide. All those with both ends coinciding constitute a single class, and these classes are related according to whether the rods in one class terminate to the right of the rods in the other class. When this is satisfied we say that the rods in the first class 'are longer than' the rods in the second.

We can now establish a mapping from the rods, via their equivalence classes, to the positive real numbers such that longer rods are associated with larger numbers. That is, letting $M(a)$ be the number corresponding to rod a, we assign numbers such that $M(a) > M(b)$ if and only if $a \supset b$, where $\supset$ represents 'is longer than', in the sense that rod a projects beyond rod b when the left-hand ends coincide. $M(a)$ is the value of the *variable* 'length' for the rod a.

By this means we establish an isomorphism between the equivalence classes and the positive real numbers, and a homomorphism between the rods and the positive real numbers.

It is a homomorphism because the equivalence classes contain more than one rod. In mathematical terms, we have established a homomorphism from the ERS denoted by $[A, \supset]$, where A represents the set of rods, to the NRS denoted by $[Re^+, >]$, where Re^+ represents the set of positive real numbers. A very important feature of this procedure is that, in general, the homomorphism will not be unique – there will be more than one mapping in which the relationships between the numbers reflect the relationships between the objects. This is an important property, and one which has led to much confusion. We discuss it in detail below.

Now, given a set of numbers which have been assigned to the objects in such a way that they preserve the 'is longer than' relationship, we can carry out statistical operations on those numbers, using the $>$ relationship, and any conclusions we reach will have empirical counterparts. We could, for example, compare the medians of the lengths of two groups of rods.

As it happens, we can go further with this example. The attribute length possessed by rigid rods has other internal empirical relationships. In particular, if our empirical concatenation of two rods involves placing them *end to end* in a straight line then we can (in principle, at least) find a third rod which, if placed next to this concatenation, has left- and right-hand ends aligned with those of the concatenated pair. That is, the concatenation of the two (shorter) rods lies in the same equivalence class as the third one. We thus have a three-component relationship between the rods, in addition to the two-component relationship above. Such three-component relationships are often written in *operation* form as $a \circ b = c$, symbolizing that $a \circ b$, the concatenation of a and b, has aligned ends with the single rod c.

Can we find an NRS with relationships which reflect both the size order relationship of the rods and this concatenation relationship? We can, of course. In particular, we can represent the relationship $\circ$ by $+$. That is, we can assign numbers to the rods to represent their lengths such that $M(a \circ b) = M(a) + M(b)$. Put mathematically, we are saying that we can find a homomorphism M from the ERS $[A, \supset, \circ]$ to the NRS $[Re^+, >, +]$. Now we can undertake statistical operations which include addition of the numbers – such as comparing the means of two groups or rods, as well as simply comparing the lengths. Again such addition operations will have empirical counterparts. Note that the introduction of the relationship $\circ$, in addition to the relationship $>$, means that the set of mappings which preserve the relationships between objects is more restricted – there are fewer mappings which preserve both the ordinality and concatenation relationships than there are which preserve merely the ordinality relationship.

The rods example is a fundamental one: length and weight measurement were used by Campbell (1920) to illustrate what he called *fundamental* measurement. They are examples of what are called *extensive* or *additive* measurements since the concatenation operation (e.g. placing two weights together on the same pan of a weighing balance) can be directly represented by addition.

It perhaps goes without saying that one must be consistent in one's choice and application of scales. I doubt that it would surprise anyone if they obtained contradictory results by using a mixture of numbers obtained by measuring in inches and centimetres. Here is a nice puzzle which slips just this sort of confusion into a problem. It is called the *two-envelopes puzzle*.

You are shown two sealed envelopes, each containing money, but with one containing twice the amount of the other. You are told you can have the money in whichever envelope you choose. You choose one (at random) but then, without looking in the first one, are given the choice of having the other one instead. Should you switch?

You might argue as follows. Call the amount in the envelope you chose x. Then, since you chose the first envelope at random, with probability 1/2 the second envelope will contain $2x$ and with probability 1/2 it will contain $x/2$. The expected amount contained in the second envelope is thus

$$\left[(2x)\frac{1}{2} + (x/2)\frac{1}{2} \right] = \frac{5}{4}x$$

This means that, on average, the second envelope will contain more than the one you first chose. So you decide to switch. However, you can now apply exactly the same argument to the second envelope, suggesting that the first will contain 5/4 times as much as the second. And so on.

This seems absurd, and a contrary argument might be that the first envelope contains (say) z units with probability 1/2 and $2z$ units with probability 1/2, so that the expected amount in the first envelope is $3z/2$. Following this argument, when the first envelope contains z the second envelope contains $2z$, and when the first envelope contains $2z$ the second contains z, so that the second envelope has probability 1/2 each of containing $2z$ and z units, yielding an expected value of $3z/2$, which is equal to the expected amount in the first envelope.

The result of the second argument certainly seems more reasonable than the infinite switching of the first argument, but what is wrong with the first argument?

Suppose the envelopes contain amounts z and $2z$. By calling the amount in the first envelope x, you are defining a measurement scale. (For example, we could define the amount in the first envelope to be 1 unit, whether it contained z or $2z$.) It is then true that, with probability 1/2 the second envelope will contain $2x$ and with probability 1/2 it will contain $x/2$, but the measurement scales in these two situations are different. That is, the basic units are different in the two cases. When the second envelope contains $2x$, the first envelope contains $x = z$. When the second envelope contains $x/2$, the first envelope contains $x = 2z$, so that x means different things in the two cases. In averaging $2x$ and $x/2$ one is averaging values from two different measurement scales. The correct result is obtained by using a consistent scale.

The consistent scale can be either that obtained by setting $x = z$ or that obtained by setting $x = 2z$. If the first envelope contains z then, by calling the amount in the first envelope x, the scale being used is the $z = x$ one. Now the second envelope contains $2z$. However, in terms of this $z = x$ scale, if the first envelope contains $2z$, then it contains $2x$. It follows that, in terms of this scale, the second envelope contains $x = z$. The average of these two possible values for the amount contained in the second envelope, *both measured on the $z = x$ scale*, is $(2x)/2 + x/2 = 3x/2$, which is equal to $(2z)/2 + z/2 = 3z/2$.

On the other hand, if the first envelope contains $2z$, then, by calling this amount x, the scale being used is the $2z = x$ one. Now the second envelope contains $z = x/2$. However, in terms of this $2z = x$ scale, if the first envelope contains z, then in terms of x it contains $x/2$. It follows that, in terms of this scale, the second envelope contains $2z = x$. The average of these two possible values for the amount contained in the second envelope, when both are measured on the $2z = x$ scale, is $(x/2)/2 + x/2 = 3x/4$, which is equal to $z/2 + z = 3z/2$ on the $2z = x$ scale.

That is, provided we adopt a consistent scale, *whichever it is*, we obtain a result of $3z/2$. In the first case, with x defined as $z = x$, this is equal to $3x/2$. And in the second case, with x defined as $2z = x$, it is equal to $3x/4$.

The initial argument, leading to the infinite regress, has used different measurement scales for the two situations: it is like measuring the content of one envelope in dollars and the other in euros, and averaging the result. It is fine to choose a unit of measurement by convention ('*this* length is taken as the unit length') in a pragmatic way, but that choice must then be applied consistently. One would not compare the weights of two groups of people by choosing one member of the first group at random to provide the unit for that group and choosing one member of the second group at random to provide a unit for the second group.

2.2.1 Axiom systems

Our aim is to represent objects and the relationships between them by numbers and certain relationships between these numbers. The relationships between the objects, of course, are in terms of a particular attribute (or attributes) of interest. This aim is expressed more formally by specifying the properties that the relationships between the objects must satisfy in order to permit representation by relationships between numbers. Such a list of properties is termed an *axiom system*, and it is clearly desirable that this should be as small and compact as possible, with little or no redundancy. *Representation theorems* are concerned with specifying the axioms which empirical systems must satisfy in order to permit representation in terms of a given numerical system. In effect, the axioms are hypothesized natural laws relating the real objects. Of course, they are idealized in the sense that any proposed natural laws are idealized. For example, one common axiom in length measurement states that whenever objects a and b satisfy $a \supset b$ then there exists an object c such that a is equivalent to $b \circ c$ (meaning that objects b and c can be combined in some way to produce a third object $b \circ c$). For any given set of rods (even an infinite set), however, this may not be the case. In any case, irreducible error in observation (perhaps things evolve over time and there is a practical limit to the size of the interval between two observations, perhaps there is uncontrollable random variation, even quantum fluctuations, perhaps our visual or instrumental acuity is insufficient to discern a difference between two objects) means that the statements about the empirical objects may not be perfect.

The earliest formal axiomatization seems to have been that of extensive structures, illustrated by the rods example above, in which the objects were required to satisfy order and 'concatenation' relationships. These permitted representation by the real numbers, with the concatenation operation being mapped to addition. However, these are not the only conditions which permit numerical assignments in which order and addition correspond to empirical relationships. Another important set of conditions leads to *conjoint measurement*, outlined below. Conjoint measurement has been promoted as of potential fundamental importance in psychology, because it derives additive quantifications from purely ordinal relationships, but it appears not to have had the impact some people expected, and its main use has been in marketing (see, for example, the applications described in Gustafsson *et al.*, 2001). Despite that, it is of fundamental importance, if only because it does provide a convincing illustration of a measurement axiomatization distinct from the classical approach to extensive measurement via concatenation.

Since representational measurement theory is all about representing the relationships between objects, we need to describe what we mean by such relationships. *Binary relationships*, relationships involving just two objects are the simplest. Formally, a binary relationship R on a set A is a subset of the Cartesian product of A with itself. That is, it is a subset of the set of ordered pairs of elements taken from A. We write aRb if two objects a

and b satisfy this relationship. There are various important properties that binary relationships may possess. They include, for all a and b in A:

Reflexivity: aRa
Symmetry: aRb implies bRa
Transitivity: aRb and bRc implies aRc
Antisymmetry: aRb and bRa implies $a = b$
Asymmetry: aRb implies a and b do *not* satisfy bRa
Negative transitivity: The relation 'not in relation R' is transitive
Completeness: for all $a \neq b$ in A, aRb or bRa
Strong completeness: for all a, b in A, aRb or bRa

Roberts (1979, p. 16) gives a nice example demonstrating that these properties depend on the set A as well as R. 'If A is the set of all people in the United States and B is the set of all males in the United States, then the relation

$$R = \{(a,b) \in A \times A : a \text{ is the brother of } b\}$$

is different from the relation

$$R' = \{(a,b) \in B \times B : a \text{ is the brother of } b\}$$

For example, R' has certain symmetry properties; that is, if $aR'b$, then $bR'a$.'

If a relationship is reflexive, symmetric and transitive, it is said to be an *equivalence* relationship. The set of objects which are related to an object a by an equivalence relationship is called the *equivalence class* containing a. It is not difficult to prove that equivalence classes are either disjoint or identical and that they partition the set A.

Order relationships are binary – we saw an example of this in the rod length example above. It is often convenient to write binary relationships in the form aRb, but one can also write them as $R(a,b)$, a form which permits more ready generalization. In the rod length example we also saw an example of a ternary relationship – a relationship involving three objects. We wrote it in operational terms as $a \circ b = c$, but we could have expressed it as $R(a,b,c)$. Higher-order relationships are also possible: for example, we might be interested in comparing the difference, in some sense, between the elements of a pair of pairs of objects – a relationship involving four objects.

Although much of our discussion will be focused on just the two relationships of order and additivity, there is no reason why further empirical relationships should not also be represented. In general, for a set of objects A, if the attribute possesses relationships $R_1, R_2, \ldots, R_n$, then we can seek to establish a homomorphism from the ERS $[A, R_1, R_2, \ldots, R_n]$ to an NRS $[T, r_1, r_2, \ldots, r_n]$, where the r_i are relationships between numbers, and T is a subset of numbers. Different relationships R_i will be represented by different r_i.

2.2.2 Order relationships

The concept of order is fundamental. Without such a concept one has only that of *different* classes of things – and not everyone accepts that so simple a structure can really be dignified with the term 'measurement'. Indeed, not everyone accepts that merely order is sufficient to permit the term measurement to be used – see Section 2.5 below. Some disciplines (such as the behavioural sciences) can make ordinal comparisons (fairly) readily but have more difficulty with other relationships, so that many of their measurement methods and statistical operations have to be built up from such comparisons. One can do this by assuming that the order relationship is the manifestation of some underlying quantitative relationship, and try to find suitable sets of numbers – using conjoint analysis or methods such as multidimensional scaling and optimal scaling (described in Chapter 3). Alternatively, one can simply restrict oneself to using only the property of order in one's modelling and analysis – as in 'distribution-free' statistical methods, of which more later.

In principle, the concept of order is simple enough. If one can place the objects on a line, from left to right, according to the magnitude of the attribute in question, so that object a is to the left of object b if and only if aRb, then the objects have an order relationship. In fact, there are several different kinds of order relationship, according to precisely what properties the relationship satisfies. These can be used to provide different kinds of ordinal measurement structures. For example:

Weak orders, in which the relationship is transitive and strongly complete.
Strict simple orders, in which the relationship is transitive, asymmetric and complete.
Strict weak orders, in which the relationship is asymmetric and negatively transitive.
Partial orders, in which the relationship is reflexive, transitive and antisymmetric.

Of central interest to us are the conditions which an empirical system must satisfy so that the order relationships between the objects is representable by the order relationship between numbers. So, let (A, R) be a binary relation. Then a starting point is the Birkhoff–Milgram theorem, which gives (necessary and sufficient) conditions under which the objects can be mapped to a numerical order representation which preserves the empirical relationship: $aRb \Leftrightarrow f(a) > f(b)$, for $a, b \in A$. Essentially, there are two conditions. The first condition is that the objects satisfy a suitable order relationship (e.g. lie in a strict weak order). The second is a condition of 'denseness' ('order-denseness'), requiring the existence of objects c in A which satisfy aRc and cRb whenever aRb. Details are given in, for example, Roberts (1979, p. 112). A similar result replaces $>$ by $\geq$ and 'strict weak order' by 'weak order'.

If we find that the objects of the empirical system do satisfy the conditions to be a weak order, then we can find a suitable set of numbers so that the order relationships between the objects are reflected by those between the numbers. But what if we find that the objects do not satisfy the axioms? This can happen for two basic reasons.

Firstly, errors of some kind may have corrupted the results. Perhaps the instrument used to order the objects is highly inaccurate (a rubber ruler). Perhaps subjectivity has crept in (as it might in preference scores) so that the result depends critically on the conditions under which ordering was done (e.g. whether the researcher slept well the night before). Secondly, and more fundamentally, perhaps the objects are not organized in a strict weak order. For example, some relationships only apply to certain pairs of objects in a set – they form a *partial order*. This often happens when objects are compared on the basis of multiple criteria. In such circumstances an alternative measurement representation must be found.

These possibilities forcibly bring home the essential unity between measurement and statistical modelling (or, indeed, scientific modelling in general). At some point we have to decide whether the departures from our chosen measurement model are sufficiently radical that the model has to be rejected, or if they can be accommodated within it by appeal to unimportant imperfections of the process of matching the mathematical model to the empirical system.

When A is finite in the definition above there is a simple practical way of choosing a set of numbers which will represent the relationship, if one exists: for an object a in A, we can simply define the corresponding number $f(a)$ as the number of elements b of A which satisfy aRb. More generally, if A is not finite (but is still countable), we can choose a set of numbers which will represent the order relationship in the following way. Assign any number $M(a)$ for the first object a. If the second object b is such that aRb then assign it a number $M(b)$ less than $M(a)$. On the other hand, if bRa, choose $M(b)$ to be greater than $M(a)$. For subsequent objects, c, if cRi for all objects i so far examined, then choose $M(c)$ to be greater than all numbers so far assigned. Likewise, if iRc for all objects i so far examined, choose $M(c)$ to be less than all numbers so far assigned. Otherwise choose $M(c)$ to be less than $M(i)$ for all objects i for which iRc but greater than $M(i)$ for all objects i for which cRi. Other such representation theorems also exist.

2.2.3 Additive relationships

The example of the rigid rods in the introduction to Section 2.2 described mathematical representations for two empirical relationships. One was order, and this has been examined in Section 2.2.2. The second was the physical concatenation of two rods: placing them end to end in a straight line. We now extend Section 2.2.2 to consider what sort of axioms an empirical system $(A, R, \circ)$ must satisfy in order that it may be mapped to a numerical system in which R is represented by $>$ and $\circ$ is represented by $+$. In particular, we want to know what are the conditions that $(A, R, \circ)$ must satisfy so that there exists a real-valued function f such that (i) a is related to b by R if and only if $f(a) > f(b)$, and (ii) $f(a \circ b) = f(a) + f(b)$; that is, we seek conditions such that there exists a function f from $(A, R, \circ)$ to $(Re, >, +)$, where Re are the real numbers. As we have already noted, attributes which do have additive representations in this manner are called *extensive* attributes.

The earliest axiom system sufficient for extensive measurement appears to be that of Hölder (1901), a refined version of the assumptions given by Helmholtz (1887). (For discussion and generalization of this system see, for example, Suppes (1951), Pfanzagl (1959), Suppes and Zinnes (1963), Luce and Marley (1969), Narens and Luce (1986), Roberts and Luce (1968), Narens (1974), and Roberts (1979).) Hölder's axioms are, however, not necessary. The following system (see Roberts, 1979, for proofs) is both necessary and sufficient:

(A1)   For all a, b, and c in A it is not true that $[a \circ (b \circ c)]\ R\ [(a \circ b) \circ c]$ and not true that $[(a \circ b) \circ c]\ R\ [a \circ (b \circ c)]$. Here $[a \circ (b \circ c)]$ represents the result of concatenating a with the concatenation of b with c. This condition thus requires that the two orders of concatenation, $[a \circ (b \circ c)]$ and $[(a \circ b) \circ c]$, are equivalent in the sense that neither has relationship R to the other. This condition is termed *weak associativity*. (Associativity will be a familiar property, since addition and multiplication are associative. However, not all operations are. In particular, averaging is not associative.)

(A2) (A,R) is a *strict weak order*, as defined in the previous section.

(A3)   For all a, b, and c in A,

$$aRb \Leftrightarrow (a \circ c)\ R\ (b \circ c) \Leftrightarrow (c \circ a)\ R\ (c \circ b)$$

These relationships tell us that the order between two objects a and b is preserved when each of these objects is concatenated with the same third object. This is a *monotonicity* axiom.

(A4)   For all a, b, c, and d in A, if aRb, then there is a positive integer n such that $(na \circ c)\ R$ $(nb \circ d)$. This is what is known as an *Archimedean axiom*, and tells us essentially that whatever the relationship between c and d, and whatever the relationship between a and b, there exists some number n such that the concatenation of that number of copies of a with c has relationship R to the concatenation of that number of copies of b with d.

This system gives the flavour of what is required to construct axiom systems for even so basic a structure as an extensive structure.

Of course, empirical data are always measured to finite accuracy, so one might think that a mapping to the real number system is not necessary. A mapping to a restricted set of numbers, using only a finite number of decimal places, might be sufficient. While this is true, mapping to the reals allows us to represent the magnitudes of attributes such as the circumference of a circle of unit radius or the length of the diagonal of a unit square (recall the problem Pythagoras encountered, mentioned in Chapter 1). In essence, we are postulating the existence of a true value for the length of the diagonal, even though we cannot measure it exactly. Our numerical representation is describing an underlying ideal reality. The alternative to mapping to the real numbers is to construct a representational system which properly represents the uncertainty inherent in practical measurement procedures. This is a perfectly legitimate alternative, and one which some believe to be more natural, though such approaches have not been as widely adopted or developed as mappings to the reals.

Not all concatenation operations can be mapped to addition. *Intensive* structures satisfy the properties that $a \circ a \sim a$ (i.e. the concatenation of two copies of object a is equivalent to a in terms of the attribute in question) and that if $a \supset b$ then $a \supset (a \circ b) \supset b$. Density is an example of an intensive structure, as is the set of real numbers along with the concatenation operation defined by weighted averaging when the (positive) weights sum to 1.

2.2.4 Conjoint measurement

The previous subsections illustrated representational measurement models for a single attribute, showing how an ordered attribute could be represented by an ordinal variable, and how concatenation relationships could be mapped to addition. However, the fundamental aim of science is typically to relate several variables to each other, rather than simply studying the properties of isolated variables. Conjoint analysis is directly concerned with relating multiple variables together. Its importance resides in the fact that it derives additive scales from merely order relationships. A primary stimulus behind its development was the fact that additive relationships in the behavioural sciences are few and far between, whereas these sciences are awash with order relationships.

We begin with a set of objects which possess several attributes and which can be ordered, for example, in terms of preference, ability, skill, or whatever. We will then seek numerical representations of the constituent attributes, and ways of combining these numerical representations, so that the order of the objects is preserved.

In fact, such situations are common. For example:

- in choosing a university, a student may be able to state a preference order for universities, while the universities might be describable in terms of location, teaching quality, and the research reputation of the faculty;
- in economics, we might be able to order baskets of commodities (see Section 8.4) according to desirability, and, of course, describe these baskets in terms of the quantity of each constituent item in the basket;

- in food design, testers may rate each of several alternative foods in terms of preference, and each alternative may be described in terms of a vector of characteristics (texture, sweetness, chewiness, brittleness, etc.);
- in marketing, one may be able to rank different brands of a product, with each product being describable in terms of its various characteristics.
- in psychophysiology, we may present stimuli which differ in intensity, frequency, and duration and ask respondents to rate them according to degree of discomfort.

Our aim will be to devise numerical scales for the constituent components of the objects, derived from the overall order relationship between objects. Call the constituent components $A_1, A_2, \ldots, A_n$. Then A is equivalent to the Cartesian product $A_1 \times A_2 \times \cdots \times A_n$, and is thus a *product structure*. Note that the A_i can be categorical or numeric, as in the examples above.

Given a product structure, our aim is to use the stated preferences between the tuples constituting A to deduce numerical scales of the constituent components such that the relationships between the elements of A are preserved. Algebraically, we have a function f which preserves the order of the elements of A, $f: A \to Re$, and we seek real-valued functions $f_1, \ldots, f_n$ such that

$$f(a_1, a_2, \ldots, a_n) = F(f_1(a_1), f_2(a_2), \ldots, f_n(a_n)) \tag{2.1}$$

where F is a real-valued function. Since this yields a system in which the different components are simultaneously measured, it is termed *conjoint measurement*. The most important special case of this is *additive conjoint measurement*, in which F is a sum of its components:

$$f(a_1, a_2, \ldots, a_n) = f_1(a_1) + f_2(a_2) + \cdots + f_n(a_n) \tag{2.2}$$

Sufficient conditions for the existence of functions f_i seem first to have been described by Debreu (1960) and Luce and Tukey (1964) (and see also Krantz *et al.*, 1971). The conditions can be written in various ways (and have various modifications). For convenience, we describe the case of two constituent components: $P \times Q \equiv A$. Then Luce and Tukey (1964) require:

(1) The binary relation (A, R) is a weak ordering on A.
(2) For all $p_1 \in P$ and $q_1, q_2 \in Q$, there exists a $p_2 \in P$ such that $(p_1, q_1) E (p_2, q_2)$ (i.e. both $(p_1, q_1) R (p_2, q_2)$ and $(p_2, p_2) R (p_1, q_1)$. Likewise, for all $p_1, p_2 \in P$ and $q_1 \in Q$, there exists a $q_2 \in Q$ such that $(p_1, q_1) E (p_2, q_2)$. This is called the *solvability* axiom.
(3) For all $p_1, p_2, p_3 \in P$ and $q_1, q_2, q_3 \in Q$, if $(p_2, q_1) R(p_1, q_2)$ and $(p_3, q_2) R(p_2, q_3)$ then $(p_3, q_1) R (p_1, q_3)$. This axiom is illustrated in Figure 2.1, where an arrow points from a to b if aRb. Thus, in the figure, the top left bold arrow points from (p_2, q_1) to (p_1, q_2). The two continuous arrows correspond to the antecedents of the axiom, and the broken arrow to its consequent. This axiom is called the *double cancellation* axiom. A more intuitive feel for it can be gained by expressing the pairs in additive terms, so that the conditions are $p_2 + q_1 \geqslant p_1 + q_2$ and $p_3 + q_2 \geqslant p_2 + q_3$. Adding these gives $p_2 + q_1 + p_3 + q_2 \geqslant p_1 + q_2 + p_2 + q_3$, so that the p_2 and q_2 terms cancel, leaving $p_3 + q_1 \geqslant p_1 + q_3$, which is the consequent above. Luce and Tukey (1964) describe the cancellation axiom as the substitute for the commutativity axiom in measurement theories based on a concatenation operation. Roberts (1979) describes a slightly different version of double cancellation, the *Thomsen condition*, which follows from the double cancellation axiom if (A, R) is a strict weak order.

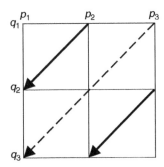

Figure 2.1 The relationship between pairs in the double cancellation axiom.

Axioms (1)–(3) imply that if $(p_1, q_1) R(p_2, q_1)$ for some $q_1 \in Q$, then $(p_1, q_2) R(p_2, q_2)$ for all $q_2 \in Q$, and also that if $(p_1, q_1) R(p_1, q_2)$ for some $p_1 \in P$ then $(p_2, q_1) R(p_2, q_2)$ for all $p_2 \in P$. If these conditions hold, the binary relation (A, R) is said to satisfy *independence*. Independence serves the crucial aim of permitting the disentanglement of the two component attributes P and Q of A. It follows from these results that we can induce weak orderings on the elements of P and on the elements of Q from the weak order on the elements of A. The order on the elements of P is obtained by defining $p_1 R_p p_2$ when $(p_1, q_1) R(p_2, q_1)$ for some $q_1 \in Q$ (and hence for all $q_1 \in Q$). Similarly, the order on the elements of Q is obtained by defining $q_1 R_q q_2$ when $(p_1, q_1) R(p_1, q_2)$ for some $p_1 \in P$ (and hence for all $p_1 \in P$). Algebraic details are given in Luce and Tukey (1964), or, slightly modified, in Krantz *et al.* (1971) and Roberts (1979).

Measurement theories based on a concatenation operation define a basic unit, and measure other magnitudes in terms of concatenated copies of that basic unit. The matching concept in conjoint measurement is given by the notion of a *standard sequence*. The sequence $\{p_i : p_i \in P, i \in N\}$ is a *standard sequence* on the first component attribute if there exist q_1 and q_2 in Q such that $\sim(q_1 E q_2)$ and, for all $i \in N$,

$$(p_i, q_1) E(p_{i+1}, q_2)$$

There is a corresponding definition for the second component attribute Q. This definition essentially matches the 'difference' between q_1 and q_2 against a sequence of 'differences' $\{p_i, p_{i+1}\}$, so that all the pairs of this sequence are, in some sense, equal, since they all 'equal' the q_1, q_2 difference. The fourth axiom then asserts that if we add repeated copies of a q_1/q_2 difference, we eventually exceed any given value. A strictly bounded sequence, for the first component, is one for which there exist p_1^* and p_2^* such that $p_1^* R_p p_i$ and $p_{i+1} R_p p_2^*$, for all $i \in N$, with a similar definition for the second component. Put another way:

(4) Any strictly bounded standard sequence (on either component) is finite.

An *additive conjoint structure* is then one which satisfies axioms (1)–(4), and Luce and Tukey (1964) show that if (A, R) is an additive conjoint structure, with $A = P \times Q$, then there exist functions $f_p : P \to Re$ and $f_q : Q \to Re$ such that, for all $(p_1, q_1) \in A$ and $(p_2, q_2) \in A$,

$$(p_1, q_1) R (p_2, q_2) \Leftrightarrow f_p(p_1) + f_q(q_1) \geq f_p(p_2) + f_q(q_2).$$

Moreover, if f_p^* and f_q^* are any two other functions on P and Q respectively, which also satisfy these conditions, then the functions are related by

$$f_p^* = \alpha f_p + \beta \quad \text{and} \quad f_q^* = \alpha f_q + \gamma$$

where α, β, and γ are real numbers and $\alpha > 0$. That is, the mappings f_p and f_q yield interval scale measurements of the component attributes P and Q. (An 'interval scale' is one in which relationships are preserved if the values are transformed by a linear rescaling: $x \mapsto ax + b$. This is discussed in more detail below.)

A system can be examined to see if the axioms required for additive conjoint measurement hold, though if the system is infinite this will require inference from a finite sample of objects. Alternatively, one can sidestep this and simply seek those additive numerical assignments which yield the best fit to the data. If the component attributes each take only a finite number of categories, this is simply a question of search over possible numerical assignments, along with isotonic regression to preserve the order of the categories of the response A. Given that real data may be expected to have error, one would not expect exact conformance to the conjoint axioms, so this is not an unreasonable approach. This is the sort of strategy which underlies conjoint analysis as used in marketing.

So far we have concentrated on *additive* conjoint measurement, where the component attributes are mapped to variables which form interval scales and which combine additively to give the overall value of the joint attribute. This is perhaps the most important type, but more general conjoint structures are also possible, taking the form of equation (2.1) above:

$$f(a_1, a_2, \ldots, a_n) = F(f_1(a_1), f_2(a_2), \ldots, f_n(a_n))$$

In general, when such functions exist, the relationship (A,R) is said to be *decomposable*.

An obvious alternative to addition is when F takes the form of a product:

$$F(x_1, \ldots, x_n) = \prod_{i=1}^{n} x_i$$

but other more complicated alternatives, involving a mixture of sums and products may sometimes be appropriate. For example, with three variables the distributive rule

$$F(x_1, x_2, x_3) = (x_1 + x_2)x_3$$

has been suggested by Coombs and Huang (1970) in the context of perceived risk, where x_1 is a quantification of expected regret, x_2 is a quantification of expected value, and x_3 is a quantification of the number of plays in a gamble. An example of an even more complicated conjoint relationship arises in relativistic combination of velocities. If x_1 is the velocity of object l relative to object m, and x_2 is the velocity of object n relative to object m, then special relativity tells us that the velocity of object n relative to object l is

$$F(x_1, x_2) = \frac{x_1 + x_2}{1 + x_1 x_2 / c^2}$$

Conjoint measurement systems in which F is a polynomial

$$F(x_1, \ldots, x_n) = \sum_k \alpha_k x_1^{\beta_{1k}} x_2^{\beta_{2k}} \ldots x_n^{\beta_{nk}}$$

are, naturally enough, called polynomial conjoint measurement systems.

Sometimes single numerical representations for the component attributes are not adequate and more complicated alternatives have to be explored. For example, one might want to supplement the simple additive model

$$(a_1, a_2) \to F(f_1(x_1), f_2(x_2)) = f_1(x_1) + f_2(x_2)$$

with an interaction term, not definable in terms of f_1 and f_2. Thus

$$(a_1, a_2) \rightarrow F(f_1(x_1), h_1(x_1), f_2(x_2), h_2(x_2)) = f_1(x_1) + f_2(x_2) + h_1(x_1)h_2(x_2).$$

2.2.5 Uniqueness of representational measurement mappings

In earlier sections we described axioms which the objects must satisfy to permit an order relationship to be representable by numerical order, a concatenation relationship to be representable by numerical addition, and complex objects to be representable by additive relationships on numerical scales of their attributes. That is, we have described behaviour which the empirical system must show in order for homomorphisms to the desired mathematical structures to exist.

Now, in general the homomorphisms from the given ERS to a particular NRS will not be unique. That is, there will typically be more than one set of numbers which are related in the same way as the empirical objects. To put it another way, a given empirical relationship R may be accurately modelled by a particular numerical relationship r using more than one numerical assignment. For example, given an assignment of numbers to the attribute length of the rods such that order and concatenation of rods map to order and addition of numbers, then an arbitrary rescaling of the numbers (changing inches to centimetres, for example) will also produce an acceptable assignment. The ordering and the end-to-end concatenation operation ∘ will be properly represented by $>$ and $+$, respectively, in both numerical assignments. The relationship between the objects in terms of the chosen attribute is preserved by both of these mappings. On the other hand, there are restrictions on the set of numbers we may use – the requirement that the numerical system be homomorphic to the empirical system enforces this. We could not replace the numbers of inches used to represent the lengths of the rods by just *any* other numbers without sacrificing the way in which the relationships between the numbers modelled the relationships between the objects. This leads us naturally to ask what numerical representations are legitimate. Note that the constraints here arise from representational considerations – further constraints, to be discussed below, may arise from pragmatic considerations. Focusing on the representational constraints, and what numerical representations preserve the empirical relationships, we can turn the question around and ask what is the relationship between legitimate numerical representations. Put another way, what are the *uniqueness* properties of the representing numerical systems?

It is clear that this question is closely related to the possibility of transforming the data. If the numbers have been assigned so that the relationship between them reflects some relationship between the objects, then we need to ensure that any transformation does not sacrifice that relationship. In the length example, different legitimate representations are related by *rescaling* transformations, but what sort of transformations are generally appropriate? Transformations satisfying the condition that they preserve the relationships are termed *admissible* or *permissible* transformations. More formally, a transformation θ mapping the NRS into itself is admissible if, for every homomorphism ϕ from the ERS to the NRS, the function composition $\theta\phi$ is also a homomorphism from A to T (with T being the appropriate subset of the number system).

In general, the notion that models must be invariant to changes which do not represent aspects of the mapping from the reality to the model is an important and powerful one. When a child plays with a toy car, the fact that it is only a fraction of the size of the real thing can be ignored, and the value of the toy is not affected by whether the fraction is 1/20 or 1/30. A statistical model of lactating cows will be just as valid, whatever the colour of the cows. The point of scientific modelling is to map from the complexity of the real world to an artificial construct which behaves in the same way as the real world *as far as the attributes under study are concerned*, and is not affected by other attributes the objects may possess. Indeed, an important component of scientific creativity is the ability to decide what attributes are important to how a system functions. The invariance of measurement models to aspects of the numerical assignment which are not part of the homomorphism from the ERS is what lies at the heart of dimensional analysis in physics (Chapter 7). It also leads us to the idea of *types* of scale.

The admissible transformations enable us to move from one legitimate numerical representation of an ERS to another. *Scale types* are defined in terms of these admissible transformations. The rescaling transformations common in physics, and illustrated for the rods example above, are one type of transformation (leading to so-called *ratio scales*). In general, however, different ERSs will be represented by different homomorphisms, which will be related by classes of transformations other than simple rescaling, so that they will form another type of scale. Note that, since there is a one-to-one correspondence between the set of homomorphisms of an ERS into an NRS and the set of mappings of the NRS to itself which preserve its structure, one can talk in terms of either mappings between the homomorphisms or mappings between alternative legitimate numerical representations. The two approaches will be equivalent. The latter approach – in terms of mappings between alternative numerical representations – seems to be the most common. It was the one adopted by Stevens (1946, 1951), who presented one of the earliest statements of scale types.

Stevens (1951, p. 23) wrote:

> Scales are possible in the first place only because there exists an isomorphism between the properties of the numeral series and the empirical operations that we can perform with the aspects of objects. This isomorphism is, of course, only partial. Not *all* the properties of number and not *all* the properties of objects can be paired off in a systematic correspondence. But *some* properties of objects can be related by semantical rules to *some* properties of the numeral series.

He went on to say that 'the type of scale achieved when we deputize the numerals to serve as representatives for a state of affairs in nature depends upon the character of the basic empirical operations performed on nature'. This led him to define the familiar four types of scale: nominal, ordinal, interval, and ratio. He characterized these 'by the kinds of transformations that leave the "structure" of the scale undistorted'. Given some numerical assignment which properly reflects the relationship between the objects in terms of the attribute under consideration, Stevens defined the scale as nominal if any one-to-one onto transformation of this assignment also leads to a legitimate numerical assignment. He defined it as ordinal if any monotonic (order-preserving) transformation led to a legitimate assignment. It was interval if any linear transformation ($x \rightarrow ax + b$) led to another legitimate assignment. And it was ratio if any rescaling transformation ($x \rightarrow ax$, $a > 0$) led to another legitimate assignment. In fact (in Stevens, 1959), he also described log interval scales, those which are invariant under transformations of the form $x \rightarrow kx^n$.

Examples of a nominal scale are the use of numbers to identify personnel in a company, and the use of numbers to identify people for tax purposes. An example of an ordinal scale is the Mohs measure of hardness of minerals, in which a set of ten minerals are ranked according to whether one will scratch another, so that any set of numbers which preserves this rank order will reflect the relationship 'harder than'. A new material can be assigned to an appropriate point on the scale by seeing which of the ten minerals it can scratch, and which it is scratched by. Another example of an ordinal scale is the assignment of numbers to represent an ordering of scents according to one's personal preference. Examples of interval scales are everyday temperature measurements (e.g. in Fahrenheit or Celsius), where the zero point and the degree size is a conventional choice, and calendar dates. The length and weight examples we have used above provide examples of ratio scales, as does the Kelvin temperature scale discussed later.

As we shall see below, modern mathematical investigations have revealed some deep insights into the possible scale types and it turns out that Stevens's types are fundamental for representations using the real numbers. Despite not having the benefit of the modern mathematical machinery, Stevens essentially got it right. This explains the persistence of the above four types, and their continued widespread use in textbooks.

For representations in terms of the real numbers, the modern mathematical classification of scale types is also based on the fact that there is a one-to-one correspondence between the set of homomorphisms of an ERS into an NRS and the group of automorphisms (structure-preserving mappings of a set to itself) of the ERS. The automorphisms are then classified using the concepts of *degree of homogeneity* (k) and *degree of uniqueness* (l) (Narens, 1981b; Narens and Luce, 1986) of the ERS.

An ERS is said to be *k-point homogeneous* if any set of k ordered elements of the ERS can be mapped by some automorphism to any other set of k ordered elements. If a structure is k-point homogeneous for all k then it is said to be ∞-point homogeneous. one-point and two-point homogeneity are the most important kinds. The simplest case is one-point homogeneity (sometimes abbreviated simply to 'homogeneity'), meaning that for any given pair of objects a and b in the ERS there is a mapping which preserves the relationships between objects in the ERS and which takes a to b. The word 'homogeneous' is used here because one-point homogeneity means that no element is special: if we single out any element for special treatment then, by a suitable automorphism, we could equally single out any other. This means that structures which do have special elements cannot be k-point homogeneous for $k \geqslant 1$. Probability is an illustration – events which are impossible and events which are certain are different from other events, so that probability structures are not homogeneous. In contrast, length is a (one-point) homogeneous structure: we could pick any length (say 1 inch) to serve a particular role (e.g. to serve as 'unit' length), but we could instead pick any other length (e.g. 1 cm) to serve the same role, simply by rescaling (we replace the usages of 1 inch by 2.54 cm).

An ERS is said to be *l-point unique* if and only if, whenever any two automorphisms agree at l points, then they agree everywhere. If there is no l for which this is true then the structure is said to be ∞-point unique. Put another way, an ERS is l-point unique if any automorphism with l fixed points leaves the ERS unchanged (i.e. is the identity automorphism). Length is clearly one-point unique since fixing the length of some object as the unit length determines the numbers to be assigned to all other objects. Temperature, as used in everyday weather forecasting, however, is not a one-point unique structure: we need to know both the zero point and the unit of measurement before we can state the temperature of any object.

Note that for an ERS with an infinite number of elements which is k-point homogeneous and l-point unique, k must be less than or equal to l. Furthermore, k-point homogeneity implies k'-point homogeneity for any k' less than k, while l-point uniqueness implies l'-point uniqueness for any l' greater than l.

The twin concepts of homogeneity and uniqueness allow us to characterize the automorphisms of a structure. Suppose that K is the largest value of homogeneity satisfied by a structure and L is the smallest value of uniqueness, and represent the pair as (K, L). Then it is possible to show that ratio scales are of type $(1, 1)$, interval scales are of type $(2, 2)$, and ordinal scales are of type (∞, ∞). Various results have also been established, especially by Louis Narens and his co-workers, about the possible scale types that can arise when mapping to the real numbers (for details, see Cohen and Narens, 1979; Narens, 1981a, b; Luce and Narens, 1983, 1985, 1986; Alper, 1984, 1985, 1987). For example, if an ERS of type $(1, 1)$ has a mapping onto the real numbers then an NRS can be found which is a ratio scale, and if an ERS of type $(2, 2)$ has a mapping onto the real numbers then an NRS can be found which is an interval scale. Moreover, it is impossible for a scale of type (M, M) to have a mapping onto the real numbers for $2 < M < \infty$. If an ERS has type (K, L), with $0 < K < L$ and $1 \leqslant L < \infty$, then the only mappings onto the positive real numbers are of type $(1, 2)$, and a discrete interval scale exists.

These results help to explain why so few scale types are used in practice – and hence why Stevens was basically right.

The above results are perfectly general. If we restrict ourselves to ERSs which include weak order and concatenation operations and which are of type (K, L), with $K > 0$ and $L < \infty$, then only types $(1, 1)$, $(1, 2)$, and $(2, 2)$ can occur. If the structure has a representation onto the positive real numbers with a continuous numerical operation, then L is finite.

2.2.6 Alternative numerical relationships

Sections 2.2.2 and 2.2.3 presented conditions which the ERS must satisfy to permit order and concatenation to be represented by particular numerical relationships and operations, and Section 2.2.4 pointed out that one could do the same for other empirical relationships. We have also now seen, in Section 2.2.5, that, even when one can represent the empirical relationship R by a given numerical relationship r, there is typically more than one way in which the assignment of numbers can be achieved (these being related by the admissible transformations). This fact, that the homomorphism from the ERS to the chosen NRS will generally not be unique, should not be confused with the fact that R may also be representable by a different r. Indeed, if r_1 is one numerical relationship which represents R for a suitable assignment of numbers, then r_2, defined by

$$x r_2 y = g[g^{-1}(x) r_1 g^{-1}(y)] \tag{2.3}$$

can also represent R, where x and y are numerical values and g is an invertible real-valued function. For example, addition of numerical length measures x and y in the rods example can be replaced by multiplication of $\exp(x)$ and $\exp(y)$. When an empirical operation can be represented by addition, then addition is by far the commonest choice, but it is not the only one which is used. A familiar statistical example indicating how addition is preferred is the use of additive structures in log-linear models.

Other examples of representing a given ERS in terms of different numerical relationships abound. One can assign numbers to represent the conductance attribute of electrical components and then the empirical concatenation operation of placing components in parallel can be mapped to the numerical operation of addition. However, one can alternatively assign numbers to represent the resistance of the components – the same physical attribute is involved, but the assignment of numbers is different. Now the 'parallel concatenation' operation will be mapped to the numerical operation $\rho_1 \oplus \rho_2 = (\rho_1^{-1} + \rho_2^{-1})^{-1}$. Similarly, velocities may be mapped to the NRS $[Re, \geqslant, +]$ in the usual classical way, or they can be mapped to $[(0, c), \geqslant, \circledR]$ where $\circledR$ is relativistic combination of velocities, given by $u \circledR v = (u + v)/(1 + uv/c^2)$. In each such mapping, since the same R is being represented by both the r relationships, the numerical representations must be isomorphic, related through the transformation g. In these three examples the isomorphisms are given by the respective $g(x)$ transformations $\exp(x)$, $1/x$, and $\tanh^{-1}(x/c)$. Note that the transformation g between NRSs determines the form of the numerical r relationship – by expression (2.3) above. In particular, only if the transformation g belongs to the family of admissible transformations will the same r relationship apply in the two NRSs.

To summarize, objects related in terms of a given attribute may be mapped to numbers in different ways (e.g. the different scalings of a ratio scale) such that the relationship between the numbers reflects the relationship between the objects. These different ways are related by admissible transformations. However, beyond this freedom of choice, there is often also more than one way to choose the numerical *relationship* itself (e.g. multiplication instead of addition), and these different ways are related by invertible transformations. The first of these choices (which numerical assignment to choose, given the numerical operation) has caused little confusion (apart from the issues described in Section 2.5). The second, however, is rather more subtle, and has occasionally led to confusion. In particular, there is the opportunity for confusion when two *different* empirical relationships can both be mapped to the *same* numerical relationship. Electrical conductance again provides an example, as is shown in the next subsection.

2.2.7 Representing different empirical relationships

We have noted above how the conductance attribute can be mapped to the real numbers in such a way that the parallel concatenation empirical operation is represented by addition. To remove ambiguity, let us refer to the physical attribute by the symbol C and the numerical assignment by c. Then we have a mapping from C to c such that parallel concatenation of C maps to addition in c. We also noted that if, instead, one chose to map the objects to numbers r, these being the reciprocals of those in c, then a different numerical operation from addition must be used to represent parallel concatenation. However, there is the possibility of representing a different empirical concatenation operation here – still relating to the same attribute C. We can, instead, choose to concatenate the electrical components by placing them end to end – in series. Our first numerical assignment (c) cannot represent this empirical operation by addition (since it represents parallel concatenation by addition, and the operations are not equivalent). Thus some other numerical operation must be found. In fact, if x and y are two numbers in c, then serial concatenation of two objects will yield a concatenation with numerical value $(x^{-1} + y^{-1})^{-1}$. However, in terms of r, this new concatenation operation, serial combination, will be represented by simple addition.

We have here two different empirical relationships (parallel and serial concatenation), both of which may be mapped to addition. Moreover, the two representations are related by a function which is its own inverse. This induces a symmetry to the situation which aggravates the confusion. (If the two representations were related by exp and log, for example, the potential for confusion would not be so great).

This particular situation, whether to use conductance or resistance measurements to represent the attribute C, has caused some controversy in measurements of galvanic skin response in psychophysiological experiments. Since the transformation between the two representations is nonlinear (r is related to c by a reciprocal transformation) relationships established in one representation may not hold in the other if they depend on linear operators. For example, as is illustrated below, it is easy to contrive situations in which the mean difference between two groups of scores is inverted according to whether one chooses to perform the calculations in terms of r or c. (That is, group A appears 'better' if r is used but worse if c is used.) The point is, however, that addition in the r and c representations corresponds to different empirical relationships – and the choice should be made on grounds of what empirical relationship one is concerned with. Lader (1970) suggests that, in the case of galvanic skin response, 'If the sweat glands are regarded as small electrical shunts through the high resistance *stratum corneum* of the skin then, as the sweat glands are essentially *in parallel*, a conductance model would be appropriate.' That is, parallel concatenation is the empirical operation which should be mapped to addition.

The empirical and mathematical structures which arise in this example also arise in other situations. Hand (1994) gave the following example (with deliberately artificial numbers, so that it is easier to see what is going on – one could contrive more realistic but more complicated examples. If the small sizes of the numbers disturb you, think of military tanks rather than cars).

A man wishes to buy a car and consults two researchers about the fuel efficiency of two different types of car. To lend force to the illustration, we shall take the researchers to be a Frenchman (F) and an Englishman (E). To compare the fuel efficiencies of the cars, samples of two cars of each of two types are taken and the fuel consumptions are measured, with the results given in Table 2.1.

The English researcher records the data in the first row of the table, finding that the two cars of type 1 ran for 1 kilometre per litre and 4 kilometres per litre respectively, producing an average of 2.5 kilometres per litre. Similarly, for type 2 cars, this researcher found that the average efficiency was 2.0 kilometres per litre. On the basis of these results, the English researcher concluded that type 1 cars were better – on average they ran for more kilometres per litre than did the type 2 cars.

The French, however, measure fuel efficiency in litres per kilometre. The French researcher thus used the reciprocals of the English researcher's raw scores. For type 1 cars this means that the two cars consume 1 litre per kilometre and 0.25 litres per kilometre respectively, producing an average of 0.625 litres per kilometre. Likewise, for type 2 cars the average is 0.5 litres per kilometre. From these results the French researcher concludes that type 2 cars are better – consuming less petrol per kilometre than type 1 cars.

Table 2.1 Fuel consumptions of two types of cars.

	Type 1			Type 2		
	Car 1	Car 2	Av.	Car 1	Car 2	Av.
E (km/l)	1	4	2.5	2	2	2.0
F (l/km)	1	0.25	0.625	0.5	0.5	0.5

The two researchers, using the same data, have reached contradictory conclusions. Concerned at this, the potential purchaser consults a pure mathematician, who points out that the essence of the contradiction arises from the fact that the averages relate to two different concatenation operations, one being concatenation of kilometres per litre and the other of litres per kilometre. If different empirical relationships are being considered (confusingly, since both are mapped to the addition of the averaging operation), it is hardly surprising that different conclusions are drawn.

From a theoretical perspective, perhaps we could stop there: the mathematician goes home, content that he has done a good job. But the poor car purchaser is not so sure. He seems now to have an explanation for why the researchers reached different conclusions, but he still does not know which is right. He still does not know which car to buy. To answer this question, we must appeal to the context from which the problem arose. Hand (1994) suggests that, 'in using a car, we are generally interested in how many litres it will take to cover a given distance (to travel from A to B) rather than how far we can travel on x litres, before we run out of petrol. That being the case, the litres per kilometre calculation will be the more appropriate (with the implication that the English are wrong!).' Alternatively, if one really cannot choose between using the data and their reciprocals as given in Table 2.1, then one is regarding the two approaches as equivalent. Since it is not possible for addition to simultaneously represent the two different concatenation operations then the empirical system being represented cannot include these operations. This leaves only the empirical order relationship being represented by the numbers. That is, 'fuel efficiency' is being measured only on an ordinal scale. If, therefore, one restricts oneself to statistical operations which give answers which are invariant to monotonic transformations (such as comparisons of medians), then the two ways of scoring the cars will yield identical answers.

Of course, distributions of ratios of positive values are often positively skewed, so that statisticians frequently take logarithms of the values before analyzing such ratios. In the petrol consumption example, this will mean that the Englishman's raw values (e, say) become log e and the Frenchman's raw values (f, say, with $f = e^{-1}$) become $\log f = -\log e$. It immediately follows from this that, on the logged data, the two analyses will yield the same conclusions. That is, if the Englishman concludes that one type of car is better than the other, so will the Frenchman will draw the same conclusion. At this point our car purchaser may set off home, feeling he now knows what type of car to buy. Provided that is, it does not occur to him, halfway home, that he does not really know what a result in units of 'log(kilometres/litre)' means. In particular, he might think, he really wanted to know which type of car led to the greater average number of kilometres/litre, not the greater average number of log(kilometres/litre), which again may not be the same thing.

Here is another illustration, also based on the reciprocal relationship. Suppose one wishes to buy stocks in a company, and that the price of these stocks fluctuates in an unpredictable way over time. We will assume that one intends to make regular purchases – monthly, say. Then there are two possibilities (at least, we will compare just two). One could either buy the same amount, x, of stocks each month, at cost c_i for the ith month, or one could spend the same amount, £c, each month, so purchasing x_i stocks at the ith month. Note that $x/c_i = x_i/c$, from which it follows that $\Sigma x_i = xc\Sigma 1/c_i$. We shall assume that, over the course of n months, we would spend the same amount whichever of the two methods has been adopted (this is not an essential assumption – it merely makes the argument easier to follow), and ask ourselves which method leads to the purchase of more stocks. The total expenditure is $nc = \Sigma c_i$, so that $c = (1/n)\Sigma c_i$. Now, the total number of stocks purchased by the first method will be nx and by the second method Σx_i. From the above, the second of these equals $xc \Sigma 1/c_i$, which is equal to $(x/n)\Sigma c_i \Sigma 1/c_i$. Since, in general, the mean of the reciprocals of a set of numbers is larger than the reciprocal of the mean (because of the convexity of the reciprocal transformation), it follows that $(x/n) \Sigma c_i \Sigma 1/c_i > nx$. That is, one will buy more stocks for the same expenditure by following the second strategy.

In this example, the objects are the monthly purchases of stocks, but two different attributes are being used to decide how much to purchase. In the first case the attribute is the number of stocks purchased in a month, and in the second it is the cost of the stocks purchased in that month. These two attributes are related, indeed they are proportional, but the relationship – the factor of proportionality – varies over time. Each of the attributes can be mapped to a numerical representation in which addition corresponds to an empirical relationship, but since the relationship between them fluctuates, addition for the numerical representation of one attribute does not correspond to addition for the numerical representation of the other attribute.

When measuring length we usually map the end-to-end concatenation operation to addition, as we have illustrated above. However, this is not the only concatenation operation involving the attribute length which can be mapped to addition. Again there is the potential for confusion if one is not aware that different operations are being used. Here is an illustration of an unusual concatenation operation for length which may appeal to statisticians (Ellis, 1966). Instead of placing two rods a and b end to end in a straight line, concatenate them by placing them end-to-end at right angles. The third rod, of length equal to the concatenation of the first two, $a \circ b$, forms the hypotenuse of the right angled triangle. Can we find a numerical assignment which will properly represent the lengths and in which addition of numbers corresponds to this concatenation operation? If so, how is this assignment related to the conventional one for end-to-end concatenation?

The answer to the first question is that, of course, we can; and the answer to the second is that numerical lengths arising from the right-angled concatenation are the squares of lengths arising from the end-to-end concatenation. A ruler graduated using squared numbers in place of the conventional graduations would yield a perfectly legitimate alternative description of space, but one in which summation of lengths corresponded to right-angle concatenation in place of end-to-end concatenation. Although this would doubtless simplify life for countless generations of students on their first encounter with Pythagoras, at first this may seem to lead to a horribly contrived description of Euclidean space. However, it is precisely what is used in statistics when variations arising from multiple sources, described in terms of variances (in squared units of measurement) rather than standard deviations, are added.

Falmagne (1985, p. 63) gives yet another concatenation operation on rods which can be mapped to addition. Let the rods define chords of a semicircle. To concatenate two rods, arrange them so that they have an end in common. The single rod corresponding to their concatenation is then the chord connecting the other two ends of this concatenated structure. If we restrict ourselves to rods such that the length of the concatenation is less than the diameter of the semicircle, then we can find a numerical assignment such that these rods can be mapped to numbers satisfying the order relationship and such that the concatenation operation maps to addition. In fact, this numerical assignment $(f(x))$ corresponds to the angles subtended by the rods at the centre of the circle, from which we can see that the relationship between this numerical assignment and the standard one involving linear end-to-end concatenation $(s(x))$ is

$$s(x) = 2r \sin\left[\frac{f(x)}{2}\right]$$

where r is the radius of the circle. If we let r increase in size, the angle $f(x)$ that a given rod subtends at the centre of the circle will decrease, so that

$$s(x) \approx 2r\left[\frac{f(x)}{2}\right] = rf(x)$$

That is, the two representations become more similar as r increases.

A much more complicated situation occurs in pharmacology. The dose response curve for a drug is a function showing the effect of the drug at different levels of dose. An 'object' here is a certain quantity of a drug, and the numerical value assigned to it is the effect of that quantity of drug. 'Effect' may be measured in various ways – number of people who survive, extent of pain relief, score on a depression scale, and so on. Now, drugs can be concatenated by taking them simultaneously. The 'object' is then the drug combination, and the numerical value assigned to this object will be the effect of taking the two drugs simultaneously. Often, however, two drugs interact with each other, so that taking them together leads to an effect greater than that one would have expected from their separate effects ('synergy') or less than that one would have expected from their separate effects ('antagonism'). The question is, how does one determine what effect one should expect from their separate effects? Or, in our terms, if the drugs do not interact, what numerical operation is appropriate through which to represent the concatenation operation of taking them together?

To explore this, let $f(x)$ be the dose response curve for drug A and $g(x)$ be the dose response curve for drug B. That is, $f(x)$ represents the effect of A when a dose of size x is taken, with a similar meaning for g. Then one straightforward possible numerical operation is addition: the drugs do not interact if the effect of the combination of dose d_a of A and dose d_b of B is equal to the sum of the individual effects of these two doses, $f(d_a) + g(d_b)$. Unfortunately, this simple additive model fails if the dose response curves are not linear, as is normally the case. To see this, consider the special case when A and B are the same drug. Under the basic premise that a drug does not interact with itself, the model for the non-interactive case should produce the effect of a dose $(d_a + d_b)$ of drug A, namely $f(d_a + d_b)$. However, in general, when $f = g$ is nonlinear, $f(d_a) + g(d_b) \neq f(d_a + d_b)$.

An alternative numerical operation is to suppose that the effect of the combination (d_a, d_b) is the equal to the effect of a dose d_a of drug A plus the effect of a dose d_b of drug B where the effect of drug B is taken as the effect of a dose d_b *on top of whatever response level was reached by* d_a of drug A. That is, the final effect is $g(g^{-1}(f(d_a)) + d_b)$. In the special case when the two drugs are the same, this reduces to the dose response curve of that drug, $g(d_a + d_b)$, as required by the basic premise.

Unfortunately, this numerical operation is also not ideal. We could, with equal justification, have defined the effect of the combination in the reverse order, as the effect of a dose d_b of drug B plus the effect of a dose d_a of drug A, where the effect of drug A is taken as the effect of a dose d_a *on top of whatever response level was reached by* d_b of drug B. This would lead to the non-interactive effect $f(d_a + f^{-1}(g(d_b)))$, and a problem arises because, in general, $g(g^{-1}(f(d_a)) + d_b) \neq f(d_a + f^{-1}(g(d_b)))$ when the curves are not identical. The numerical operation is not well defined.

Somehow we need to capture the fact that the drugs have their effects simultaneously. We can do this if we assume that the rate at which the response changes with increasing dose depends on the current level of response. So, for example, if the current level of response was y, and drug A alone were being used, then the rate at which the response would change would be $f'[f^{-1}(y)]$. Likewise, if drug B alone were being used, the response would change at rate $g'[g^{-1}(y)]$. Now suppose that the drugs are combined in proportions α of drug A and β of drug B. Suppose also that a dose d_c of this combination has been given such as to induce response $y = h(d_c)$. Then, if there was no interaction, the rate at which the response to the combination would change with increasing dose would be

$$h'[h^{-1}(y)] = \alpha f'[f^{-1}(y)] + \beta g'[g^{-1}(y)]$$

This choice of numerical operation has all the properties we require. It can alternatively be expressed as

$$h^{-1}(Y) = \int_0^Y \left\{ \frac{\alpha}{df^{-1}(y)/dy} + \frac{\beta}{dg^{-1}(y)/dy} \right\}^{-1} dy$$

Examples of such numerical operations, and further discussion of the ideas leading to this approach, are given in Hand (1993b, 2000).

The last few subsections have been concerned with the freedom of choice we have in selecting the numerical assignment and the numerical relationship so that the relationships between numbers reflects the relationships between objects. This has necessitated the discussion of transformations between different numerical assignments. These transformations were constrained by the need to faithfully represent the empirical relationship. However, in statistics, other transformations are also often used. We shall discuss the relationship between statistical operations, statistical statements, transformations and measurement scales in more detail in Section 2.4.

2.2.8 'Classical' measurement

Adams (1979) provides a review of pure representational measurement theory, and then points out some weaknesses (see Adams, 1966, for details). He also notes:

> For the Greeks the things measured and their measures (line segments) were scientifically on a par, to be described in a 'seamless web of theory' in which the basic assumptions of measurement are made explicit. Modern theorising treats the measures (real numbers) as elements of an essentially different realm (pure mathematics) whose principles are not to be regarded as 'empirical' or as part of empirical science.
>
> (Adams, 1979, p. 221)

Michell (1986, 1990) has explored this point in detail, developing what he terms 'classical measurement theory'. He called it 'classical' because, according to him, traces may be found in the works of Aristotle and Euclid and it was 'developed during the Middle Ages and the Scientific Revolution and sustained the practice of measurement until at least the beginning of [the twentieth] century'. According to Michell's classical theory, measurement addresses the question of 'how much' of a particular attribute an object possesses and thus only refers to attributes which are 'quantitative'. This fundamentally quantitative perspective is also one which has been espoused by Rozeboom (1966) (his 'usage 3': ' "measurement" is *assessment of quantity*'). Rozeboom (1966, p. 225) notes that in Campbell's restricted representational approach variables are taken to be quantitative 'if and only if their causal implications for some binary comparison among physical concatenations of X's arguments have an additive scale representation', but then argues that this does not

completely resolve things because it leaves unanswered the question of 'why this particular [representational] structure should be singled out for special attention'. We might
answer, with the benefit of nearly 40 years of hindsight, that nowadays other kinds of structures are also widely used, but this does not detract from the force of Rozeboom's question
regarding why extensive systems are so important.

Michell also defines a quantitative attribute as one whose values satisfy ordinal and additive
relationships. Again, this sounds very representational, but Michell sought to distinguish his
classical perspective from the representational viewpoint by stressing that it is the *attribute*
which has these properties, and not the objects. The behaviour of a set of *objects* may or may
not reflect the quantitative nature of the attribute in question: the behaviour of objects is a function of their other properties as well as the attribute in question. Rozeboom (1966) had earlier
argued in the same way, for example commenting on the meaninglessness of the notion
of concatenating objects in terms of the 'height of a person' attribute (if we concatenate
two 6 foot tall men, do we get one 12 foot tall man?). Rozeboom thus remarked (p. 227)
'[Campbell's] error has been to regard "units" as *objects* (i.e. as the bearers of properties) and
their concatenation as a *causal* process unfolding in time; whereas understood properly, the
units comprised by a quantity *P* are component *properties* of which *P* is an *analytic* aggregate'.

While the distinction between comparative and concatenable properties of objects and
of attributes is noteworthy, I think it is something of a red herring: the properties of an
attribute are necessarily inferred by observations on the behaviour of objects. A physical
concatenation operation between objects certainly provides evidence for the quantitative
nature of an attribute, but the lack of such an operation does not mean that the attribute is
not quantitative. Evidence for the assertion that an attribute is quantitative may be found in
other ways. Conjoint analysis is an example, of course, and Michell cited the example of
temperature: objects possessing temperature do not satisfy a concatenation relationship
(in terms of that attribute) and yet this attribute is generally regarded as quantitative. Also:

> For example, length and density are both taken to be quantitative variables (i.e. to
> possess additive structure), although the kinds of observations supporting each are
> quite different. In the case of length, the behaviour of rigid rods directly reflects the
> hypothesised additive structure. In the case of density the evidence is less direct,
> being via the relation of density to mass and volume.
>
> (Michell, 1990, p. 13)

Other examples may be found at physical extremes – for example, astronomical distances
cannot in any practicable sense be concatenated.

According to Michell's 'classical' perspective, the hypothesis that an attribute is quantitative is a scientific hypothesis just like any other. Measurement then involves the discovery
of the relationship between different quantities of the given attribute. The key word here is
'discovery'. He argues that, whereas the representational theory *assigns* numbers to objects
to model their relationships (and an operational or pragmatic approach *assigns* numbers
according to some consistent measurement procedure), the classical theory *discovers* pre-
existing relationships. By definition, any quantitative attribute has an associated variable.
I would say that the representational approach chooses numbers in such a way as to model
pre-existing relationships (between objects in terms of their attributes), so that whether one
describes this as 'assigning' or 'discovering' is a matter of taste rather than substance.

Developing a measurement procedure according to the classical theory requires relating
the hypothesized quantitative attributes to observable quantities within some theoretical

framework. The hypothesized quantitative attributes can then be measured by virtue of their relationships. This seems to me to be a good description of indirect representational measurement procedures. Here the hypothesized attributes, as well as their quantitative nature, are all a part of the theory being studied. Rasch's (1977) notion of *specific object-ivity* might be regarded as a good example of this framework. Rasch gave an example in which observed scores on a test are described by a Poisson model (discussed further in Chapter 3). The parameter of the Poisson model can be viewed as an underlying measure of ability for a given test. Rasch then showed that this model permits parameters to be sep-arated into two sets, one describing comparisons between the abilities of individuals (inde-pendent of which test is used) and one describing comparisons between test difficulties (independent of which subject is assessed). (It is this separation which leads one to believe that, for example, ability is an intrinsic property of the individual.) The link between the observed scores and the parameters is stochastic and nonlinear – quite complicated, as Michell suggested it might sometimes be.

According to Michell, classical theory measurements are always real numbers: if one has been able to measure them, the numbers which have resulted satisfy all the properties required for arithmetic manipulation, so that one can manipulate them using any statistical operation. This is as true for latent variable scores – measures of a hypothesized underlying quantitative attribute – as it is for straightforward observables such as length or weight. It is also true for measures such as preference scores – they are held to be measurements of a quantitative preference attribute, though with measurement error and possible bias, which may indeed be nonlinearly related to the attribute's value. Such bias, nonlinearity and meas-urement error can be investigated by refining the theory in which the preference scale is embedded – by relating the scores to other variables – and by using subtle statistical methods. The classical theory is predicated on the condition that measurement is necessarily concerned with 'quantitative' attributes. Modern representational theory includes this as a special case.

Michell (1990) developed his classical perspective, and in particular attempted to distin-guish it from the representational approach, in some detail. For example:

> [Representationalism] is motivated by the belief that numbers are not part of the fur-niture of the universe. Its aim is to explain how, given this premise, numbers find their way into the practice of measurement. Whatever the philosophical motivations for this view about numbers (and the motives are many and varied), it is mistaken. There is no place in science for entities thought to be outside of space and time or for 'convenient' fictions. When we speculate in science, our speculations are about real things located in the same spatio-temporal universe as ourselves.
>
> (Michell, 1990, p. 24)

This is certainly one viewpoint, though not one held by everyone. Others hold that both the empirical and the formal have a role to play in science. Stevens (1968, p. 856) himself, for example, described science as presenting

> a two-faced, bipartite endeavour looking at once toward the formal, analytic, schem-atic features of model building, and toward the concrete, empirical, experiential obser-vations by which we test the usefulness of a particular representation. Schematics and empirics are both essential to science, and full understanding demands that we know which is which.

Kyburg (1984, p. 253) also developed a particular form of representational approach: 'Most approaches to measurement that have been suggested in recent years have taken the process of measurement to be the assignment of *numbers* to objects and events. I have suggested that the value (or interval of values) assigned to an object or event by measurement is a magnitude (or interval of magnitudes), rather than a number.' Thus, instead of assigning the number '2' to the length, in feet, of an object, he assigns the magnitude '2 feet' to the object. His book develops the consequences of this approach.

2.3 Pragmatic measurement

In the representational approach to measurement, empirical relationships between objects, observed in terms of a particular attribute or attributes, are modelled by corresponding relationships between numbers. The attributes are supposed to exist because one can actually observe the relationships between the objects in terms of the attributes in question. The trick is to characterize these relationships and to match them up to equivalent numerical relationships. This is fine when the relationships between objects in terms of the attribute can be unambiguously studied – in classical physics, as taught in schools, for example. However, it is not so fine when things are not so clear. Dawes and Smith (1985, p. 509) remark: 'It is not uncommon for psychologists and other social scientists to investigate a phenomenon at great length without knowing what they're talking about.' If we cannot define what we are talking about, and if we cannot describe it sufficiently precisely that we will recognize it when we see it, how then can we study it, and how can we measure it? There is clearly difficulty in formulating a mapping from something so ill defined to a representing numerical system. On the other hand, we do need to measure and manipulate such measurements all the time. We need to monitor the growth of the economy, the content (or discontent) of customers, the quality of life of patients, and the complexity of a computer program. Perhaps this appears especially difficult for the social and behavioural sciences, but in fact it is also true for other sciences. If one looks back at early attempts to formulate precise definitions of probability, of the efficiency of computer programs, even of basic physical concepts such as gravity, one sees that things were not always as clear as they may seem now.

The solution to the conundrum implicit in the quotation from Dawes and Smith above lies in the concept of *pragmatic* measurement. This is the aspect of measurement, chosen on grounds distinct from the faithful representation of the empirical system, which is needed to lead to a unique numerical result when a measurement is taken. At its extreme, instead of taking the attribute and the relationships between objects as given, requiring only an adequate formal representation, in pragmatic measurement models the measurement procedure entirely *defines* the variable being measured. Thus, representational aspects aside, the variable 'length' may be *defined* as the number of segments covered when a marked ruler is laid by the side of the object to be measured. The 'weight' of an object may be *defined* as the number of unit weights required to balance the object on a weighing scales. In place of representation playing the key role, the procedure and definition are motivated by the need for a well-defined measurement procedure, so that the resulting numbers should be of practical value in some sense. Of course, most measurement situations are less extreme than this, and the pragmatic component is only a part of the measurement process.

The extreme pragmatic approach is closely linked to the philosophical position of *operationalism*. This defines scientific concepts in terms of the operations used to identify or measure them. Operationalism was developed by the Nobel prize winning physicist Bridgman as a response to the dramatic changes which physics had experienced around the turn of the century. He focused on the measurement procedure and the use made of the result: 'the concept is synonymous with the corresponding set of operations' (Bridgman, 1927, p. 5). The approach was adopted by Dingle (1950), who summarized it thus:

> Formerly science was regarded as the study of an external world, independent of the observer whose experiments and observations were simply means of finding out how the world was constructed and by what laws its behaviour was governed. The emphasis has now shifted from the nature of the world to the operations of experiment and observations. These are no longer regarded as more or less arbitrary means of discovering the already established order of nature, but rather as affording primary data for rational study; and any world that we may contemplate is no longer an independent existence whose nature demands or determines them, but rather a logical construct, formed and shaped and modified so as to afford a true picture of the relations which the observations exhibit.

Thus, an attribute is defined by its measuring procedure, no more and no less, and has no 'real' existence beyond that. In operationalism the attribute and the variable are one and the same. This approach thus defines 'a measurement [as] any precisely specified operation that yields a number' (Dingle, 1950, p. 11). This principle overcomes the problem of ambiguity or uncertainty in the underlying concepts: Boring (1945) illustrated it nicely when he described intelligence as being 'what [intelligence] tests test.'

Although they are related, the operational and pragmatic perspectives on measurement are not identical. The essence of pragmatism is that a deliberate choice has been made, on grounds external to the system being represented. You can vary this choice on various grounds, including simple convenience. Although both this and the operational perspective see the measurement process as defining the attribute being measured, the operational viewpoint does not generally include explicit notions of 'convenience' arising from external sources.

Michell (1990, p. 23) describes the operational approach as being 'against speculation' and hence not scientific. He argues that this approach (p. 24) 'involves the deliberate confusion of *what* is being measured with *how* it is being measured'. That is certainly one way to put it. But one might alternatively have said that it 'involves the explicit recognition that *what* is being measured is a function of *how* it is being measured'.

From our perspective, operationalism is an extreme viewpoint, rejecting the need for any underlying reality which is being modelled. In contrast, the position adopted in this book is that most measurement procedures have both representational and pragmatic aspects. The representational part does model some underlying reality, while the pragmatic part is purely operational, and is simply an arbitrary aspect of the process of taking the measurement – arbitrary in the sense that other measurement procedures may adopt alternative pragmatic constraints. The choice of pragmatic constraint is a conscious choice of the person taking the measurement, and is made on grounds quite separate from any properties of the empirical relational system being modelled. Of course, one might reasonably aim to adopt pragmatic constraints which are widely used, so that one can communicate effectively with others concerned with investigating the same phenomenon.

One way of thinking about the difference between representational and pragmatic aspects of measurement is that, since the former seeks to *represent* or *model* empirical relationships, it is about *understanding* the substantive domain of investigation. In contrast, pragmatic measurement is concerned with *prediction*. Accurate prediction can be achieved without any understanding of the underlying mechanism (witness someone who can drive a car well without any understanding of how it works). Confusion between these two aims is widespread, and is probably promoted by the unfortunate adoption within statistics, described in Chapter 1, of the term 'model' to denote both descriptive and iconic models.

Of course, to be useful, any numerical assignment procedure, even those based entirely on pragmatic rather than representational principles, has to be well defined. Arbitrariness in the procedure will reflect itself in ambiguity in any conclusions drawn from the measurements. This is one reason why problems arise in the social and behavioural sciences, where, inevitably, measurement procedures are complex. Complete specification of the procedure is often difficult or impossible and different researchers may use the same name for variables that actually have subtly different pragmatic aspects, leading to different conclusions. Of course, this issue does not solely arise in the human sciences. The Mohs scale and the Brinell scale (Chapter 7) are both measures of hardness. The former is determined on the basis of whether a mineral will scratch or be scratched by others. The latter is based on the size of an indentation produced in a specimen. In pragmatic terms, since the definition of the concept lies in the measurement procedure, it is not a cause for concern that different procedures lead to different conclusions, but rather an indication that more refined theory needs to be developed to cope with the results of two different measures, one measuring 'hardness (Mohs)' and the other 'hardness (Brinell)'.

This issue of different definitions (i.e. different pragmatic aspects) for a given variable name is sometimes confused with the fact that there may be different ways of measuring an attribute in the representational theory. We can, for example, measure length by using rigid rods or by using the time for light to transit from one point to another or by using a strain gauge. From a representational perspective, the measurements are underpinned by a complex physical theory which shows them to be different descriptions of the same underlying attribute. From the operational or extreme pragmatic perspective, however, the different numerical variables are simply distinct types of 'length'. Our perspective is that the extreme pragmatic perspective is untenable in this case, and that a strong representational interpretation is more appropriate. However, a less extreme example is given by two questionnaires, each seeking to tap into some attitude to a proposition but using different questions. Here a significant pragmatic aspect is necessary: the 'complex psychological theory' showing how and why the different variables are different descriptions of the same underlying attribute probably does not exist. The different variables are slightly different because of the different pragmatic aspects. Of course, if, in a supposedly purely representational model, two alternative measurement methods lead to consistently different results, then this is an indication that the ERS is not as straightforward as was thought. The distinction between mass and weight provides an illustration.

We described above how Stevens generalized the (representational) work of Campbell, pointing out that empirical relationships other than simple concatenation operations might be mapped to representing mathematical structures, so that measurement scales beyond ratio scales also existed. This led him to describe his four basic types of scale. However, Stevens (e.g. 1946, 1951) also more generally described the process of measurement as 'the assignment of numerals to objects or events according to rules' and remarked that 'provided a consistent rule is followed, some form of measurement is achieved' (Stevens, 1959). Indeed, the first of these statements has been quoted countless times and has been used as a justification for all sorts of 'measurement' procedures. Stevens (1959) himself remarked that the only procedure excluded as a type of measurement by this definition was 'random' assignment.

At first sight, the flexibility in Stevens's 'numerical assignment according to rules' description of measurement seems to have lost contact with the constraints imposed by faithfully representing empirical relationships between objects. It seems, on the face of it, to be an operational definition – measurement is the result of the procedure one adopts to assign numbers to the objects – and some authors (especially psychologists, who were particularly concerned about the nature of the data on which their work was based; see Section 2.5) seem to have taken this as indicating Stevens legitimizing operational approaches. I think this is a misinterpretation of Stevens's intent. Time after time, he stressed the role of the empirical relationship being modelled, not merely in his earlier publications on this topic, but also in his later ones, after he had had more time to digest things. Thus:

> To the British Committee, then, we may venture to suggest by way of conclusion that the most liberal and useful definition of measurement is, as one of its members advised, 'the assignment of numerals to things so as to represent facts and conventions about them'.
>
> (Stevens, 1946, p. 680)

Later we have:

> Measurement is possible in the first place only because there is a kind of isomorphism between (1) the empirical relations among objects and events and (2) the properties of the formal game in which numerals are pawns and operators the moves ...
>
> (Stevens, 1951, p. 1)

and, in the same work,

> A rule for the assignment of numerals (numbers) to aspects of objects or events creates a *scale*. Scales are possible in the first place only because there exists an isomorphism between the properties of the numeral series and the empirical operations that we can perform with the aspects of objects ... The type of scale achieved when we deputize the numerals to serve as representatives for a state of affairs in nature depends upon the character of the basic empirical operations performed on nature.
>
> (Stevens, 1951, p. 23)

Still later:

> Most measurement involves the assignment of numbers to aspects of objects or events according to one or another rule or convention ... It has proved possible, however, to formulate an invariance criterion for the classification of scales of measurement ... The theory proposes that a scale type is defined by the group of transformations under which the scale form remains invariant ... The admissible transformations defining a scale type are those that keep intact the empirical information depicted by the scale. *If the empirical information has been preserved, the scale form is said to remain invariant. The critical isomorphism is maintained.*
>
> (Stevens, 1968, p. 850; emphasis added)

To me, it seems pretty clear from this that by 'rule' he meant a mapping which preserves some kind of empirical structure. The details of the structure preserved determine the scale type. The rule might be a highly prescribed one, such as 'take this rod as consisting of the unit, and map other objects to numbers using end-to-end concatenation'. Or it might be a very loosely prescribed one, such as 'randomly pick objects, one after the other, and assign consecutive integers to them'. While this latter is a legitimate rule, it clearly preserves very little structure in the empirical system, and so can only lead to weak conclusions. In fact, of course, it leads to only a nominal scale.

Stevens explicitly notes that one can choose to model different empirical relationships, according to one's choice of mapping. As we have seen before (Stevens, 1951, p. 1), 'measurement is the assignment of numerals to objects or events according to rules' but 'the fact that numerals can be assigned under different rules leads to different kinds of scales and different kinds of measurement. The rules themselves relate in part to the concrete empirical operations of our experimental procedures which, by their sundry degrees of precision, help to determine how snug is the fit between the mathematical model and what it stands for.' I interpret the 'in part' of this last sentence as saying, in my language, that there is a pragmatic component in (almost) all measurement procedures. Measurement procedures which do not enforce a 'snug fit between the mathematical model and what it stands for', that is, those which do not define a rigorous isomorphism between the mathematical model and what it stands for, require a larger pragmatic component to bridge the gap.

Michell (1999) distinguishes between 'external representation', which is that which isomorphically represents some structure of the empirical system (and is what we normally mean by 'representational' in the measurement context), and 'internal representation', which occurs 'when the attribute represented, or its putative structure, is logically dependent upon the numerical assignments made, in the sense that had the numerical assignments not been made, then either the attribute would not exist or some component of its structure would be absent.' (Michell, 1999, p. 166). This seems very close to what we are terming the pragmatic aspects of measurement. However, it is not quite the same: 'simply to presume that a consistent rule for assigning numerals to objects represents an empirical relation possessing such properties is not to discover that it does; it is the opposite' (Michell, 1999, p. 168). In pragmatic measurement we presume no such thing: our pragmatic constraints are external to and additional to any constraints on measurement arising from relationships within the empirical system. Michell also refers to 'ambiguous representation', in which 'it is not known whether or not the attribute or its putative structure exists independently of the numerical assignments made'. This is *not* the same as a mixture of our representational and pragmatic aspects.

Michell then argues that Stevens is seeking to extend the concept of representationalism, from simply modelling the relationships in the empirical system, to also modelling relationships arising from the operations performed in taking measurements on the objects. An operationalist, after all, would assert that only the operations and their results have meaning. In support of this interpretation, Michell cites Stevens (1951, p. 23), when he says 'scales are possible in the first place only because there exists an isomorphism between the properties of the numeral series and the empirical *operations that we can perform* with the aspects of objects'. The italics are Michell's, to emphasize the supposed operational interpretation. I do not agree with this interpretation. After all, Stevens then goes on to say (as we have already noted in Section 2.2.5) that 'this isomorphism, of course, is only partial. Not *all* the properties of number and not *all* the properties of objects can be paired off in a systematic correspondence. But *some* properties of objects can be related by semantical rules to *some* properties of the numeral series.' There is nothing here about purely operational aspects of the measurement process: it is all solely representational.

Before we leave the subject of Stevens's approach to measurement, it is worth remarking that his writing is beautiful. His papers are worth reading for their eloquence and elegant turns of phrase, whatever one's perspective on his theories.

Of course, as far as the pragmatic aspect of measurement goes, in order for useful discussions to take place, and for progress to be made, everyone involved must use the same conventions to discuss the phenomena. That is, the same pragmatic measurement procedures must be used. If they are not, then apparently contradictory conclusions can be reached. This is illustrated by questionnaires which can elicit contrary responses by means of subtly different wordings of the questions. Furthermore, the pragmatic measurements which will be most useful will be those in which the measurement sits properly and effectively in a theoretical web of relationships with other variables – that is, those which yield effective *predictions*. (Those which yield ineffective prediction are presumably not used, at least in good science.)

As with the representational approach, techniques for constructing pragmatic measurements fall into two classes: those that focus on single variables (direct methods) and those that define a variable in terms of others (indirect methods). Examples are given in Chapter 3. Pragmatic measurement procedures often arise in complex situations, where there is an underlying attribute to be measured, but where it is very difficult to define this adequately. Intelligence provides a classic example. We have a general notion of what we mean when we say that someone is more intelligent than someone else, but quantifying this is difficult. That is, it is difficult to tie this ill-formed notion, via a clear and well-formed theory, to particular variables which can be measured. For this reason, intelligence tests are indirect, consisting of multiple items, each loosely related to the underlying attribute of interest, and such that the sum of the scores is taken to be a measure of 'intelligence'. This sum, the IQ, serves as an indirect pragmatic measure of intelligence. We shall have more to say about this example in the next chapter and in Chapter 5 (on psychological measurement). We note parenthetically here that, although intelligence is generally measured by indirect methods, there have been some suggestions that it might be measured simply in terms of reaction time. In this case reaction time would be regarded as a proxy variable directly measuring intelligence. Quality of life (Chapter 6) is most commonly defined as a latent construct in terms of other variables which can be directly measured (the items in a quality of life questionnaire) but attempts have also been made to measure it directly (by asking a single question about the quality of life of the respondent). Given the many distinct aspects to quality of life, it is clear that it is essentially a pragmatic construct (whether direct or indirect) whose precise meaning depends critically on the particular question(s) used and the way they are combined to give an overall score.

Since extreme pragmatic measurement models contain no notion of mapping from an empirical reality to a numerical representation, there is no notion of relationships to be preserved. This means that there are no restrictions (apart from consistency – the same mapping has to be used each time it is applied) on the transformations which the resulting numerical representation may be subjected to: the results of any such transformation would define an alternative, equally legitimate, pragmatic measurement procedure. Of course, it would be a different procedure, yielding a different measurement from the original (untransformed) measurement – so perhaps we should give the resulting variable a different name. For example, suppose the attribute weight is measured by the operational procedure of using a spring balance, with the results expressed in kilograms. We might call the resulting variable *wt1*. Alternatively, we could, equally legitimately, use an electronic weighing instrument, which gave results in log-transformed kilograms – variable *wt2*.

The fact that there are no restrictions on the transformations which may be applied to purely pragmatic measurements means that there is no intrinsic notion of scale type in such measurements. If we regard a thermometer measurement of temperature as simply yielding a pragmatic measurement then we can equally regard the log transform or any other transformation of the temperature as a legitimate pragmatic measurement. In particular, in marked contrast to the representational viewpoint, it does make sense to say that the temperature on a day in June was twice the temperature of a day in January. This is so because transforming the temperature (e.g. from Fahrenheit to Celsius) yields a different measurement of temperature – and one would have no reason for expecting the statement to preserve its truth value when transferred to this different measurement. (To make clear which variable one was talking about, one would explicitly specify the particular measurement scale used and state whether the measurement was in degrees Fahrenheit or Celsius – unless the context made it clear.) Having said this, certain assignments of numbers will be more useful than others since they will lead to more convenient descriptions of relationships between variables. It is simpler to state that pressure is proportional to volume than that log(pressure) is proportional to the cube root of volume plus 10, both statements being true for certain specific scales. Of course, all of this applies only to entirely pragmatic measurements. If there is a representational aspect to the measurement, then there are restrictions on legitimate transformations.

The fact that, if one is adopting an entirely pragmatic perspective, there are no empirical constraints to be modelled means that there are no automatically defined sets of variables which are equally legitimate as representations. Under the representational approach, as described in Section 2.2, the variable 'weight' is defined so as to model certain empirical relationships between objects, and there is only freedom in assigning numbers up to the choice of unit. We cannot log-transform the numbers without destroying the fact that they represent the concatenation property by addition. Such constraints do not apply to purely pragmatic measurements, so there are no such things as admissible transformations or sets of variables which provide equally legitimate representations. However, such sets are sometimes defined in terms of the *invariance of statistical statements to transforming the variables*. For example, we might find that comparing the arithmetic means of the *wt1* weights of two groups of people led to the mean of group 1 exceeding that of group 2. Such a comparison would be invariant to linear transformations. We might thus use the statistical statement to determine the class of legitimate transformations, and hence use this as a definition of the scale type of the variable in question. This possibility is discussed in more detail in Section 2.4.1.

Other terms are also used for purely pragmatic measurement. Dawes and Smith (1985) use the term *nonrepresentational*, and Suppes and Zinnes (1963) use the term *pseudopointer* measurement for what appears to be essentially the same thing. Luce *et al.* (1990) described *index measurement* as the use of proxy variables which are understood and which are easily measurable to act as indicants of others (not to be confused with the notion of *index numbers*). They gave the example (Luce *et al.*, 1990, p. 323) of measuring 'hunger' using 'amount of food ingested, initial rate of ingestion, force exerted to overcome a restraint to reach food, percentage reduction in normal body weight, time since last food ingestion, etc.'. Equally, though, and with as much justification, we could use log(amount of food ingested) as a measure of hunger. This has as much empirical justification as simple amount of food ingested. So, while, in representational terms, the amount of food ingested may be measured on a ratio scale, when regarded as a direct measure of hunger it is inappropriate to regard it as a ratio scale, and it is at best ordinal. A measurement may be representational in its own right, but pragmatic in the use to which it is put: reduction in body weight, or weight of food ingested, are both measures of weight, and so have a clear representational derivations, but used as a measure of hunger they become merely pragmatic. Alternatively, amount can be regarded as a purely pragmatic measurement of hunger: 'amount' defines 'hunger' (we might call it 'type 1 hunger') when measured in this way. Since the different measurement procedures for measuring hunger (amount versus rate of ingestion versus log(amount) and so on) may yield different conclusions from one's research, if they are regarded as equally legitimate measures, this is indicative that deeper theoretical work relating them is needed.

Superficially similar situations arise with the use of *pointer* measurements in the physical sciences. For example, the use of the extension of a spring to measure weight. However, as Luce *et al.* (1990) pointed out, there are ratio scale representations of weight and of length and there are theories (Newton's laws and Hooke's law) connecting the two so that using 'a spring to measure weight directly is valid according to this theory'. Expressed another way: we can put markers on the scale of length of extension which correspond to the values of weight contained within the theory, but nothing equivalent can be done in the hunger example: whatever markers we put on the length scale define the extent of hunger. The point is that the weight example has a clearer and more obvious representational component.

Adams (1966) has criticized the representational approach, and produced a model very similar to the pragmatic approach, though one which still relates to attempts to understand an underlying reality in the same way as does the representational approach. To Adams (1966) measurements were merely indicators (though they may be more or less effective) of the underlying phenomenon. His motivation was that 'the proponents of [the representational] approach have neglected to consider what it is that measurements are made *for*, and in so doing have been led to conclusions as to what measurement *ought* to be which are in serious disagreement with what scientists do.' He started from the premise that things like IQ are measurements and then considered what sort of measurement theory justifies this. This looks rather like the pragmatic approach – he is saying that IQ is that aspect of intelligence which is measured by IQ tests, and that IQ provides an objective and replicable (subject to change over time and random variation, of course) measure of something closely related to intelligence. This is certainly very different from the approach of the British Association committee described above, which started from the premise that measurement theory was (a restricted subset of) representational measurement theory and then pointed out that psychological 'measurement' did not conform – and hence was not measurement. To Adams, measurements provided 'systematic, objective indices of phenomena'. This indicial nature is explicit in areas such as economics and psychological rating scales. Laws of measurement connect the phenomenon under investigation with the results of making the measurement, but these laws need not be exactly satisfied for measurement to be useful (and, indeed, may not be exactly formulable: IQ is useful, but stating exactly (in representational terms) how it relates to *intelligence* is impossible, at least in part because there is no clear definition of intelligence). It follows that one should not ask whether a measurement procedure yields a *true* measure of the quantity but, rather, how good an indicator it is of the phenomena it is supposed to give information about. The ultimate test will be how well it does in modelling empirical relationships with other variables. Recall the comment from Cronbach and Gleser (1957) mentioned in Section 1.3.

Perhaps, as a cautionary note about extremes of pure pragmatic measurement, we can quote from Krantz *et al.* (1971, p. 32), who say:

> A recurrent temptation when we need to measure an attribute of interest is to try to avoid the difficult theoretical and empirical issues posed by fundamental measurement by substituting some easily measured physical quantity that is believed to be strongly correlated with the attribute in question: hours of deprivation in lieu of hunger; skin resistance in lieu of anxiety; milliamperes of current in lieu of aversiveness, etc. Doubtless this is a sensible thing to do when no deep analysis is available, and in all likelihood some such indirect measures will one day serve very effectively when the basic attributes are well understood, but to treat them now as objective definitions of unanalyzed concepts is a form of misplaced operationalism.

2.4 Measurement and statistics

2.4.1 Meaningfulness

The representational aspect of measurement hinges on a homomorphism between the empirical and numerical systems. Moreover, as described above, since the homomorphism will generally not be unique, there will be alternative legitimate numerical representations of the empirical system. Transformations between these representations are the admissible transformations. Statistical computations may be performed on the results of any of the homomorphisms, and the numerical results will carry identical substantive import. The question then arises, what if the numerical conclusions disagree? The classic example is of calculating the arithmetic mean of data measured on an ordinal scale. The class of admissible transformations for such data is that of monotonic strictly increasing transformations. As is well known, however, the order of the means of two groups can usually vary according to the transformation employed. For example, imagine two sets of three objects in which the attribute is measured on a purely ordinal scale. If one collection yields the three measurements 1, 2, and 6, while the other yields 3, 4, and 5, then the mean of the first three is $(1 + 2 + 6)/3 = 3$, while that of the second three is $(3 + 4 + 5)/3 = 4$, so that the second group has the larger mean. However, since the scale is purely ordinal, any order-preserving transformation is admissible, and will yield an equally legitimate numerical representation. In particular, a transformation of $(1, 2, 3, 4, 5, 6)$ to $(1, 2, 3, 4, 5, 12)$ would still satisfy the order relationships between the objects. Now, however, the first collection of objects has mean $(1 + 2 + 12)/3 = 5$, while that of the second collection is still 4.

Stevens's (1946, 1951) suggested solution to this problem was to restrict the statistical manipulations according to the scale type: only those manipulations which were invariant to admissible transformations were legitimate.

Unfortunately, such an approach has its problems – in particular, how to define exactly what is meant by a statistic being invariant (after all, change inches to centimetres and the *value* of a mean changes). Such issues can be overcome, but closer examination of this invariance approach shows that it is not the statistic *per se* which causes the difficulties, but the use to which that statistic is put – the interpretation made of that statistic or the statements made about it. Recognizing this, Adams *et al.* (1965), in an important paper, shifted attention from statistics *per se*, to statements made about them. A statement is defined as being *empirically meaningful* relative to a measurement scale if and only if its truth value is invariant over admissible transformations of that scale. This is the definition of 'meaningful' adopted here.

So, for example, a statement that one mean, $\bar{x}$, is greater than another, $\bar{y}$, $\bar{x} > \bar{y}$, is not generally empirically meaningful for ordinal scales: the order of the means can (usually – we shall consider a special case in a moment) be inverted by a suitable ordinal transformation, as in the illustration above. It is, however, empirically meaningful for interval or ratio scales: the statement retains its truth or falsity under all admissible transformations for these two scale types (these being linear and rescaling transformations, respectively).

Conversely, although the statement $\bar{x} = \bar{y}$ is empirically meaningful for interval scales, one cannot infer that therefore it is always legitimate to use means with such scales: the statement $\bar{x} + \bar{y} = \bar{z}$ is not empirically meaningful with interval scales. Similarly, the statement $\bar{x} \times \bar{y} = \bar{z}$, all components being measured on the same variable, is not empirically meaningful even with ratio scales.

We see from these examples that whether or not it is legitimate to use an arithmetic mean depends on the statement being made. The legitimacy is not simply a property of the statistic and the measurement scale, but is a property of the nature of the statistical statement being made. It is the *context* in which a statistic is used which determines whether it is legitimate or not to use that statistic, and not merely the scale type. Sometimes a statistic is said to be *appropriate* relative to a statement and a scale if and only if the statement is empirically meaningful (invariant over admissible transformations) relative to the scale.

Since admissible transformations are transformations to alternative equally valid representations of an underlying empirical system, the concept of an admissible transformation has no relevance to the pragmatic aspect, since this does not refer to an underlying empirical system. Consequently, the notion of empirical meaningfulness is meaningless as far as the pragmatic aspect goes. (However, as was mentioned above, one can define a notion of scale type for pragmatic measurements, in terms of the transformations which preserve the truth value of the statistical statement in question. This is discussed below.)

Invariance over admissible transformations is an attractive definition for a statement to be empirically meaningful, but does it encapsulate all we want? Weitzenhoffer (1951) suggested that in fact the real criterion of 'meaningfulness' should be that a relationship could be expressed in terms of the relationships of the ERS under consideration. He argued that, in principle at least, one should be able to arrive at substantive conclusions merely from consideration and manipulation of the objects themselves – but that this is generally so unwieldy as to be impracticable, hence the use of numerical representations and mathematics. Adams *et al.* (1965, pp. 118–119) also said:

It is worth noting that the association between empirical meaningfulness and intrinsic definability outlined here suggests an alternative way of characterising empirical meaningfulness which is to an extent independent of considerations of numerical measurement and admissible transformations. That is, a formula may be described as empirically meaningful relative to a system of measurement (more generally, any precisely formulated empirical theory) just in case it expresses a relation over the objects of the theory which is intrinsic in the sense of being definable in terms of the empirical operations and observations on which the measurement theory is based. It is a matter of conjecture that this criterion of meaningfulness is in fact more fundamental than that which defines it in terms of invariance under admissible transformations.

While this is clearly appealing, elucidating exactly what is meant by saying that a relationship is 'definable in terms of other relationships' is not easy. (That there is a difference is illustrated by an example of Luce *et al.* (1990, Section 22.5), showing that 'definability' is narrower than invariance.) The subtlety of the issue may be illustrated by the following example.

Consider a single ordinal scale and the statement $\bar{x} < \bar{y}$ for data from it. This is meaningless in terms of the 'definability' criterion since arithmetic mean is not defined for ordinal scales – there is no empirical operation corresponding to addition. Also, in general, it is meaningless in terms of the invariance criterion since arbitrary monotonic transformations will change its truth value. But now suppose that we find that its truth value *is* invariant for some particular data set. For such a data set it would appear to be meaningful in terms of the invariance criterion but meaningless in terms of the definability criterion. However, closer examination shows that for such data sets the y sample stochastically dominates the x sample. (This can be seen as follows. Consider the subclass of monotonic increasing transformations consisting of the step functions defined by $g(z) = 0$ for $z < Z$ and $g(z) = 1$ otherwise, with z ranging over the entire range of the data. Then $\bar{x} < \bar{y}$ on *all* such transformed scales implies that the proportion of the x points lying to the right of Z is less than the proportion of the y points lying to the right of Z, for all Z.) One could adopt the premise of this example, that $\bar{x} < \bar{y}$ for all monotonic increasing transformations, as an operational definition of one group being 'less than' another group, even though arithmetic mean is not defined in the ERS. In fact, of course, this is exactly what the Mann–Whitney–Wilcoxon two-sample test statistic does, estimating the probability that a randomly chosen member of one sample is larger than a randomly chosen member of the other sample.

In what follows the invariance definition is adopted as the defining characteristic of meaningfulness, but two points are worth making. Firstly, given an NRS representing an ERS, any arbitrary relation defined on the NRS corresponds to a matching one on the ERS. However, general arbitrary relations, defined by exhaustive listing (i.e. defined *extensionally*) of the sets of objects satisfying them, are not of interest. What are of interest are relations defined by means of a simple mathematical formula (*intentional definitions*) on the NRS (see Luce *et al.*, 1990, Chapter 22). Secondly, in statistics we are often concerned with statements about *aggregates* of objects. (Of course, these aggregates may subsequently be used to make inferential statements about the behaviour of individuals, but that is a different matter, as is the fact that the objects may be aggregates themselves.) In particular, our aim is to make descriptive and comparative statements about aggregates. Now, aggregates of objects are not the same as objects. The properties of aggregates and the relations between aggregates differ from those of and between individual objects. For example, an aggregate of objects can have (a property measured by an interval variable which has) a leptokurtic distribution, but the value of (this variable for) individual objects and pairs of objects cannot have such a distribution. In general, though, aggregates inherit their properties from individual objects.

For some situations it is easy to show analytically that all of the legitimate homomorphisms will lead to the same truth value for a statement. In others it is possible to show it, conditional on the observed data, by exploring the admissible transformations using numerical methods. An important subclass of the latter type of problem arises with categorical ordinal data with only a few categories. In one clinical trial that I encountered, the data had four response categories (none, mild, moderate, severe) and the question was whether there was an interaction in a study in which each subject was exposed to a 2×2 cross-classification of factors. In such a situation, all monotonic increasing transformations can be explored by fixing the two most extreme categories (at 0 and 1, say) and letting the two intermediate categories range over this interval (preserving their order, of course).

One might also reasonably expect summated rating scales, which are very popular in the behavioural sciences, to reflect a stronger empirical property than mere ordinality, though not so strong as that in interval scales. Therefore one might not be prepared to countenance *arbitrary* monotonic increasing transformations. One consequence of this is that a relationship between two means computed from such a scale may well have invariance properties within the subclass of transformations one is prepared to consider. For example, we may find that $\bar{x} < \bar{y}$ *for all the transformations we consider reasonable*, if not for all monotonic increasing transformations. In essence we have identified a set of data configurations lying between those for which one group does not stochastically dominate another and for which it does. This is the sort of situation described by Abelson and Tukey (1959) when they said: '*the typical state of knowledge short of metric information is not rank-order information*; ordinarily, one possesses something more than rank-order information'.

Pure representational measurement starts with objects, finds relationships between them in terms of *attributes*, and maps the attributes to numbers (which are values of *variables*), and we may then go on to make a statistical statement about those numbers. There are constraints on what statistical statements can be made as a consequence of the requirement that the truth values of the statements must remain the same under all admissible mappings. The nature of these mappings determines the scale types. In contrast, pure pragmatic measurement starts with objects and maps the objects to numbers (which are values of *variables*) using some operation (which could, of course, be the same operation as was used in the representational approach), and we could then go on to make a statistical statement about those numbers. This statistical statement will have a truth value which is invariant to certain transformations and *we could use that class of transformations to define the scale type of the variable*. Of course, this means that the scale type will depend on the statistical statement – 'the scale type of data may be determined in part by the questions we ask of the data' (Velleman and Wilkinson, 1993a, p. 70). In a sense this makes the scale type the choice of the researcher.

Let us take a concrete example. I have a collection of rocks. I hang them, one at a time from a spring. To each resulting extension of the spring I assign a specific number, in such a way that larger numbers correspond to greater extensions. In this process, the rocks are the objects, the quality of causing an extension of the spring is the attribute, and the number assigned is the value of the variable to be associated with that rock. The ordered nature of the numbers reflects the ordered magnitudes of the attribute. I have deliberately *not* included any attempt to represent any notion of concatenation in the NRS, so any other similarly ordered sets of numbers would provide an equally valid reflection of the ordered magnitudes of the attribute. The representational aspect of this measurement procedure has thus resulted in a family of sets of numbers, which preserve the order. Now, of course, to report the results of the study, I have to pick a member of this family; I have to pick a particular set of numbers. This choice is the pragmatic aspect. It might be made on a purely arbitrary basis, or some other external considerations might lead to a particular choice (e.g. that a particular rock, of which there is a similar one in all such rock collections worldwide, is given the value 1). The fact that some statements will be invariant (empirically meaning-ful) over the pragmatic choice – they will be invariant over transformations preserving the order of the numbers assigned to the rocks – leads to the representational scale type of the variables. An example of such a meaningful statement is that the median value of the variable for one group of rocks is greater than the median value for another group.

However, having chosen a particular numerical representation, that is, one of the members of the family of sets of numbers, I can manipulate them in any way I like – in the words of Lord (1953), 'the numbers don't remember where they came from, they always behave just the same way, regardless'. Given the particular choice of numbers, we do not have to restrict our considerations to those statistical statements which have invariant truth values under alternative legitimate numerical assignments – once we have crystallized our choice, there are no others. So, consider an arbitrary statistical statement using our particular choice of numbers. Take, as an example, that the arithmetic mean of the measured variable for the rocks in one group is larger than the mean for another group. Now, this statistical statement has a truth value which is invariant to linear transformations (but not to arbitrary monotonic increasing transformations). Any alternative set of numbers, related to the original ones by a linear transformation, will yield the same result. Thus one might describe the variable as being of *interval* scale type. But, and this is a crucial point, *this notion of an interval scale hinged on the statistical statement*. Had we chosen a different statistical statement (e.g. a comparison of medians) then invariance of its truth value would have been maintained under a different set of transformations (ordinal ones in this example). While this has been sug-gested as a way of defining scale type, it seems to me to be of limited practical value.

Adams (1966) has a more relaxed perspective on meaningfulness. He suggests (p. 148) that meaningfulness depends on the usefulness of the measurement procedure:

> it is sometimes said that *differences* in Mohs scale hardness measures are not mean-ingful, and that taking the mean of Mohs scale hardness measures of all of the rocks in a class is not appropriate. I conjecture that, to the extent that we have intuitive feel-ings about meaningfulness and appropriateness, we tend to judge these operations and the statements made about them to be meaningful, significant, appropriate, etc., according as they are informative about things of concern to us. I think that we intui-tively regard differences and means of Mohs scale hardness measures as insignificant simply because (so far as we know) they are not reliable indicators of anything besides themselves: i.e., these quantities [*sic*] are not derived measures. If we were to discover, say, that differences in Mohs scale hardness were highly correlated with some other relation between rock specimens besides scratching, then this would immediately confer on differences a significance or 'meaningfulness' which they are not now known to have.

As I will now demonstrate, the empirical meaningfulness concept arising from invariance under admissible transformations leads to a powerful tool for investigating phenomena. I shall begin with the important paper by Luce (1959). On p. 84, he remarks that

> In physics one finds at least two classes of basic assumptions: specific empirical laws, such as the universal law of gravitation or Ohm's law, and a priori principles of

theory construction, such as the requirement that the laws of mechanics should be invariant under uniform translations and rotations of the coordinate system. Other laws, such as the conservation of energy, seem to have changed from the empirical to the a priori category during the development of physics. In psychology more stress has been put on the discovery of empirical laws than on the formulation of guiding principles, and the search for empirical relations tends to be pursued without the benefit of explicit statements about what is and what is not an acceptable theory. Since such principles have been used effectively in physics to limit the possible physical laws, one wonders whether something similar may not be possible in psychology.

Following this idea, Luce described several theorems which restricted the classes of possible relationships between pairs of variables. He based these theorems on the following principle (Luce, 1959, p. 85):

A substantive theory relating two or more variables and the measurement theories for these variables should be such that:

1. (*Consistency of substantive and measurement theories*) Admissible transformations of one or more of the independent variables shall lead, via the substantive theory, only to admissible transformations of the dependent variables.
2. (*Invariance of the substantive theory*) Except for the numerical values of parameters that reflect the effect on the dependent variables of admissible transformations of the independent variables, the mathematical structure of the substantive theory shall be independent of admissible transformations of the independent variables.

To me, when I first heard of this idea and saw its implications, it seemed remarkable, lending extraordinary power to the search for scientific laws. Some examples are given below. In mathematical terms, they are consequences of the solutions of four functional equations called Cauchy equations (Cauchy, 1821; Aczél, 1966). The formulations of the Cauchy equations below are based on Roberts (1979, p. 159):

(I) Let Re' represent either the real numbers, Re, or the positive real numbers, Re^+. Suppose that a continuous function $\psi: Re' \rightarrow Re$ satisfies

$$\psi(x + y) = \psi(x) + \psi(y) \qquad (2.4)$$

for all x and y in Re'. Then ψ has the form

$$\psi(x) = cx$$

for some real constant c.

A proof of this is straightforward and proceeds along the following lines. For example, when $Re' = Re^+$, it follows immediately from (2.4) that $\psi(nx) = n\psi(x)$ for all positive integers n. Now let $x = (m/n)y$ with m and n positive integers. Then

$$\psi\left(\frac{m}{n}y\right) = \psi(x) = \frac{1}{n}\psi(nx) = \frac{1}{n}\psi\left(n\left(\frac{m}{n}y\right)\right) = \frac{1}{n}\psi(my) = \frac{m}{n}\psi(y)$$

By letting $y = 1$ we obtain

$$\psi\left(\frac{m}{n}\right) = \frac{m}{n}\psi(1)$$

Putting $\psi(1) = c$ means that $\psi(r) = cr$ for all rational numbers r. Taking limits on both sides means that the expression also holds for all reals r.

(II) Let Re' represent either the reals or the positive reals. Suppose that a continuous function $\psi: Re' \rightarrow Re$ satisfies

$$\psi(x + y) = \psi(x)\,\psi(y)$$

for all x and y in Re'. Then either ψ is identically zero, or

$$\psi(x) = e^{cx}$$

for some real constant c.

(III) Suppose a continuous function $\psi: Re^+ \rightarrow Re$ satisfies

$$\psi(xy) = \psi(x) + \psi(y)$$

for all positive real numbers x and y. Then

$$\psi(x) = c \ln x$$

for some real constant c.

(IV) Suppose a continuous function $\psi: Re^+ \rightarrow Re$ satisfies

$$\psi(xy) = \psi(x) + \psi(y)$$

for all positive real numbers x and y. Then either ψ is identically zero or

$$\psi(x) = x^c$$

for some real constant c.

These theorems permit one to constrain the forms of relationships between pairs of variables in putative laws relating them. For example, let us consider the relationship ψ between two ratio scales, $f(a)$ and $g(a)$, each mapping to Re^+, $\psi: f \rightarrow g$. The first part of Luce's principle above requires that changing the units of one of the variables (i.e. transforming it by an admissible transformation) will not change the form of the relationship, but will simply result in a change of units to the other variable. So, when f and g are related by the function ψ,

$$g(a) = \psi[f(a)]$$

we have that

$$\psi(rf(a)) = \phi(r)\psi(f(a)) = \phi(r)g(a)$$

for some function ϕ and some positive real number r.

Now, choosing a such that $f(a) = 1$ gives

$$\phi(r) = \frac{\psi(r)}{\psi(1)}$$

from which

$$\psi(rf(a)) = \frac{\psi(r)}{\psi(1)}\psi(f(a))$$

From this it follows that

$$\ln\left(\frac{\psi(rf(a))}{\psi(1)}\right) = \ln\left(\frac{\psi(r)\psi(f(a))}{\psi(1)\psi(1)}\right) = \ln\left(\frac{\psi(r)}{\psi(1)}\right) + \ln\left(\frac{\psi(f(a))}{\psi(1)}\right) \qquad (2.5)$$

If we define a function θ such that $\theta(z) = \ln(\psi(z)/\psi(1))$ then (2.5) can be expressed as

$$\theta(rf(a)) = \theta(r) + \theta(f(a))$$

which is the form of the condition in (III) above. It follows that

$$\theta(z) = c \ln z$$

so that

$$\ln\frac{\psi(z)}{\psi(1)} = \ln z^c$$

and hence

$$\psi(z) = \psi(1)z^c = \alpha z^c.$$

In this derivation we have glossed over various conditions (e.g. that $\psi(1) > 0$), but these can be added without difficulty.

This seems almost magical. It tells us that, without any empirical knowledge of the system being modelled, if two variables x and y are both ratio scale variables, then any relationship between them must have the form $y \propto x^\beta$. For example, knowing simply that the length l of the side of a square and its area A are ratio scales means that we know that $A = \alpha l^\beta$ for some constants α and β, that the volume V of a sphere of radius r has the form $V = \alpha r^\beta$, and so on. It also implies that if a set of variables, $v_1, \ldots, v_p$, all have ratio scales, then the relationship between them has the form

$$c = v_1^{\beta_1} \times v_2^{\beta_2} \times \cdots \times v_p^{\beta_p}$$

Such relationships are illustrated and explored further in Chapter 7.

Luce (1959) also presented other such relationships. For example:

If $f: A \to Re^+$ is a ratio scale and $g: A \to Re$ is an interval scale, and if f and g are related by a continuous function ψ, $g(a) = \psi[f(a)]$, then either $\psi(z) = \alpha \ln x + \beta$ or $\psi(z) = \alpha z^\beta + \gamma$.

If $f: A \to Re$ is an interval scale and $g: A \to Re^+$ is a ratio scale, and if f and g are related by a continuous function ψ, $g(a) = \psi[f(a)]$, then ψ is a constant.

If $f: A \to Re$ is an interval scale and $g: A \to Re$ is an interval scale, and if f and g are related by a continuous function ψ, $g(a) = \psi[f(a)]$, then $\psi(z) = \alpha z + \beta$.

Unfortunately, Luce's principle was not received with universal acclaim: 'Luce's Principle has met with something less than enthusiastic acceptance in all quarters … the Principle's methodological status is dubious at best' (Rozeboom, 1962, p. 542). Both Luce (1959) and Rozeboom (1962) note that there are laws which do not conform to the principle. An example is the law of radioactive decay, which takes the form $q = ae^{-bt}$, which has q a quantity of mass, and hence a ratio scale, and t a time interval, and hence a ratio scale. This being the case, presumably 'there must exist additional criteria by which we can discriminate between laws which fall under the Principle's sovereignty and those which do not' (Rozeboom, 1962, p. 544). The key to this (Rozeboom, 1962; Luce, 1962) seems to be that, in cases which do not adhere to Luce's results, the independent variable is multiplied by a parameter which has units the reciprocal of the units of the independent variable, so rendering the product dimensionless. Dimensionless objects are, of course, not subject to the restrictions explored above. The conclusion seems to be that Luce's 'Principle' does not solve all the problems, but it is certainly a powerful tool in the right circumstances. Luce (1962) retracts the use of the word 'principle'. These issues are explored in more depth in Luce (1990) and Narens (2002, Chapter 6).

2.4.2 Alternative measurement scales

Sometimes alternative scales, which superficially appear to have the same admissible transformations but which are not themselves related by an admissible transformation, can be found for an attribute. This happens when there are subtle differences in the empirical system being represented. Here is an example from Luce *et al.* (1990).

Given a set of objects A, and adopting the representational approach, suppose that the attribute and certain relationships, including an order relationship $\supseteq$, can be represented by an interval scale variable. Call a particular numerical assignment S_1 (e.g. if the attribute in question is temperature, then S_1 might be temperature measured in degrees Celsius). Now, since the scale is interval, the statement that $\bar{x}_1 > \bar{x}_2$ for the mean temperatures of two groups of objects has a truth value which is invariant for all admissible transformations. This means, for example, that the statement would be true when temperature was measured in degrees Fahrenheit if and only if it was true when measured in degrees Celsius.

Now, however, suppose that a mapping from A to Re, also representing the $\supseteq$ relationship by $\geqslant$, can be found such that the numbers assigned to the objects in A are normally distributed. (If A is finite, we could assign normal scores to the elements of A.) Such an assignment (S_2, say) is also invariant up to linear transformations in the sense that any other numerical assignment preserving $\supseteq$ and such that the numbers in A are normally distributed is related to the numbers in S_2 by a linear transformation. So, again, the objects in A can be represented by an interval scale variable: again the statement that $\bar{x}_1 > \bar{x}_2$ for the mean temperatures of two groups of objects has a truth value which is invariant for linear transformations. Not, note, for 'admissible' transformations – there is no empirical relationship beyond ordinality being represented and which could make linear transformations the 'admissible' set.

We thus have two numerical assignments, both of which preserve the truth value of the statement $\bar{x}_1 > \bar{x}_2$ under linear transformations. This means that anyone who adopts the same assignment procedure as that leading to S_1 (i.e. a procedure that preserves the empirical relationships between the objects) will obtain the same truth value for this statement as was obtained using S_1. Similarly, anyone who adopts the same assignment procedure as that leading to S_2 (namely, assigns numbers so that they are normally distributed as well as preserving the order relationship) will obtain the same truth value for this statement as was obtained using S_2. Unfortunately, however, there is no reason to expect the numbers assigned by the two processes to be linearly related: the statement $\bar{x}_1 > \bar{x}_2$ might be true under assignment S_1 but false under assignment S_2.

The numbers in the first assignment process were chosen so that the relationships between them represented the relationships between the objects. That is, the NRS was homomorphic to the ERS. In contrast, in the second assignment process, the only empirical relationship (which was deliberately) preserved by the mapping was order. This means that there is extra freedom in the second assignment. We chose to remove this arbitrariness by requiring the numbers to follow a normal distribution – but this, in itself, was arbitrary – we could have chosen some other distributional form. (And, indeed, the relationship between the means which results when a normal distribution is adopted need not be the same as the relationship resulting when another distributional form is adopted.) It means that, in the second process, the chosen numerical assignment has a pragmatic component – an additional constraint adopted for pragmatic reasons, reasons distinct from those of faithfully reflecting the ERS.

The fact that there may be two or more alternative, nonlinearly related representations of a given set of objects, both possessing the same scale invariance properties, has led to some confusion. Anderson (1961), for example, remarked that 'possession of an interval scale does not guarantee invariance of interval scale statistics'. This was discussed in a stimulating paper by Velleman and Wilkinson (1993a) and further in Hand (1993a, 1996) and Velleman and Wilkinson (1993b). As pointed out in Hand (1993a), Anderson's (1961) example of whether to measure duration or speed, of which Velleman and Wilkinson (1993a) said 'both are valid interval scales, and yet statistics computed on one form may be quite different from those computed on the other', tells only part of the story. The full story was given in the fuel efficiency example in Section 2.2.7 above. If one is interested in speed then numbers preserving the relationships between speeds can be chosen and these will preserve the interval scale structure of speed. The same is true for duration, if one is interested in duration. However, speed and duration are not the same things – the empirical relational structures being modelled are different, with different internal relations – so it is no surprise that analyzes of the two corresponding numerical relational systems may yield different conclusions. Care must be taken to ensure that different alternative representations are describing the same empirical operation. If the researcher expects them to yield identical conclusions, it suggests that the researcher believes the two variables, speed and duration, are tapping the same underlying attribute, and it is this which is the real object of study, and not either of speed or duration. If this belief is correct, then the numerical assignments can only be ordinally related to the empirical system (they are only ordinally related to one another). To put it another way, the representation is representing only the ordering of the objects, and nothing more, so that only statements which are invariant to ordinal transformations should be made. One must not fix the statistical statement while changing (transforming) the data without being aware that this changes the statistical model and may change the meaning of the statement, in just the same way that fixing the data and changing the statistic may change the model and the meaning of the statistical statement.

2.4.3 Transformations and statistics

The discussion in the preceding subsection does not mean that the researcher should not experiment (so that 'as a good scientist he is willing to entertain the possibility that what he thinks might not be the way the world really is'; Velleman and Wilkinson, 1993b) by transforming the data (in this example, by analyzing both duration and speed). Simply that one should be aware of the implications of such transformations. Generalized linear models (McCullagh and Nelder, 1989) represent something of an advance in this context. They establish a relationship between the conditional mean and a linear function of the covariates: the response variable is not transformed, and so remains on the original scale.

If the data have been measured on some scale such that the relationship between numbers reflects the relationships between objects, then adopting a transformation outside the class of admissible transformations is risky. But such transformations are often adopted for statistical reasons – for example, to achieve approximate normality or to stabilize variances. Indeed, many are based on subjective assessment of the distribution of the data. The trouble is that, after a non-admissible transformation, the mathematical relationship between the numbers which reflects the relationship between the objects will have changed. If, before the transformation, the mathematical operation of addition modelled a physical operation of concatenation, then, after the non-admissible transformation, some other mathematical relationship will model it. If one continues to use addition, one is no longer representing that concatenation operation – see Section 2.2.7. Given that one wants to compare the arithmetic means of two groups, then comparing the arithmetic means of the logarithms of the measurements would be incorrect (in fact, of course, in terms of the raw data one would then be comparing the geometric means). Of course, this is predicated on the assumption that it *is* the arithmetic mean of the groups one really wants to compare. One has to be very clear about what is the real question under investigation (for a discussion of such issues, see Hand, 1994). The US Food and Drug Administration, the regulatory authority which oversees applications for new medications, stresses the caution which needs to be exercised: 'Unnecessary data transformation should be avoided. In the event that a data transformation was performed, a rationale for the choice of data transformation *along with interpretation of the estimates of treatment effects based on transformed data should be provided*' (FDA, 1988; emphasis added).

To illustrate, suppose one wants to compare the effects of two diets on the weight of calves. A basic experiment to permit such a comparison will involve giving one diet to one calf and the other diet to another, and comparing the gains in weight. This has the disadvantage that it does not allow for the uncontrollable variation between calves – some grow faster than others and we would not be certain that any apparent difference between the two diets could not be attributed to other differences between the two calves. To control for this, we instead give the first diet to several calves and the second to several (different) calves, and compare the total weight gain of the two groups. By such a process one hopes, based on solid properties of probability, that the uncontrollable differences between the natural growth rates will be diluted relative to the effect of the diets (one would be very unlucky if all of those calves with the faster intrinsic growth rate happened to fall in one group). Of course, direct comparison of the total weights of the two groups is not sensible except in the special case that the groups have the same numbers of calves, so one normally standardizes by the numbers in each group – and uses the arithmetic means. Various tests can then be used. In this example, with arithmetic mean being the focus of interest, it would be wrong to use median, geometric mean, or some other 'average'. Similarly, it would be wrong to (nonlinearly) transform the data and then use arithmetic mean. (As Frank Yates said (Michael Healy, personal communication) 'the farmers are not being paid by the log(kg)'.) In particular, it would be wrong to transform non-normally distributed data to normality and perform a t-test on the result since the test would not be comparing arithmetic means on the original scale, which is the scale which maps the empirical concatenation relationship ('putting several calves on the

Table 2.2 Proportions of women developing depression. (Brown and Harris, 1978).

x_1	x_2	
	Event	No event
No intimacy	0.32	0.03
Intimacy	0.11	0.01

scales simultaneously') to addition. That is, the question being explored, the hypothesis being tested, would not be the one to which an answer was required.

As another example, Keene (1995) describes a crossover study with the primary end point being gastric half-emptying time, in minutes. Comparison between treatments A and B in the study (by a *t*-test?) is non-significant. However, Keene points out that the standard deviation increases with group mean, and says this 'is suggestive of the need for a log transformation'. An analysis with log-transformed data leads to a treatment effect which 'approaches significance'. The trouble is that the two analyses are answering different questions and can give results which not only disagree in their degree of significance but are even contradictory. For example, with artificial numbers just to illustrate, consider two groups of only two subjects, the first group yielding values 1 and 9 and the second values 2 and 6. The means of the groups are 5 and 4, respectively, so that the first group takes longer. However, after (natural) log-transformations the means are 1.098 and 1.243, so that the second group appears to take longer. One can get back to the original units of measurement by antilogging the final result, but this does not resolve matters. This result has been achieved by, effectively, multiplying together the raw scores, an operation which does not correspond to an aspect of the empirical system. Addition, on the other hand, makes perfect empirical sense. (Consider subsequent prescriptions of the medicine, yielding a concatenated gastric emptying time for the complete set of episodes.) The point is that one has to be very aware of the question one is trying to answer, and to relate any proposed transformations to that question.

Of course, causes are often multiplicative in their effects. When this is the case the analysis can be simplified by log-transforming the model to turn it into an additive one. The distinction between multiplicative and additive effects has sometimes led to confusion. Table 2.2 (from Brown and Harris, 1978) shows, for a sample of women from south-east London, the proportions developing depression over a one year-period cross-classified by the presence or absence of intimacy with a husband or boyfriend (x_1) and whether or not they experienced a severe life event in this period (x_2). Brown and Harris (1978) used an additive model and noted a clear interaction: the difference between the values in the upper row is 0.29, while that in the lower is 0.10, a marked difference. However, Tennant and Bebbington (1978) used a multiplicative model and obtained a ratio for the values in the upper cells of 10.7 and an almost identical ratio for the values in the lower cells of 11.0: apparently there is no interaction. Which of the two analyses is correct depends on one's aims, and on the type of question one wishes to answer.

2.4.4 Measuring change

In the weight of calves example, we have made the basic assumption that it was the difference in weight gains attributable to the diets which was of interest. Assuming that the amount of money a farmer receives for an animal is proportional to the gain in weight, then the weight gain will tell us how much profit the farmer will make (I am ignoring the cost of raising the animal, so that we do not get too far from the essence of the example!). The difference in weight gains will tell us the differences in profits which result from using the two diets. It can be argued instead, however, that the difference in *percentage* gains is the primary interest (where percentage gain is 100 times the increase in weight divided by the original weight). After all, the farmers will be concerned with the rate of return on their initial investment. This is not essentially different from the previous question, except now the arithmetic means of the percentage gains will be compared. Again, using any average of the percentage gains other than the arithmetic mean would be inappropriate, as would transforming the data prior to using the arithmetic mean.

An equivalent, of course, is simply to use the ratio of final to initial weights. Like the percentage weight gain, this has the advantage of being unit-free so that comparisons can be made with changes in other variables measured in different units. However, if used to measure change, it has an asymmetry which can be disturbing. If the pound increases in value by 25% against the dollar, then the dollar decreases in value by only 20% against the pound. The ratio of final to initial values is also not additive because of the changing baseline. For example, two successive increases of 10% lead to an increase of 21%, not 20%. Törnqvist *et al.* (1985) recommend the use of the log of the ratio of final to initial values, on the grounds that it is the only measure of change which is:

(i) symmetric, $K(x,y) = -K(y,x)$;
(ii) additive, $K(x,y) = K(x,z) + K(z,y)$; and
(iii) normed, $dK(x,x)/dx = 1$, to force K to behave approximately like $x/y - 1$ when $x \approx y$.

These extra conditions are (axiomatic) pragmatic conditions, derived externally to the empirical system being represented.

In comparing two weight gains, one might go even further and argue that it is the *ratio* of percentage gains which is really of interest. This tells us the relative rate of return of one diet compared to the other. Whether or not this is appropriate will depend on the context. In the calf weight example, it is all very well to know that one diet is twice as profitable as the other (which is what the ratio of percentage gains tells us), but perhaps it is more useful to know that one diet will earn 5% more profit on the basic investment than the other (the difference in percentage gains), or that it will earn £100 per animal more than the other.

2.5 Measurement, statistics and meaningfulness: a twentieth-century controversy

The struggle to understand the nature of measurement, and in particular its relationship to statistics, is very much a twentieth-century story. This section describes an aspect of this struggle: the amount of freedom one has to apply arbitrary statistical manipulations to data. It illustrates the difficulty researchers have getting to grips with ideas which, with the advantage of some decades of hindsight, seem clear enough. We begin with a paper by one of the fathers of classical psychological test theory, Frederic Lord (1953). In this entertaining paper, Lord describes two groups of American football players (freshmen and sophomores), each of whom is assigned a number by a machine. When both groups had received their numbers, which were supposed to have been assigned at random, the freshmen complained that theirs were too low.

The professor responsible for the machine which handed out the numbers believed that numbers on less than an interval scale should not be added (and hence that their means and standard deviations should not be calculated – they are 'meaningless' in the sense discussed above) and he sought the advice of a statistician. Much to the professor's consternation, the statistician, regardless of the fact that the numbers presented to him lay on a purely nominal scale, nevertheless went ahead and conducted a location test. The calculations led the statistician to conclude that the two sets of numbers were highly significantly different. The statistician claimed that it was legitimate to carry out the test since 'the numbers don't remember where they came from, they always behave the same way, regardless'. This phrase, 'the numbers don't remember where they came from', has served as a clarion call for those who wished to justify operating on data in arbitrary ways. The paper concludes with the professor accepting the statistician's approach, and calculating means, etc., as if the numbers lay on an interval scale.

The paper itself is not completely unambiguous in its statement of the moral of this little tale, but was in Lord (1954): 'The conclusion is that nominal and ordinal numbers (including test scores) may be treated by the usual arithmetic operations so as to obtain means, standard deviations, and other similar statistics from which (in certain restricted situations) correct conclusions may usefully be deduced with complete logical rigor.'

Numbers, of course, do have the necessary properties to permit these manipulations, and the conclusion arrived at – that the mean of one group of numbers is lower than that of the other – is fine. But all that means is that the freshmen have been assigned numbers lower on average than the sophomores. Indeed, as far as differences between the two sets of numbers go, one can perform any mathematical manipulations and conduct any tests one likes. If any such operation establishes a difference between the two sets of numbers then a difference between the two sets of numbers has been established.

However, the numbers were assigned arbitrarily ('at random'), and were in no sense a measure of any ability or property of the players – the size of the numbers does not signify anything. The calculation, and the conclusion that one set of numbers is on average lower than the other is correct, but it implies nothing about the football players. The statistician is right in the sense that the conclusion he has drawn about the two sets of numbers is valid (one set has a smaller mean than the other set), but no inference may be made from this about any property of the players. Although any numerical operations may be performed on the numbers to legitimately draw conclusions about the numbers as numbers, when regarding the numbers as measurements of some underlying property, only those operations may be performed which correspond to some empirical operation between the objects being measured. In this case, the numbers do not correspond to any empirical property – so there are no empirical operations and hence no legitimate numerical operations. Numerical manipulation may lead us to draw conclusions about the numbers which have no counterpart in the empirical system. Of course, the complaint of the freshmen may still be regarded as legitimate: they had been assigned numbers which were smaller, on average, than those assigned to the sophomores, even if those numbers did not represent any property of the players.

In summary, we must separate the numbers and the conclusions which we may wish to draw about them from the system being modelled and the conclusions we may wish to draw about it. Any numerical operation and relationship is legitimate for the former, but for the latter only those corresponding to empirical counterparts make sense.

This seems clear enough now, but a lengthy exchange took place in the scientific journals, in which various authors strove to make the distinction between the two uses of the numbers. We illustrate with extracts from some of these articles.

Burke (1953) opines that 'the use of the sample mean and standard deviation does no violence upon the data, whatever the properties of its measurement scale'. However, he also comments that 'the classification of a scale as additive depends upon the presence or absence of a certain correspondence, expressed in sets of axioms, between the numbers of the scale and the objects to which they refer – with, of course, appropriate ordering relations and combinative operations for each', before explicitly separating the statistical computations from the interpretation: 'The statistical technique begins and ends with the numbers and with statements about them. The psychological interpretation given to the experiment does take cognizance of the origin of the numbers but this is irrelevant for the statistical test as such.' The implication is that statistics is about manipulating the numbers while interpretation, meaningfulness, and so on are within the domain of (in this case) psychology. Unfortunately, Burke then confuses things by concluding with an example in which 100 measurements are taken on each of two sticks and the two means computed and compared by a t-test. He says: 'A moment's reflection will show that we are not at all concerned with the additive nature of the scale for length ... We make no interpretations whatever about adding 100 sticks together – there are only two sticks.' But surely this is not right: implicit in the entire exercise is the notion of an imaginary population of identical sticks, measured with error, and for which the length attribute has the appropriate concatenation property.

Senders (1953), who was later to be severely criticized by the opposing camp (Kaiser, 1960), takes issue with Burke's conclusions. She says:

> When numbers have been assigned to objects according to some stated rules these numbers can indeed be manipulated in any way we desire. But once the manipulations have been completed and the tests made, a dilemma arises. If we have performed operations on the numbers which we could not perform on the objects, we must choose between two interpretive procedures. We must either assume that (a) what is true of the numbers is also true of the objects, or (b) what is true of the numbers is not necessarily true of the objects.

To address (a) Senders then gives two examples showing how adding numbers which have been assigned to represent a merely ordinal relationship can lead to contradictions. And, regarding (b) she says:

> If we accept conclusion b, on the other hand, we are in an even more ridiculous position. Though our statistical procedures may have been perfectly justified and our interpretations correct when considered strictly in relation to the numbers, we can make no interpretation about the properties of the objects to which the numbers have been assigned. As psychologists, we can draw no conclusions about responses, organisms, or behaviour, but only about numbers.

She concludes:

> Since psychologists are presumably more interested in the behaviour they describe with numbers than in the numbers themselves, they will learn more if their statistical

techniques correspond with the properties of the set of numbers as a measurement scale than if these properties 'have no effect upon the choice of statistical techniques for representing and interpreting the numbers.'

Bennett (1954) objects to Lord's (1953) conclusion on the grounds that the assigned numbers are merely serving in a nominal way: 'to make the mean meaningful, we must insist upon some cardinal aspects of lowness to highness in respect to our 2, 6, and 9. When such is the case, we have imposed a lot more meaning than just football-player identification. We *have* cardinality.' He concludes: 'Our Professor X had best re-retire; his helpful statistical friend had best return to his TV set.'

Behan and Behan (1954) comment that 'if the empirical procedure does not contain an operation that corresponds to one of the mathematical operators, then that operator may not be used to manipulate the numbers obtained. The use of an operator not provided for in the empirical procedure results in numbers that bear no relationship to possession of the quality measured', and they earlier say 'while we are *able to* add, multiply, divide, take the square root, etc., of nominal numbers, it is an exercise in arithmetic manipulation only', having pointed out that the random assignment of numbers to the players is using them in a nominal way only.

Boneau (1961) defends his argument (in Boneau, 1960) in favour of 'the use of the *t*-test in many typical psychological situations where there are measurement considerations'. He says:

Note that we make absolutely no assumptions about any underlying dimension and we deal only with the numbers assigned by the measurement operation. We, and statistical tests, are concerned only with this operationally defined manifestation of the underlying dimension. The statistical test cares not whether a Social Desirability scale measures social desirability, or number of trials to extinction is an indicator of habit strength. It does not even care whether the measuring scale is monotonically related (or unrelated) to the underlying dimension.

But, towards the end of his note, he says: 'Certainly one cannot ignore the problem of measurement. It would seem to make a difference to psychology whether or not (and in what sense) the numbers we assign by means of an intelligence test are related to the underlying concept, but the problem is a measurement one, not a statistical one.' If we had adopted some rule in our measurement procedure, which related values of an attribute to numerical values, and if, for example, a *t*-test concluded that there was a difference between the mean values of two groups, then we could confidently conclude that there was some kind of empirical difference between these groups. However, unless the numerical assignment rule was based on a representation of a property which mapped to addition, then it would be difficult to decide to what sort of empirical difference this mean difference corresponded.

Anderson (1961) is clear about what he views as the irrelevance of measurement scales. Commenting on Siegel's (1956) assertion that the variables involved in a *t*-test must have been measured on at least an interval scale, he says that in cases (when comparing the locations of two groups) where equinormality holds: 'This statement of Siegel's is completely incorrect. This particular question admits of no doubt whatsoever. The *F* (or *t*) test may be applied without qualm. It will then answer the question which it was designed to answer: can we reasonably conclude that the difference between the means of the two groups is real rather than due to chance?' He cites Lord (1953) in support of this position and goes on to say that the conclusion is also valid in situations where equinormality does not hold.

Anderson's attitude seems to be that while meaningful statistical statements will (by definition) be invariant under admissible transformations, such transformations are not the only ones that a researcher might wish to consider, so that notions of invariance under such transformations are irrelevant. The reason why a researcher might wish to consider other scale changes is that the scales being used to produce scores are simply indicants for the underlying psychological variables (or, I would say, have a substantial pragmatic component). Thus, 'it is evident that an indicant may be an interval scale in the customary sense and yet bear a complicated relation to the underlying psychological dimensions. In such cases, no procedure of descriptive or inferential statistics can guarantee invariance over the class of scale changes which may become necessary' (Anderson, 1961, p. 312). One of his examples involves the choice between time intervals and speed as the response variable in an experiment. Both of these are at least interval scales and yet analyses of them may yield contradictory results. He remarks that 'evidently then, possession of an interval scale does not guarantee invariance of interval scale statistics' (p. 311). Unfortunately, I think this example is poorly chosen, since it simply confuses two different empirical types of relationship (that involving speed and that involving duration). The doubt about the nature of any empirical constraints does not arise from the difficulty of identifying them.

In discussing Anderson (1961), Adams *et al.* (1965, p. 124) say that

> what he is asserting is that the validity of a statistical test depends only on the assumptions of a statistical model with respect to a distribution of numbers – and not on properties of the measuring scale. We agree with Anderson and with Burke (1953) on this conclusion with respect to the statistical test as such. However, it is obvious that the statistical hypothesis is a statement about statistical operations, and as such may be empirically meaningful or meaningless, depending on the nature of the statement and the scale type. Therefore, although it is true that the statistical model does not explicitly mention the type of scale, the practice of ignoring scale type in making statistical tests could lead to the formulation of empirically meaningless hypotheses, and hence to statements about the test which are not empirically significant.

(See also Adams *et al.*, 1964.) Adams *et al.* (1965, p. 122) also comment that

> a second limitation of the theory of appropriate statistics is that it would not appear to be applicable to systems of measurement for which there are not clearly defined sets of permissible transformations. For example, this would appear to rule out test scores and other measurements of this kind ... The importance of this limitation cannot be over-stressed. Test scores are the sorts of measurements over which disagreement most often arises concerning questions of appropriateness. We believe that the failure to provide a precise definition of permissible transformations for such scales is not just a defect in our formulation – the imprecision of the concept in these applications is inherent, and no deeper analysis can discover what the permissible transformations 'really are' in these cases. This has an important bearing on the controversy over the validity of Stevens' theory of appropriate statistics, since much of this has centred around measurements whose classes of permissible transformations are undefined.

Is this not a recognition of what we would call the pragmatic aspects of measurement?

Gaito (1960) rather muddies the water. He quotes from Kempthorne (1955) 'the fact which has been noticed by most statisticians, that the level of significance of the analysis of variance test for differences between treatments is *little affected by the choice of a scale of measurement for analysis*' (Gaito's italics). This is a rather odd statement. Certainly one can carry out ANOVA on numbers on any scale. But the result will only have empirical meaning if the statistical operations and statements involved have corresponding empirical relationships. The level of significance is completely unaffected by the choice of measurement scale: but whether it means anything is not. Gaito goes on to stress the *statistical* assumptions (such as normality) underlying the ANOVA manipulations and conclusions.

He picked up the issue again in Gaito (1980) in response to a spate of statistics texts which adopted the view that scale types precluded the use of certain statistical techniques (describing one of them as reaching 'the height of absurdity'). He felt that the authors of these books 'apparently do not read the statistical journal literature, inasmuch as a number of articles on this topic showed clearly that measurement scales are not related to statistical techniques ... Scale properties do not enter into any of the mathematical requirements for the various statistical procedures. I have not known of any mathematical statistician who agreed with the Stevens misconception.' This seems a little strong to me. Stevens's argument was not about requirements for the statistical procedures in general, but about requirements for them if one wanted them to mean anything about the empirical system being studied. Gaito then went on, quite explicitly, to separate the meaning of the numbers (issues of measurement scale and the domain of the researcher – such as a psychologist) from the statistical operations which might be applied, concluding: 'In mathematical statistics literature one will not find scale properties as a requirement for the use of the various statistical procedures. This requirement was merely a figment of the imagination of a number of psychologists because of a confusion of measurement theory and statistical theory. Statistical procedures do not require specific scale properties.' This is right. But presumably psychologists are interested in more than drawing conclusions about sets of numbers – the point made by Senders (1953).

Townsend and Ashby (1984) took Gaito (1980) to task, asserting that assumptions about distribution type, parameter values, and so on 'must be taken in addition to the constraints of measurement'. And they say, referring to Lord's (1953) football numbers example:

> The outrage the freshmen suffered because of their minuscule numbers was an outcome of an invalid external interpretation, having absolutely nothing to do with the measurement scale or the way the numbers were assigned ... Perhaps a generous interpretation of the above contention would be to question whether a statistical analysis is ever of interest without reference to an empirical system.

They wearily conclude: 'It seems unlikely that uncompromising antagonists to the measurement approach will be convinced by the present reasoning, or any other for that matter.'

Gaito struck back in Gaito (1986), although his section 'Statistical conclusions vs empirical conclusions' seems to be an explicit recognition of the difference between the purely statistical conclusions (relating only to the numbers and their distribution) and the empirical conclusion. He quotes Binder (1984, p. 475):

> investigators computed means and standard deviations with IQs, correlated IQs with many other variables (some of which were nominal), and tested hypotheses involving the IQ with analysis of variance. What resulted was rich, empirical knowledge, a theoretical structure that matches any other structure in the social sciences for predictive

usefulness ... The point is that important empirical advances were made by procedures that were said to be inappropriate by Stevens, Siegel, and the others.

Of course they were. But IQ measurement procedures contain a large pragmatic component: there are arbitrary choices made in developing such measurement instruments,
beyond those constraints implied by the empirical system. This does not prevent them from
yielding useful predictions, correlations, and relationships with the results of other measurements. Binder's statement that 'what resulted was rich, empirical knowledge, a theoretical structure that matches any other structure in the social sciences for predictive
usefulness' does not lie in contradiction to measurement theory when one recognizes the
pragmatic components of the measurements he is describing.

In a review of the controversy, Stine (1989, p. 153) says: 'The origin of the numbers used
in a statistical analysis is central to the interpretation of the analysis. For the scientist, what
counts is that there is an empirical analog to the numerical results, a situation that will
occur only when the analog is viewed with the appropriate scale', and (p. 154) 'When all
of the arguments are considered, concern for the empirical origin of one's numbers would
seem central to the successful conduct of science.' True, but one should not forget the role
of pragmatic constraints in measurement. Even if one is not clear about precisely what
empirical operations are being preserved by one's 'rule for assigning numbers to objects',
one might nevertheless have a measurement which is powerful for predictive purposes. A
great deal of therapy and social policy, not to mention other areas, depends on this.

MacRae (1988) in a section entitled 'Scale type is not an intrinsic property of data' gives
an example of

> a study in which we count the number of food pellets dispensed to an animal as a
> reward in a learning experiment. The number we record is on an absolute scale of
> pellet count. It may well be on something close to a ratio scale of food quantity, but
> if the pellets vary appreciably in size then it is not even an interval scale of food quan-
> tity ... From a more psychological point of view, it is plausible to view each recorded
> number as a measurement on an absolute scale of number of rewards ... Whether or
> not the number is on an ordinal, an interval or a ratio scale of amount of reward can-
> not be determined except in the context of an even stronger theory of the rewarding
> effect of successive food pellets.

Surely the ambiguity arises here because of the weak link between what we want to measure (reward) and what we do measure (some aspect of the quantity of food pellets). Total
weight of food will be measured on a ratio scale, but it will only have a weak (we might
reasonably assume ordinal) relationship to the amount of reward. We can interpret it in a
purely representational mode – but then, because of the weak link and lack of clear empirical operations between different magnitudes of reward, we must at best regard it as an
ordinal scale. Conversely, we can interpret it as having a pragmatic aspect – and take number of food pellets, or weight of food, or any of the other measures we like, as a measure of
(and definition of how we are measuring) reward.

The step from scales defined purely in terms of representations of empirical relationships
to those used in practice is described in Stevens (1951, p. 26):

> Most of the scales used widely and effectively by psychologists are ordinal scales. In
> the strictest propriety the ordinary statistics involving means and standard deviations

ought not to be used with these scales, for these statistics imply a knowledge of something more than the relative rank order of data. On the other hand, for this "illegal" statisticizing there can be invoked a kind of pragmatic sanction: in numerous instances it leads to fruitful results.

Hence our 'pragmatic' aspect.

The great statistician John Tukey presented a variant on this point of view. Abelson and Tukey (1959), comment that declaring a scale ordinal, and so restricting oneself to non-parametric statistical techniques designed for the analysis of rankings, is 'to restrict the flexibility of statistical analysis severely and unnecessarily'. They illustrate by saying (p. 416):

> If you say your non-metric information is rank order and nothing more, then you implicitly acknowledge the possibility of a 'true' sequence of the form (a,a,a,a,b) … If you say that this possibility is inconceivable, then you really have more non-metric information than mere rank order … Here lies the heart of the situation. Quite commonly, when we say we only know rank order, we actually know more than this, but don't know how to express what else it is that we know.

Tukey makes the same point in Tukey (1986a): 'the typical state of knowledge short of metric information is not rank-order information; ordinarily, one possesses something more than rank-order information.' And (Tukey, 1986a, p. 244): 'The limitations discussed by Luce [Luce (1959, p. 84)] do not control which statistics may "sensibly" be used, but only which ones may "puristically" be used.' He goes on to say that, though the Luce and Stevens proscriptions on what analyses may be used seem cogent and logical, they fail for two main reasons: '(1) A lack of adequate recognition that knowledge is approximate not precise. (2) A lack of appreciation that all useful conclusions are not fundamental.' One might regard our 'pragmatic' aspect of measurement as an attempt to bridge the gap implied by point (1). As we said above, the pragmatic aspect crystallizes the measurement operation, yielding actual numbers.

Point (1) leads (Tukey, 1986a, p. 246) to

> the failure to recognize that many scales, such as the early scales of temperature, are approximate interval scales … Large differences in variability between two populations of temperatures remain large on all scales … Many assignments of scale values to ordered classifications, assignments which may be either equally spaced or more carefully or appropriately chosen, produce approximate interval scales … But the approximation is still there; arithmetic means and measures of spread can be very useful, provided they are interpreted with proper caution.

Gardner (1975) also supported the position that many measurements used in practice have scale types somewhere between interval and ordinal. He gave as an example 'summated rating scales' in which the score is a sum of a set of responses to questions (each of which might be, for example, binary or ordinal in its own right).

Baker *et al.* (1966) carried out a practical investigation of the effect, on the conclusions of a *t*-test, of perturbing the intervals between consecutive values in a nonlinear way. From one perspective, this is simply an exploration of the robustness of the *t*-test. Their conclusion was that for inference (especially if the sample sizes in the two groups were equal and if the tests were two-tailed) the results were robust to the perturbations, but for descriptive statistics they were less so. Labovitz conducted similar studies. Labovitz (1967) took a four-category ordinal response variable and assigned score distributions ranging from equally spaced to highly skewed (e.g. 0, 1, 2, 10). He summarized the results in Labovitz (1970) as 'the monotonic scoring systems produce largely similar point-biserial coefficients, t-tests, and critical ratios'. Of course, such conclusions are only valid if one believes the empirical relationship between objects to be 'more than ordinal' in the sense that there are limits to the amount of (ordinal) distortion one is prepared to contemplate. By choosing a particular numerical coding for the data, one is adopting a particular pragmatic constraint. The claim arising from these studies is that the results are fairly invariant over the particular constraint adopted.

Not everyone was happy with this. Champion (1968) wrote: 'When a person has ordinal data, it is ordinal data regardless of how close he would like it to approach interval data. Nothing save violating assumptions will bring his data and the higher-level statistical techniques into "harmony".' Morris (1968) agrees that sometimes it may be appropriate to assume that (for example) intervals between categories are equal, so that an interval scale can be used, but argues that this is not always the case: 'It is likely to be much more difficult to compare the interval between the categories "agree strongly" and "agree" with the interval between the categories "agree" and "no opinion".' It seems to me that these criticisms ignore the restriction that Baker *et al.* (1966) and Labovitz (1967) make, that their data are *more than* ordinal (in the sense that arbitrarily large nonlinear transformations are not allowed), to the extent that the recorded numbers are slight deviations from an unmeasurable underlying interval scale. Labovitz (1968) replies to Champion and Morris, but does not stress this point. In any case, Labovitz was not to give up easily. He went on, in Labovitz (1970), to extend his 1967 results on the degree of error in treating an ordinal variable as interval by exploring the relationship between an ordinal prestige ranking and the suicide rate for 36 professions, using 20 different (order-preserving) scoring systems to assign numbers to the prestige ranking. Two of these alternative scoring systems were the original NORC prestige score and the simple ranks score (1 to 36), and the other 18 were generated randomly (subject to the order restriction). Labovitz assessed the agreement between the resulting sets of scores using Pearson product moment correlation coefficients. All 190 of these were above 0.90 and 157 of them were 0.97 or above. He also looked at the correlations between each of these 20 scales and suicide rate. For the original prestige rankings this was 0.11 and for the 20 alternative scoring systems it ranged from 0.09 to 0.15. Of these results he says (p. 520): 'The similarity in predicting an outside variable is extremely high ... Given some degree of unreliability in occupational prestige and suicide data, and the rather crude measurement procedures, these results substantiate the point that different systems yield interchangeable variables.' Of course, the value of these results depends on the extremeness of the differences between the alternative systems, a point which he acknowledges, illustrating with an extreme case which effectively binarizes the scores. Mayer (1970) tackles him on this:

> all he has shown is that the Pearson r is fairly stable with respect to non-linear monotone transformations on the numbers assigned to ranks ... I fail to see how the author can justify treating ordinal data as interval data in all cases. Clearly multivariate techniques such as multiple regression, discriminant analysis, or even the simple Hotelling's T^2 test can not be used with arbitrarily assigned ranks ... By treating an

ordinal variable as an interval variable, the researcher is declaring that its underlying distribution is meaningful. It is unclear, yet crucial to know, how to decide which distribution an ordinal variable possesses ... Unfortunately, many procedures which are based on statistics that are invariant under both scale and location changes are extremely sensitive to departures from the underlying assumptions.

Schweitzer and Schweitzer (1971) also comment on the robustness of r to arbitrary monotonic transformations, supporting Mayer's implicit contention that the choice of this measure to support Labovitz's arguments was perhaps misleading.

Vargo (1971) was also unhappy with Labovitz's arguments, concluding that 'the paper's methodology does not produce the asserted findings'. He also commented that 'the monotonicity constraint, the large range differences (more than two orders of magnitude), and the large number of occupational items in the original data set all combine to force a free random number generation process to be a quasilinear transformation process ... What Labovitz demonstrated in a particular case is the power of monotonicity in correlation measurement. The significance is numerical, not methodological.'

Labovitz responded to these various criticisms in Labovitz (1971), but we will not go into the details here. In Labovitz (1972) he presents his relaxed view of measurement scales, arguing that measurement scales are not (always?) clearly defined in terms of the empirical systems, but that we should be willing to contemplate the possibility that our data have arisen from other scale types. This notion has more force in the social sciences (Labovitz's area) than the physical sciences. Labovitz (1972, p. 15) says:

In experimenting with numbers and techniques, the system should be kept open, because closure may be premature. Especially in a young science, we should apply several techniques and use several different ways of expressing 'it' in numbers. We should not say, for example, that marital status is nominal, but that marital status has been treated as nominal: and perhaps we should ask 'would it be helpful to treat it as ordinal or interval?'

Other authors have also adopted a position similar to that of Labovitz (1972). For example, Borgatta (1968): 'As a scientist, I would think, one would have to be most concerned with improvement of his measures ... The answer [that science ordinarily seeks] may be imperfect as found, but the task is one of successive approximations. It is not one of avoiding answers because they will not be completely accurate representations of the "real" relationships.' He concludes: 'Our motto might become: "Some parametric estimate is better than no parametric estimate".' This is further reinforced in Borgatta and Bohrnstedt (1980, p. 160):

An ordinal scale is appropriate if we assume that only the properties of 'greater than' and 'less than' define an underlying latent construct. We doubt this is the case for most variables of interest to social scientists. As it seems to us that most constructs are conceptualized as continuous and can be thought of as reasonably distributed in the population using a bell-shaped curve as a model, we see no reason not to analyze the manifest data using parametric statistics, even though they are imperfect interval-level scales.

Velleman and Wilkinson (1993a, p. 68) (further discussed in Hand, 1993, and Velleman and Wilkinson, 1993b) say: 'Experience has shown in a wide range of situations that the application of proscribed statistics to data can yield results that are scientifically meaningful, useful in making decisions, and valuable as a basis for further research.' However, as we have discussed above, there is no such thing as a proscribed statistic. There are only meaningful and meaningless statistical statements. By all means carry out a nonlinear monotonic transformation of a ratio scale, and analyze the data using whatever statistics you like, but be aware that a statistical statement about the arithmetic mean of the result may not also be true for the arithmetic mean of the original data. Velleman and Wilkinson (1993a) later say 'Good data analysis … is a general search for patterns in data that is open to discovering unanticipated relationships. Such analyses are, of course, impossible if the data are asserted to have a scale type that forbids even considering some patterns … A scientist must be open to *any* interesting pattern.' But this is surely wrong. The data type does not 'forbid considering some patterns'. The relationship between the data and what they describe merely reflects the fact that certain kinds of statistical statements may have truth values which are not invariant under some transformations. Whether or not a transformation is sensible to contemplate depends on the question one is trying to answer.

2.6 Parametric versus nonparametric statistics

Any mismatch between theoretical predictions and observations can be explained in two ways: either the theory is inadequate or there is measurement error (or, and probably more usually, both). Implicit in representational measurement is a theory about the objects: that they are related (in terms of some kind of behaviour – the attribute) in certain ways that form the relationships of the ERS. So, for example, we might assume that the objects are ordered and satisfy some concatenation relationship, despite the fact that we can establish this only for a finite number of (sets of) objects. The question of establishing relationships for the ERS is tightly bound up with the problem of induction – the core of statistical inference itself. If the assumed relationships do not hold, or hold only approximately, then we should expect the predictions and inferences drawn from our numerical calculations not to hold or to hold only approximately. But such approximations will also appear if the measurements are not perfectly reliable – measurement error manifests itself most clearly in the fact that repeated measurements of the same attribute of the same object (using the same measuring instrument) can yield different values. It is a ubiquitous aspect of measurement in all areas. As such, one might argue, it should be integrated into the theoretical structure describing measurement. Attempts in this direction have been made by, for example, Falmagne (1979, 1980).

In situations where the theory describing an empirical system, or relating that system to a numerical system, is poorly formulated, or where the domain is of such complexity that fundamental theory is impossible (the social and behavioural sciences may well fall into this mould), the pragmatic aspect of measurement will play a larger role. 'Engineering' in the domain in question (mechanical engineering, social engineering, software engineering, or whatever) can take place, with some effective results, not because one 'understands' what is going on, but because one's pragmatic measures have some predictive content. Then, as understanding matures, perhaps in some domains deeper representational perspectives on measurement can be developed. This has already occurred in classical physics. Of course, it may never occur – perhaps the behavioural sciences are of such intrinsic complexity that one is forever bound to measurement which is largely pragmatic.

The debate about measurement scales and statistics has been tightly woven with the choice between parametric and nonparametric statistical methods. Nonparametric or distribution-free methods generally rely only on ordinal properties of the data, often achieving this by working with ranked data. The use of only ordinal properties means that it is not necessary to make distributional assumptions. It also means that one requires only weaker measurement scales. Of course, the use of weaker measurement scales means that one is necessarily restricted to weaker descriptive statistics. As a consequence, much of the comparative discussion of parametric versus nonparametric tests has been in terms of significance or hypothesis tests. Sometimes, working with ranks means that the results are conditional on the given data, for example by means of randomization or permutation tests, instead of based on the theory of sampling from a population. Put another way, the switch from parametric to nonparametric tests may change the hypothesis being tested, so care is needed. Some prefer parametric tests on the grounds of their greater power, but the difference is often not very large: under assumptions of equinormality, rank order tests are almost as powerful as parametric tests.

Anderson (1961) cites Siegel's book on nonparametric statistics (Siegel, 1956, p. 19): 'The conditions which must be satisfied ... before any confidence can be placed in any probability statement obtained by the use of the t test are at least these: ... 4. The variables involved must have been measured in *at least* an interval scale'. And, as we have noted above, Anderson then goes on to say: 'This statement of Siegel's is completely incorrect. This particular questions admits of no doubt whatsoever. The F (or t) test may be applied without qualm. It will then answer the question which it was designed to answer: can we reasonably conclude that the difference between the means of the two groups is real rather than due to chance?' However, as we have also seen above, this statement of Anderson's is ambiguous and it is this that lies at the root of the controversy above. There is ambiguity in whether the 'two groups' in question are the two groups of numbers or the two groups of objects from which the numbers were obtained. Without the measurement-theoretic constraints, a difference between the two sets of numbers may have no implication for a difference between the two sets of objects.

3

The process of measurement

... we must remember that measures were made for man and not man for measures.

Isaac Asimov (1967, p. 143)

3.1 Introduction

Now that we have established what we mean by measurement, 'all that remains' is to describe processes by which we may measure things. This, of course, will turn out to be an exercise at least as elaborate as deciding what we mean by measurement. Methods for measurement, and the ease with which we can apply those methods, will depend on the nature of the things being measured, the balance of the representational and pragmatic aspects, the objective of the measurement, whether the procedure can be direct or has to be indirect, and so on. Differences such as these explain why there are so many procedures.

In their seminal work on representational measurement, Krantz *et al.* (1971, p. 32) distinguish between measurement and scaling:

> The work on fundamental measurement representations, which is relatively recent in the behavioural sciences, contrasts with an older research field known as psychometrics and scaling theory. Most of the psychometric literature is based on numerical rather than qualitative relations (e.g. matrices of correlation coefficients, test profiles, choice probabilities), although there is a tradition, which has recently grown considerably, focusing on ordinal relations. This work aims to represent such relations by numerical relations, mostly of a geometric nature, that are more compact and more revealing than the input data ... Most of these scaling procedures assume the validity of the proposed model and produce a best-fitting numerical representation of the data, whether or not the assumed model is really appropriate.

From our perspective, it is not so much a matter of 'whether a model is really appropriate', by which we take Krantz *et al.* to mean whether or not the numerical relationships in the model properly reflect empirical relationships between the objects under study, but the extent to which pragmatic constraints are used in defining the model form. Thus, from our perspective, scaling methods are simply kinds of measurement procedures. The explicit recognition that real measurement procedures involve pragmatic aspects seems to resolve the difficulty implicit in the statement (Krantz *et al.*, p. 33): 'In scaling aptitude, intelligence, or social attitudes, test scores or numerical ratings are usually interpreted as measures of the attribute in question. But in the absence of a well-defined homomorphism between an empirical and a numerical relational structure, it is far from clear how to interpret such numbers.' The numbers are values of the attribute which is being defined by the measurement procedure.

All of this means that there is a tremendous variety of measurement procedures, ranging from the very simple to the exceedingly complex, from the almost purely representational to the almost purely pragmatic, from direct to indirect, from univariate to multivariate. This chapter describes some of these.

3.2 Direct scaling

An example of what is probably the simplest direct measurement procedure is the classic example of straightforward representational measurement of the physical attribute 'length'. Here, one picks a particular object to serve as a unit length and determines the length of any other (larger) object by the number of duplicates of the unit length required to match it. It is also easy to devise strategies for partitioning the unit length into equal segments (e.g. halves) so that fractional lengths can be measured. Of course, the practicality of this procedure may leave much to be desired, so, as with money, universal 'unit lengths' have been adopted, and copied onto rods called rulers. This is discussed at some length in Chapter 7.

This direct procedure, based on the notion of concatenating duplicate objects, is widely used in physics: attributes such as weight, time, charge, velocity, volume, and so on can all be measured in this way. Of course, the fact that they *can* be measured by such a process does not mean that they have to be measured by such a process. Even length and distance nowadays are measured by much more sophisticated indirect processes. An example is the measurement of a room's dimensions by an estate agent using a laser range finder. Time, here, is measured directly (by concatenation of duplicates of a very tiny time unit) and this is combined with the known velocity of light to give the distance traversed.

Length measurement, by the concatenation process, is almost entirely a representational measurement strategy, requiring only the choice of unit to be made on external grounds. In contrast, another simple direct measurement procedure has a major pragmatic component. This is the visual analogue scale or the semantic differential.

There are variants of visual analogue scales, but a common format consists of a horizontal 10 centimetre line, with its end points clearly marked, and with a word at each end indicating the extremes of the attribute to be measured (hence 'semantic differential'). For obvious reasons, this is called a *bipolar* scale. Respondents indicate their position between these two extremes with a mark on the line. The length, in suitable units, from one end of the scale to the mark constitutes the value of the attribute. Millimetres are a common unit. Note that length is, yet again, playing a key role. From a rigorous representational perspective, scores from a single subject might be regarded as ordinally (and only ordinally) related, with scores from different subjects having no obvious relationship. However, it is common to regard the length as having more than merely ordinal properties. This is, of course, acceptable from a pragmatic perspective.

It is clear from this brief description that a number of pragmatic decisions are made in constructing such an instrument. Perhaps chief of these is the choice of descriptive terms for the end points, but another is having an appropriately standardized instruction or question introducing the scale and what the respondent is expected to do.

McCormack *et al.* (1988) list the merits which have been claimed for visual analogue scales. These include simple construction, straightforward administration and scoring, suitability for frequent and repeated use, ease of explaining to respondents, speed of completion, and the fact that they require very little training to administer.

Variants of this basic idea include using numbers or other descriptive terms to indicate intermediate points. Some have argued against this, since it can lead to clustering of responses (e.g. Aitken, 1969; Scott and Huskisson, 1976; Huskisson, 1974). However, another common approach takes this idea further, and abandons the definition of the scale solely in terms of the end point descriptors, instead giving ordinally related labelled categories. Thus, for example,

- I am very confident in the Government's management of the crisis.
- I am fairly confident in the Government's management of the crisis.
- I am not very confident in the Government's management of the crisis.
- I am not at all confident in the Government's management of the crisis.

These can often be simplified to a single statement with different categories according to strength of response:

- better than usual
- same as usual
- worse than usual
- much worse than usual.

If discrete categories are presented, then one must decide how many to have. Common choices are five and seven, but fewer are used in special cases (e.g. mild, moderate, severe, in medicine) and occasionally more are used. People often avoid the extreme positions, so reducing the effective number. If an odd number is used, the middle one will typically be a position representing ambivalence or neutrality, and sometimes an even number is used to force people away from sitting on the fence.

Although, from a representational perspective, the scale points here have solely an ordinal relationship, it is common to code them as integers – an additional pragmatic constraint on the definition of the concept being measured. More advanced methods, some of which are outlined below, seek to find optimal numerical values for the scale points, in terms of some criterion. These criteria generally come from considerations beyond the empirical system being modelled (e.g. being based on correlations between sets of observations) and so are just as much pragmatic constraints as is the simple choice of integers.

There are various other topics which, one could argue, should be covered in a discussion of direct scaling. One example is unidimensional scaling. Unidimensional scaling begins with a matrix of pairwise similarities between the objects, and finds the linear configuration (the single dimension) which yields between-object distances which best match those given similarities. This is a direct approach if the distances are given initially. However, if the distances are deduced from other attributes then it is clearly an indirect method. In any case, the ideas underlying such methods will be easier to explain once we have introduced some other concepts. We therefore leave their discussion until Section 3.4.

A second topic which might well have been included in this section is a discussion of those scaling methods which aggregate direct ratings from a number of raters to produce a scale value. For example, we could ask ten judges to assess the position of Prime Minister Tony Blair on a visual analogue scale stretching from extreme left wing to extreme right wing. We then have the problem of deciding how to summarize the ten values into a single score. Clearly each individual judge is providing a direct assessment, and it is merely a matter of aggregating them appropriately. However, now consider a situation in which a single judge rates an object on ten different scales, all thought to be related in a similar way to the underlying attribute of interest. The aim is to extract, from the ten values, the value of the underlying attribute – an indirect measurement model. But the problem is structurally identical to the one with ten judges. Of course, one may postulate a different kind of relationship between the scores of the judges and the 'underlying variable' in each of the two situations, but that is a matter of the model one chooses to adopt, and does not reflect any fundamental difference between the two situations. For these reasons, we have chosen to discuss such methods as indirect methods.

3.3 Indirect scaling

The previous section measured the attribute of the objects directly, in the sense that no intermediate measurements or calculations involving other attributes were needed. In many situations, however, direct measurement of the attribute in question is not possible, and indirect strategies, in which its value is deduced from its relationship to other attributes, which can be directly measured, must be used.

Many different indirect measurement methods have been developed, and are discussed in the following subsections. Their differences essentially lie in the nature of the information collected about the relationship between objects. Much of the research on scaling methods has been reported in the behavioural science literature (e.g. van der Ven, 1980), so the literature often speaks of 'subjects' rating 'objects'. We will often adopt this terminology below, but note that it is not intended to be as restrictive as it might imply. As noted at the end of the previous section, the multiple subjects rating a particular object could simply be ratings of different manifest (directly observable) attributes rated by a single subject. Thus, we could rate people in terms of their height, weight, girth, wrist circumference, and hat size, with our aim being to condense these down into some measure of 'person size'.

At an extreme, one might collect explicit numerical information, from each of a number of subjects, on how they would position the objects along a continuum representing the size of the attribute being investigated. For each individual subject, this could be a direct measurement process – simply ask the subjects how they think the objects are related. Or it could be a more elaborate indirect approach in its own right: each measure of similarity could be based on summarizing a set of direct assessments. We say more about this in Section 3.4 below. However these ratings are obtained, it is unlikely that all of the subjects would give identical numerical assignments, so some method needs to be developed to synthesize the scores into a single measurement scale. A variety of ways of doing this are considered below, but note that the issue of synthesis is a general one, which applies whatever the level of information collected, if different subjects can provide conflicting or inconsistent data or if multiple direct measurements, requiring integration, are collected.

Less demanding than requiring full numerical values would be to collect information on how each subject ranks the objects in terms of the attribute in question. Such rankings might be on scales spanning two adjectives (bipolar scales, mentioned above) or might be in terms of, for example, a scale which is anchored at only one end (a unipolar scale). An example of the latter would be a rank ordering in terms of preference.

Less demanding still would be just to ask each subject to make comparisons between pairs of objects, in terms of the attribute in question. For example, to state whether object A is heavier, more attractive than, or brighter than object B, for each pair $\{A, B\}$. Of course, there is a possibility that different subjects may not give the same ordering for each pair, and even that a single subject may give a non-transitive result (preferring A to B, B to C, and C to A). This cannot happen (at least, for an individual subject) if entire rankings are requested: the very request that someone should place all the objects in a rank order contains, implicit within it, a model of the underlying empirical system. The ranking is *representing* the *belief* that the empirical objects have an ordinal relationship in terms of some attribute, and such a ranking precludes the possibility that a non-transitive relationship could exist. Of course, unidimensional scales, by definition, assume a transitive relationship between their points. It follows that, for us, if we insist on a unidimensional scale, intransitivities represent inconsistencies, and an effective way of synthesizing the raw data needs to be developed to overcome such conflicts. Quite what is the right way will depend on how one views the inconsistencies.

Even less demanding than pairwise comparisons would simply be to ask each subject to assign each object to one of two classes. For example, one might ask each subject to take an examination consisting of a series of questions of different degrees of difficulty, so that each subject would get each question right or wrong. In principle, if the difficulty of the questions and the ability of the subjects are regarded as lying on a continuum, then the questions and subjects can be assigned to points on this continuum which would preserve the difficulty and ability order. The procedure results in an ordinal assignment, for each person splitting the questions into two classes, with the given person's ability lying between the two classes.

Yet another type of information which could be collected would be to compare *pairs of pairs* of objects. One could, for example, ask each subject to state if A was more similar to B than C was to D, for sets of objects $\{A, B, C, D\}$.

We can see from this that there is a wide variety of ways (indeed, perhaps an infinite number of ways) that one can seek information about the position of subjects and objects on the postulated underlying univariate continuum describing the magnitude of the attribute in question. However, at several points in the paragraphs above we have stressed the need to develop methods to synthesize the results from several subjects in a way which will overcome any differences between them. As we shall illustrate, early methods of scaling started from the perspective that the scale was perfect (no inconsistencies) and that where conflicts did arise it represented a weakness of the representation. Early methods of scaling were therefore developed which could easily handle perfect results, but required some rather awkward and *ad hoc* adjustments for imperfect results. In fact, of course, imperfect scales are the norm in real life: perfectly scalable results are rare. This is not to criticize the early researchers. Firstly, it was necessary to go through their developmental phases to reach our current position; and secondly, we now have computational tools of which they could not have dreamt.

So how should conflicts and inconsistencies be handled? This partly depends on how one regards such occurrences. One strategy is to say that the unidimensional representation is inadequate, and that something more elaborate is needed. For example, human beings differ in both height and weight, and no unidimensional measure of size can completely capture both aspects of this variation. Methods of multidimensional scaling, essentially extensions of some of the methods we describe here, develop this approach. However, this book is about developing single measurement scales for attributes, so we need an alternative strategy. One such alternative is to acknowledge that a complete unidimensional representation of a multidimensional data distribution is not possible, and to adopt the perspective that we are merely trying find the single best-fitting unidimensional representation. That is, to acknowledge that the definition of the variable being measured must include some pragmatic component. Height and weight are correlated, after all, and while we will have tall thin people and short fat people, we can fairly reasonably represent the bulk of the population on a size scale (the first principal component of these two measures, say, or the sum of the two measures, in specified units). This, of course, is what we do when we construct models in general. The famous adage of George Box comes to mind that 'all models are wrong, but some are useful'. The pragmatic constraint means that we are all talking about the same thing, and a good choice of pragmatic constraint will mean that the resulting variable is useful in some way.

If we are going to seek the unidimensional model which best summarizes the data, we need to be clear about what we mean by 'best summarizes': we need a measure of goodness of fit to the model. The dominant modern approach is to construct a model in which random measurement error or random subjects effects ('individual differences') account for any inconsistencies. The parameters of such a model, as well as the points represented the values of the objects and the scores of the subjects, can then be estimated in various ways. The most common single approach is to describe the measurement error or between-subjects effects in terms of probability distributions, with likelihood being the most popular goodness-of-fit criterion (and, hence, maximum likelihood being the most important modern fitting method). However, other authors (e.g. Gifi, 1990) have developed rather more *ad hoc* methods based on optimizing other measures of goodness of fit. We illustrate these various ideas below.

To illustrate the flavour of the early methods, we shall describe just five of them, showing how they work with perfectly scalable data. The five methods we shall consider are:

Thurstone scaling. Here, information on the rank order of items is obtained from a set of 'judges', and this is summarized to yield individual item scores. These are then used to score new subjects.

Likert scaling. A subject's score is the total score from a set of items, each thought to be related to the attribute of interest, and each scored (for example) on a categorical scale from one to five.

Guttman scaling. Here, information is obtained on how each subject assigns objects to one of two classes, implicitly representing an order going from one class, through the subject's value to the other class. The classic example is that given above, in which the objects are the individual questions in a test and the two classes are 'answered correctly' or 'answered incorrectly'; the subject's ability value lies between the difficulty levels of the questions in the two sets.

Coombs scaling. Here, each subject gives information on their rank order of the objects, in increasing (say) order of the magnitude of the attribute in question.

Bradley–Terry models. Here each subject compares two objects, deciding which is greater on the attribute in question.

The description of scaling methods above, in terms of the degree of refinement of the data collected about the relationships between objects, is one way to approach a taxonomy of such methods. We think it is a useful one, because of the potentially unlimited different kinds of information. However, Coombs (1964) presented another, more elaborate, 'theory of data'. It distinguishes between *dominance* relationships and *proximity* relationships. A dominance relationship enforces an order on a pair of elements. For example, we might say that a subject dominates a particular examination question if they can answer it correctly, or that one sound dominates another if it is louder than the other. A proximity relationship give a measure of similarity or difference between two elements. A very simple such relationship would be whether two elements were or were not indistinguishable in terms of the attribute in question. A more sophisticated proximity relationship would be a measure of how similar they were – for example, the proximity of different variables in a data set could be measured by their correlation.

Orthogonal to the dominance–proximity dichotomy, Coombs also makes the *single-set–multiset* distinction. A comparison between two sounds, to see which was the louder, would be comparison between elements from a single set (the set of sounds). An individual's response to an examination question would be a multiset comparison, with individuals (subjects) and questions (objects) being the two sets.

We can also distinguish between *cumulative* scales, in which a subject who can correctly answer or who agrees with a given item will also be able to correctly answer or will agree with all simpler or less extreme items (as in a Guttman scale – see below), and *differential* scales, in which a subject will give correct answers to or will agree only with those items which are similar in some sense to his or her ability or attitude. In general, in *summated* or *summated rating* scales, a subject's score is the sum of the individual scores of each of a set of items.

3.3.1 Thurstone scales

A large initial set of items is created, each consisting of a statement describing the attribute to be measured. For example, if the attribute is 'attitude towards one's direct line manager', then the statements might include 'effective in all aspects of their work', 'good at organization,' 'sometimes does not interact well with their staff', and so on, so that the complete set of statements covers the range from positive to negative. Duplicates, ambiguous items, and so on are eliminated. Then each of a large group (where 'large' could mean as many as 50 or more) of 'judges' independently sorts the items into (say) 11 groups, according to the positions of the items on the favourable/unfavourable axis towards the topic in question. The groups are supposed to be as 'equally spaced' as possible. Note that the judges are not asked to give their views, but are intended to say how they believe the statements should be interpreted in terms of favourableness towards the topic in question.

For each judge, each of the groups is then scored 1 to 11. In general, not all items will be assigned to the same rank of group by all judges, so that each item will receive a distribution of n scores, where n is the number of judges. Items for which this distribution has a large dispersion are discarded, because clearly there are substantial differences of interpretation for such items. Each of the remaining items is assigned a value equal to the median of the scores assigned to it by the judges. From this remaining set a subset (of, say, 20) is selected so that their values are distributed across the range of attitudes.

A subject's score on such an instrument is simply the mean of the values of those items with which he or she agrees.

3.3.2 Likert scales

In a Thurstone scale, each item has a position (determined, by the scale construction process, as the median of the positions chosen by the judges involved in creating the scale). A subject's score is then the mean of the positions of the items he or she endorses. In contrast, in a Likert scale, each item has a range of possible responses, and the final score is the sum (or mean) of the individual responses. Typically, in Likert scaling, each item is scored on a five-point categorical scale, ranging from one extreme to the other (e.g. strongly disagree, disagree, undecided, agree, strongly agree), with these categories scored from one to five. To avoid response set problems (Section 5.2) it is a good idea to phrase some questions positively and some negatively. To construct such a scale, a large pool of items thought to be indicators of or related to the attribute in question is administered to a sample of subjects drawn from the same distribution as those on whom the scale is ultimately to be used. (Contrast this with the use of judges in Thurstone scale construction.) Items which do not correlate highly with the total score are rejected.

The fact that a Likert scale score is a sum (or mean) of component scores means that any given final score can be made up from different profiles of scores. If the components are all tapping aspects of the same attribute, then this is not a problem. If, on the other hand, in fact multiple attributes underlie the set of component scores, then it could be problematic. One way to think about this is to interpret the final score as the best-fitting univariate approximation to the multiple-attribute truth. In general, the representational foundation of Likert scaling is rather weak, being based only on the fact that each item has a monotonic relationship with the attribute to be measured, so using scores for the attribute categories and aggregating these by summing them, without any solid theoretical reasons for these choices, might be regarded as a pragmatic choice. Later in this chapter, we provide a more solid representational foundation via explicit latent variable models. Factor analysis, in particular, is such an extension of Likert scaling.

The General Health Questionnaire, described in Chapter 6, is an example of a Likert scale.

3.3.3 Guttman scaling

In a selection examination for a school or college, we might want to measure the arithmetic ability of the applicants. To do this, we typically ask them a set of arithmetic questions of varying difficulty. Some people will be able to answer most of these questions without any problem, while others will falter at even the simplest of the questions. If we had a difficulty scale for the questions, and assuming that, whenever someone correctly answered some question then they also correctly answered all the easier questions, then we could also position people on this scale (not precisely, but certainly in the interval between two questions: the hardest one they could answer correctly and the easiest one on which they failed). The scale would thus serve as an ability scale for people, as well as a difficulty scale for questions. Of course, this also works in reverse. If we were able, a priori, to position people on an ability scale, and if the questions were such that a person of a given ability would correctly answer all questions up to a certain difficulty level and none beyond that, then we could also place each question on this scale.

The difficulty with using this strategy for deriving ability scores for people is that we do not know the difficulties of the questions a priori. Somehow, we need to derive the ability scores and the difficulty scores simultaneously. This is exactly what Guttman scaling (Guttman, 1941, 1944) seeks to do: it seeks to position both sets of objects (people and questions in this example, though, of course, the two sets of objects involved could be other things) on a unidimensional continuum on which both people and questions can be located in such a way that a person lies to the right of all items he or she answered successfully and to the left of all others.

The relationship between objects, one from each set (people or questions) is a dominance relationship. Let $i > j$ represent the fact that subject i successfully answers question j. Then our aim is to choose scores $s(i)$ and $s(j)$ for subject i and question j respectively on our unidimensional continuum such that $s(i) > s(j)$. If each subject has attempted each question, then there are $S \times Q$ constraints which must be satisfied by our choice of numerical scores (where S is the number of subjects, and Q is the number of questions).

Another way of representing this system may make it clearer. We illustrate with an (unrealistically small) example. Suppose that five people, labelled A to E, have taken a test which has four questions, labelled 1 to 4. The results of the tests are given in Table 3.1, with a row for each person and a column for each question, and where a 1 signifies that a question was answered correctly and a 0 that a question was answered incorrectly. Thus, for example, candidate E answered question 2 correctly, but was wrong on all of the others. Observation of the matrix in Table 3.1 shows that we can reorder the rows (people) so that if a person gets any question right, then all rows (people) above that person also get all the question right. This reordering is shown in Table 3.2. Since person B got every question right, this person appears as the top row of the table. Likewise, person E got fewest questions correct and appears as the last row.

Observation of Table 3.2 shows that we can now similarly reorder the columns in such a way that if a person gets any question correct, then they also get every question to the left of this correct. For example, question 2 was correctly answered by everyone, so this appears as the leftmost column. Such a reordering is shown in Table 3.3. Note the triangular structure of the matrix in Table 3.3. We also note that we could also have achieved Table 3.3 by first reordering the questions, as in Table 3.4, and then reordering the people.

This process allows us to position points representing both people and questions on a common continuum, satisfying the dominance property that the position of subject i, $s(i)$, is to the right of the position for question j, $s(j)$, whenever subject i correctly answers question j (i.e. whenever $i > j$). Such an arrangement is shown in Figure 3.1. Note that the order of persons C and A is arbitrary, since they have exactly the same response pattern.

The result of this exercise has been a linear configuration of points, such as that shown in Figure 3.1. But note that only the *order* of the points is determined by this exercise. Our representational arguments can take us no further in terms of assigning particular numbers to the points. It follows that only statements invariant to monotonic increasing transformations are meaningful. Of course, we can go further than this if we impose some pragmatic constraints. For example, we could arbitrarily assign the number 1 to the point corresponding to question 2, the number 2 to the point corresponding to question 4, the number 3 to the point corresponding to question 3, and the number 4 to the point corresponding to question 1.

Table 3.1 Scores of candidates.

Person	Question			
	1	2	3	4
A	0	1	0	1
B	1	1	1	1
C	0	1	0	1
D	0	1	1	1
E	0	1	0	0

Table 3.2 Reordering of people.

Person	Question			
	1	2	3	4
B	1	1	1	1
D	0	1	1	1
C	0	1	0	1
A	0	1	0	1
E	0	1	0	0

Table 3.3 Reordering of people then questions.

Person	Question			
	2	4	3	1
B	1	1	1	1
D	1	1	1	0
C	1	1	0	0
A	1	1	0	0
E	1	0	0	0

Table 3.4 Reordering of questions before reordering people.

Person	Question			
	2	4	3	1
A	1	1	0	0
B	1	1	1	1
C	1	1	0	0
D	1	1	1	0
E	1	0	0	0

2   *E*   4   *A* = *C*   3   *D*   1   *B*

Figure 3.1 Ordering of questions and people satisfying the dominance relationships.

New people can easily be positioned on the scale simply by finding the toughest question which they answered correctly (or, equivalently, counting the number of questions they answered correctly). We can then, pragmatically, score any person by the mean of their two neighbouring question scores.

The construction of the Guttman scale is based on the property that if someone correctly answers some question, then they correctly answer all easier questions (or, equivalently, that if some question beats a person, then all harder questions beat that person). Put another way, this scaling procedure assumes that the questions have a fixed difficulty order and that the people have a fixed ability order. It assumes that no mistakes are made (that no random variation intrudes), and that the relationships really can be described in terms of a single dimension. Thus, in our example above, the only possible response patterns to the four questions are (0000), (1000), (1100), (1110), and (1111). In general, with Q questions, there will be just $Q + 1$ possible response patterns. Patterns other than the indicated $Q + 1$ patterns reflect violations of the basic Guttman property. If, as is very likely in practice, other patterns are found, how should they be dealt with? How do we score someone whose response pattern is (1010)? In general, the total number of possible patterns is 2^Q.

There are various approaches to this problem, including:

(i) constructing probabilistic variants of Guttman scaling;
(ii) finding the ordering which has the smallest number of inconsistencies – that is, accepting that the representation is not correct but nevertheless seeking the best-fitting single continuum in some sense, even though we believe it is not a 'correct' representation of the underlying structure generating the data;
(iii) accepting that the representation is not correct, that the points cannot be aligned on a single continuum, and then either replacing questions until such a representation can be achieved or extending the model to have multiple dimensions (though this may defeat the object).

'Best-fitting' in (ii) might be defined in various ways. Indeed approaches (i) and (ii) could be regarded as defining criteria through which the quality of fit is measured. Other criteria can be based on error counts: Guttman defined a *coefficient of reproducibility*, c_r, as

$$c_r = 1 - \frac{N}{I \times S}$$

where N is the total number of inconsistencies, I is the number of items, and S is the number of subjects. Or, to put it another way, c_r is the proportion of responses which are consistent with the proposed scaling.

As we will see, many of the scaling methods described later in this chapter can be regarded as attempts to cope with violations of the Guttman assumptions in the ways indicated above, though they are seldom presented in that way nowadays.

In the above basic non-probabilistic scaling model, we assumed that if a person knew the answer to a question then they always answered correctly, and if they did not know the answer then they always answered incorrectly. Probabilistic models relax this by assuming that if a person knows the answer they answer correctly with some probability p and if they do not know the answer they answer correctly with probability q (and hence incorrectly with probability $1 - q$). Usually, of course, $p > q$. Then, given any proposed order of the points representing questions and people, one can easily work out the likelihood of obtaining the observed data dominance matrix (the matrix indicating which questions each person

Table 3.5 Scores of candidates who do not conform to a Guttman scale.

Person	Question			
	2	4	3	1
A	1	1	0	0
B	1	0	1	1
C	1	1	0	0
D	0	1	1	0
E	1	0	0	0

answered correctly). The maximum likelihood scale would then be the order which maximized this likelihood.

The details of the probabilistic model will depend on how one models p and q. A very simple model could just have fixed values of p and q for all people and all questions. For example, one might assume that people who knew the answer to questions had a probability of $p = 0.9$ of giving the correct answer (the remaining 0.1 being attributed to carelessness) while people who did not know the answer had a probability of $q = 0.2$ of guessing the correct answer. For example, suppose that the data matrix was as shown in Table 3.5. If the (unknown) true dominance order was $2E4AC3D1B$, then person E would have a probability of p of getting the right answer to question 2, and $1 - q$ of getting the wrong answer to each of questions 1, 3 and 4. Likewise, person B would have a probability of p of getting the right answer to each of questions 1, 2 and 3, and a probability $1 - p$ of getting the wrong answer to question 4. Assuming that all answers given by all people are independent, then the overall probability of obtaining the data matrix in Table 3.5 if the order $2E4AC3D1B$ is correct is

$$P = p^2q^2 \times p^3(1 - p) \times p^2q^2 \times p^2(1 - p)q \times p(1 - q)^3$$

$$= p^{10}q^5(1 - p)^2(1 - q)^3$$

We can similarly calculate the probability that the data matrix in Table 3.5 would have been obtained under the assumption that any other ordering was correct. Inverting this relationship, these probabilities are the likelihoods of the orderings, given that this data matrix was observed. We can then choose that ordering which maximizes this likelihood.

This example is very simple, in that it assumes p and q are the same for all pairs of person and question. Any number of elaborations, better matching the real circumstances of the test, are possible. For example:

- If the test is a multiple-choice test, with different questions having different numbers of possible answers, we could let $q = 1/a$, where a is the number of possible answers for a given question. This would model random guessing across the a answers.
- We could let the probability of a person correctly answering a question depend not merely on the dominance relationship (i.e. whether the point representing person i is to the right of the point representing question j in the order) but on the distance between the points. Thus, for example, in the order $2E4AC3D1B$ we could construct a model such that p was a function such that the probability that person D correctly answered question 2 was greater than that they correctly answered question 3. Indeed, using this model one can easily combine p and q. Let r represent the probability that a given person answers a

given question correctly. Then if the point representing a person is to the right of the point representing the question, we will have $r > 1/2$, while if the point representing a person is to the left of the point representing the question, we will have $r < 1/2$, and r will depend on the distance between the points. We explore models such as this in Section 3.5.4, where $r = f(t - b)$, some function of the ability value t of a person and the difficulty value b of a question. Guttman scales are such models with $r = 1$ when the point representing a person is to the right of the point representing the question, and $r = 0$ otherwise.

Probabilistic variants are one approach to developing scales from data which do not exactly conform to dominance relationships. They are based on the implicit assumption that random error has induced the anomalies. An alternative is to suppose that those items which show anomalies are in fact not measuring the same thing: they cannot be conveniently positioned on the unidimensional scale one is trying to construct. In this case, one might discard those items, replacing them by alternatives which better conform to the objectives – and which show fewer anomalies. Item analysis is discussed further below. Note that this strategy of selecting only those items which conform to the unidimensional attribute model means that the validity of this model is based on how subjects have responded to the items. This is different from the case of Thurstone scaling, where the validity (or otherwise) of the model arises from the efforts of the judges: if they produce widely dispersed scores for any individual item, then that item will be dropped.

Evaluating individual questions in this way is tantamount to testing whether the questions conform to the unidimensional model: it is a test of the representational theory. If the departures are substantial, one may be unhappy that the theory is a good one: that the objects are related in the postulated way. If this is the case, one might abandon the model. Alternatively, as we said above, one might accept the departure (e.g. if it is not too large), acknowledging that the unidimensional model is not perfect, but nevertheless accepting it as a good enough approximation to reality for practical purposes.

Of course, there is no reason why various strategies should not be adopted simultaneously. One might, for example, fit a probabilistic model while still discarding those items which show significant departures from this model.

Guttman scales can also be constructed for more complicated situations in which each individual question permits an ordered set of responses. For example, we might have permitted a 'don't know' response, intermediate between the 'right' and 'wrong' responses considered above.

3.3.4 Coombs scaling

Guttman scaling represented the dominance relation existing between particular objects and subjects in terms of some attribute. In the opening example, the point representing a person was positioned to the right of the point representing an arithmetic question if the person could correctly answer the question. The only information represented by this scaling procedure was a binary response given by each subject to each question. Coombs (1950, 1964) went further than this, and described a more elaborate approach based on representing the rank ordering subjects give to stimuli. So, for example, the subjects are asked to arrange the objects or stimuli in order of sweetness, in order of perceived loudness, or whatever. We can think of the dominance relation of Guttman scaling as a coarsening of the ranking used in the Coombs method to be described below. It is as if each subject can identify that a certain

1 EABDC
2 EADBC
3 CDEAB

B . . . . . | . . . | 1 E . 2 . | . | . A . | . D . | . . . | . | . . . 3 . . . C
     BE   BA BD   EA ED BC AD   EC   AC DC

Figure 3.2 A simple example of Coombs unfolding method.

set of objects have a low (or high) rank, but cannot clearly distinguish between them and so merely assign them a value 'low' (or 'high'). Of course, one can (and we discuss this below) develop even finer gradations of distinction between the objects: one could, for example, request that subjects actually position the objects on a numerical scale.

Figure 3.2 provides a simple illustration of Coombs 'unfolding'. Suppose that three subjects rank objects A, B, C, D, and E. Subject 1 ranks them in order given by the first ruler in the figure. That is, subject 1 has rank order EABDC. Similarly, subject 2 has rank order EADBC, and subject 3 has rank order CDAEB. The last ruler shows both objects and subjects positioned so that the order of distances from a given subject to each object matches that subject's order of the objects. For example, the last ruler shows that objects have increasing distance from subject 1 if we take them in the order EABDC – the same as this subject's ordering of the objects.

Coombs described a formal method of moving from the individual subject rankings to the overall ordering of objects. He called his method *unfolding* because the subjects' rankings can be obtained from the overall ranking by 'folding' it at the subject's position. The 'subject's position' on the scale is called the subject's *ideal point* because no other point on the scale can be closer for that subject. Thus, for subject 2, for example, if we fold the last ruler over on itself at the ideal point for subject 2, the point indicated by 2, we find that the object letters fall into the right order for this subject.

The ordering of the items given by a subject is often called that subject's *I* scale. The scale which results from the unfolding process is often called the *J* scale, standing for *joint*, as it locates both subjects and items.

The last ruler in this example also shows the mid-points between the objects' positions. If we move along the scale, whenever we cross over one of these mid-points our order for the objects changes. For example, to the left of mid-point EA subjects are closer to E than to A, but to the right of this point they are closer to A than to E. At the far left of the scale, the ordering is BEADC, so, as we move to the right and cross mid-points, the preference order will change, first to EBADC (as we cross BE), then to EABDC (as we cross BA), and so on, finally arriving, at the other end of the scale, at the order CDAEB. This last order is the reverse of the first one – and note that only these two orderings have this property. No other pair of orderings are the reverse of each other (because they all involve folding the overall scale at some internal point). This fact tells us the order of points on the *J* scale: it is the order of either of the subject orderings which are reflections of each other. It does not matter which of these two we take.

Starting from one of these orderings, we can now use the individual subjects' *I* scales to order the subjects. If two subjects have the same order of objects apart from an interchange of one adjacent pair (XY, say), then these two subjects lie next to each other on opposite sides of the XY mid-point. To one side of this mid-point X is ordered above Y, and to the other side Y is ordered above X. Subjects 1 and 2 in Figure 3.2 illustrate this. Subject 1 has ordering EABDC and subject 2 has ordering EADBC. As we move along the scale, subject 1 will be located next to subject 2, separated by the mid-point BD. The first two rulers in the table show how the order of objects B and D differ between these two subjects.

It is clear from the last ruler shown in the figure that, if there are I objects, then the number of distinct I scales we can observe is $I + 1$ (or less if two objects coincide on the scale, but we will ignore this case here). If $I + 1$ distinct I scales are observed, and if these orderings are such that they can be themselves ordered, then this gives the order of objects and subjects on the J scale. If the data have arisen, without error, from such a scale, then this will be possible. We will discuss the case when it is not possible below.

We have now reached the situation of having ordered the objects and subjects, but there is more metric information available from the order of the mid-points. For example, suppose that we know that the object order is ABCD and that mid-point AD lies to the left of mid-point BC. This means that $(A + D)/2 < (B + C)/2$, so that $B - A > D - C$, so that we have an ordering of the distances between objects, as well as an ordering of objects themselves.

Further details of these kinds of procedures, along with algorithms for applying them, are given in Coombs (1964). A detailed example is given in Long and Wilken (1974). However, we will not pursue them further because they hinge around the assumption that there are no inconsistencies in the data. Instead, we will move on to discuss more sophisticated methods which can allow for mismatches between the data and the model. (*Multidimensional scaling* methods (Cox and Cox, 2001; Davidson, 1983; Gower and Hand, 1996) extend the idea of fitting a unidimensional model to the situation where there are multiple underlying dimensions, and then seek to locate the objects in that space in such a way that the distances between them, the rank order of distances, etc, are preserved. In this book, however, we are concerned with finding the best-fitting *single* dimension.)

3.3.5 The Bradley–Terry model

Several of the above models position both subjects and objects on the same continuum. The Bradley–Terry model (Bradley and Terry, 1952) is a particularly simple scaling approach which ranks only one of the two modes. Suppose our aim is to rank objects, and suppose that we can make pairwise comparisons between objects (e.g. we could ask a series of judges to rate pairs of objects). Let object i have (true unknown) score s_i, $i = 1, \ldots, m$, and model the probability that a randomly selected judge rates object i as superior to object j as $\pi_i/(\pi_i + \pi_j)$, with $\pi_i = e^{s_i}$. The likelihood function is easily constructed from this as

$$L = \prod_{i \neq j} \left(\frac{e^{s_i}}{e^{s_i} + e^{s_j}} \right)^{n_{ij}} \tag{3.1}$$

where n_{ij} is the number of judges who rate object i as superior to object j. From this, we can estimate the scores s_i.

In fact, rewriting (3.1) as

$$L = \prod_{i \neq j} \left(\frac{1}{1 + e^{s_j - s_i}} \right)^{n_{ij}}$$

we will see that this has a very similar structure to the Rasch model to be described below. In the Rasch model (after Georg Rasch, 1901–1980) items are tested by judges to form a dominance relationship, whereas here items are compared with other items.

3.4 Modern unidimensional scaling methods

In Section 3.3, we suggested that, at an extreme, the information provided about the relationships between objects might be numerical estimates of the similarities or dissimilarities between objects. The term *proximity* is sometimes used to indicate a general measure of such similarity or dissimilarity. A proximity measure is a function from the set of pairs of objects to the real numbers, typically taken to be non-negative. Usually the dissimilarity of the *i*th object with itself, δ_{ii}, the minimum possible dissimilarity, is taken to be zero. Likewise, similarities are often scaled so that the similarity of an object with itself, the maximum possible similarity, is unity, $s_{ii} = 1$. The nature and properties of similarity and dissimilarity measures have been explored in great depth: see, for example, Gower and Legendre (1986), Baulieu (1989), Gordon (1999) and Cox and Cox (2001). It is, of course, simple to invert similarity measures to yield dissimilarity measures, and vice versa. For this reason, most of the following discussion will only be in terms of dissimilarity measures.

General proximity measures can be obtained in a variety of ways. For example, one could simply ask subjects to give their subjective measures of the similarities between objects. This could be pairwise: 'say, on a scale from 0 to 10, with 0 being identical and 10 being maximally different, how similar you think objects A and B are', for some set of pairs. Or it could be using multiple objects simultaneously: 'position the objects on a continuum from 0 to 10, so that similar objects are close to each other'. The second approach forces consistency within subjects, while using pairs permits subjects to show inconsistencies. Classic examples of such subjective ratings are the similarities between Morse code signals and similarities between the tastes of different whiskies.

As a general point, we remind the reader that, although we are using the word 'subjects' here, the proximity ratings could be obtained by other means: instead of subject 1 we might use a measure of weight, instead of subject 2 a measure of height, and so on. As when the ratings are produced by different subjects, such direct ratings are likely to be inconsistent between ratings. When ratings are obtained by particular direct measurements (such as weight and height in this example), we will often have actual values for individual objects, not merely proximities for pairs of objects. In this case, the data matrix could first be contracted down to yield overall proximities for pairs of objects, or the different component scales could be regarded as providing different information about the proximities.

Once one has dissimilarity measures between objects, one seeks a configuration of points (in our case a unidimensional configuration), one point for each object, such that the distances between points match as closely as possible the dissimilarities between the corresponding objects. That is, if the distances between points are denoted by $\{d_{ij}\}, i, j = 1, \ldots, n$, then we seek to minimize a loss function defined as some aggregate measure of the difference between the dissimilarities δ_{ij} and the distances $d_{ij}, i, j = 1, \ldots, n$.

If there does exist an *exact* unidimensional representation of the dissimilarities δ_{ij}, that is, if $\delta_{ij} = d_{ij}$, it can be found as follows. Suppose the position of the *i*th object (i.e. the attribute value of the *i*th object) is s_i. In fact, of course, given only the interpoint distances, we need to choose an overall location constraint if we are to be able to find a unique solution. We choose to let $\mathbf{1}^T \mathbf{s} = 0$ – that is, the mean of the positions of the points is set to zero. Then the squared distance between the *i*th and *j*th points is given by

$$d_{ij}^2 = (s_i - s_j)^2 = s_i^2 + s_j^2 - 2s_i s_j \qquad (3.2)$$

In matrix form, with $\mathbf{D} = \{d_{ij}^2\}$, $\mathbf{s} = (s_1, \dots, s_n)^T$, and $\mathbf{G} = \text{diag}(s_1^2, \dots, s_n^2)$, this is

$$\mathbf{D} = \mathbf{G11}^T + \mathbf{11}^T\mathbf{G} - 2\mathbf{ss}^T \tag{3.3}$$

Now

$$(\mathbf{I} - \mathbf{11}^T/n)\mathbf{G11}^T(\mathbf{I} - \mathbf{11}^T/n) = \mathbf{0} \tag{3.4}$$

$$(\mathbf{I} - \mathbf{11}^T/n)\mathbf{11}^T\mathbf{G}(\mathbf{I} - \mathbf{11}^T/n) = \mathbf{0} \tag{3.5}$$

and

$$(\mathbf{I} - \mathbf{11}^T/n)\mathbf{ss}^T(\mathbf{I} - \mathbf{11}^T/n) = \mathbf{ss}^T \tag{3.6}$$

(this last using the constraint that $\mathbf{1}^T\mathbf{s} = 0$) so that

$$(\mathbf{I} - \mathbf{11}^T/n)\mathbf{D}(\mathbf{I} - \mathbf{11}^T/n) = -2\mathbf{ss}^T \tag{3.7}$$

from which

$$-\frac{1}{2}(\mathbf{I} - \mathbf{11}^T/n)\mathbf{D}(\mathbf{I} - \mathbf{11}^T/n) = \mathbf{ss}^T \tag{3.8}$$

This means that the vector of positions of the points, $\mathbf{s}$, is given by the eigenvector of

$$\mathbf{M} = -\frac{1}{2}(\mathbf{I} - \mathbf{11}^T/n)\mathbf{D}(\mathbf{I} - \mathbf{11}^T/n) \tag{3.9}$$

If there is more than one non-zero eigenvalue, then an exact univariate representation is not possible. If there is not an exact representation in one dimension there may be such a representation in $p < n - 1$ dimensions. In fact, the double centring transformation of (3.7) can be generalized as follows. In place of s_i let the p scores for the ith point be $\mathbf{s}_i = (s_{i1}, \dots, s_{ip})$, define $\mathbf{S}^T = (\mathbf{s}_1^T, \dots, \mathbf{s}_n^T)$, and, as a further generalization, replace $\mathbf{D}$ by $\mathbf{D}_w$, with ijth element $d_{wij}^2 = (\mathbf{s}_i - \mathbf{s}_j)^T\mathbf{W}(\mathbf{s}_i - \mathbf{s}_j)$, where $\mathbf{W}$ is a symmetric positive definite matrix (which, in a still greater generalization, may vary from subject to subject). Equation (3.7) then generalizes to

$$(\mathbf{I} - \mathbf{11}^T/n)\mathbf{D}_w(\mathbf{I} - \mathbf{11}^T/n) = -2\mathbf{SWS}^T$$

Solving this for the components of $\mathbf{S}$ gives the point vectors representing each object. The case with $\mathbf{W} = \mathbf{I}$ was described by Torgerson (1952, 1958), and is often termed the classical metric scaling method.

In this book, however, we are concerned with unidimensional representations, so that if there is not an *exact* unidimensional representation we will seek the best-fitting one. It can be shown (Gower, 1966) that the best-fitting single dimensional representation is given by the first eigenvector of the matrix in (3.9).

Our aim, when seeking the 'best' univariate representation, is to minimize a measure of the difference between the interpoint distances $d_{ij} = |s_i - s_j|$ and the observed similarities δ_{ij}, i, $j = 1, \dots, n$. In the multidimensional case, distance between points can be defined in many ways, but in the unidimensional case one is more constrained. One important measure is

$$S_1 = \sum_{ij}\left(\delta_{ij}^2 - |s_i - s_j|^2\right)$$

It can be shown that the set of scores which minimizes S_1 are the scores on the first eigenvector of $\mathbf{M}$ above. Note that points can always be found in a space of $n - 1$ dimensions such that the interpoint distances d_{ij} equal the δ_{ij} exactly, and that as we project down into fewer than $n - 1$ dimensions, so the interpoint distances d_{ij} decrease. This means that the terms $(\delta_{ij}^2 - |s_i - s_j|^2)$ in S_1 are always positive.

Another important way to define the distance between two points s_i and s_j is as $d_{ij} = |s_i - s_j|$, so that another obvious loss function is

$$S_2 = \sum_{i=j} \left(\delta_{ij} - |s_i - s_j| \right)^2$$

As one might suspect, the function S_2 has many local minima, posing something of an optimization challenge. Various algorithms for finding minima have been proposed (e.g. Guttman, 1968; Hubert and Arabie, 1986, 1988; Simantiraki, 1996). Given the large number of local minima, and the power of modern computers, stochastic optimization methods, such as simulated annealing, have appeal. To test the accuracy of location of individual points, one can look at the individual measures $(\delta_{ij} - |s_i - s_j|)^2$, expressed as a fraction of the total deviation S_2.

The function S_2 can be generalized in various ways. For example, we could make general continuous monotonic transformations of the dissimilarities, yielding the criterion

$$S_3 = \sum_{i<j} \left(g(\delta_{ij}) - |s_i - s_j| \right)^2$$

Hubert et al. (1997) explored the special case of $g(x) = x + c$ for the univariate case, and a common transformation in the more general multivariate case is $g(x) = bx + c$. Methods which use a parametric transformation for g are sometimes called *metric scaling* methods.

So far, in this section, we have considered the case of inter-object dissimilarities being given as numerical values. However, this is the extreme case (of maximum information) described at the start of Section 3.3. Often less extensive information is given: perhaps, for example, only a rank ordering of the objects is given. In such situations *isotonic regression* is often used for the g transformation. This constrains the values $g(\delta_{ij})$ to preserve the order of the δ_{ij}, so that, for all i, j, r, s,

$$\delta_{ij} < \delta_{rs} \Leftrightarrow g(\delta_{ij}) \leqslant g(\delta_{rs})$$

or

$$\delta_{ij} \leqslant \delta_{rs} \Leftrightarrow g(\delta_{ij}) \leqslant g(\delta_{rs})$$

For the general multivariate case, these ideas were introduced in Shepard (1962a, b) and Kruskal (1964a, b), and more recent reviews are given in texts on multidimensional scaling (e.g. Cox and Cox, 2001) or multivariate statistics (e.g. Krzanowski and Marriott, 1994; Everitt and Dunn, 2001). Because of the nature of the function g here, such methods are often termed *nonmetric scaling* methods. Ramsay (1982) has stressed the merits of ensuring continuity in the transformation, and has proposed the use of spline methods, though Gower (1982) is less convinced.

Given a dissimilarity matrix, all of the above methods will produce a scaling of the objects. Often, however, one has such matrices from more than one subject, and these are unlikely to be identical. Some way has to be found to aggregate the results from such exercises. A common approach to this is that described by Carroll and Chang (1970). We saw above that the best-fitting univariate representation of the dissimilarity matrix, when 'best' is defined in terms of the measure S_1, is given by the first eigenvector of the matrix

$$\mathbf{M} = -\frac{1}{2}(\mathbf{I} - \mathbf{11}^T/n)\mathbf{D}(\mathbf{I} - \mathbf{11}^T/n)$$

The corresponding matrix for the ith subject, $i = 1, \ldots, m$, is

$$\mathbf{M}_i = -\frac{1}{2}(\mathbf{I} - \mathbf{11}^T/n)\mathbf{D}_i(\mathbf{I} - \mathbf{11}^T/n)$$

where $\mathbf{D}_i$ is the $\mathbf{D}$ matrix for the ith subject. Carroll and Chang (1970) then suggest fitting a model $w_i\mathbf{ss}^T$ to $\mathbf{M}_i$, where $\mathbf{s}$ is common to all the $\mathbf{M}_i$ matrices, and where w_i is a weight specific to the ith individual. (In fact, Carroll and Chang describe the more general multi-variate representation.) The w_i allow differences between subjects to be represented, so that the overall model is an individual differences model. The model has been generalized in various ways – see, for example, Takane *et al.* (1977).

In the above, we have described our aim of locating objects on a univariate continuum as a special case of multidimensional scaling. If the objects did differ solely in terms of a single attribute, then one might expect that a multidimensional scaling solution would show this, detecting that more than one dimension was unnecessary. However, researchers have observed that, even when a single attribute is present, often two dimensions are identified as apparently necessary. In particular, it has been observed that the points representing objects often lie on a curve – leading to the phenomenon being termed the *horseshoe* effect. An early classic illustration of this appears in Kendall (1971). Kendall described the attempt to order 59 tombs according to the time at which they were created, using an incidence matrix describing the presence or not of 70 types of artefacts in the tombs: tombs close together in time would be expected to have substantially overlapping sets of artefacts. One explanation for this is that there is a ceiling effect: once the origins of the tombs are more than a certain time apart, the tombs have very little in common, so that their dissimilarities will tend to take the same (maximum) dissimilarity value. Another explanation is that larger dissimilarities are generally associated with greater inaccuracy (see the discussion of Weber's law in Section 5.6).

If much of the above seems rather *ad hoc*, note that S_2 can be obtained from a probability model. Suppose that the attribute value of the ith object x_i is drawn from a normal distribution $N(\mu_i, 1)$, $i = 1, \ldots, n$. Then the distance between the ith and jth point is distributed as $f(\delta_{ij}) \sim N(|\mu_i - \mu_j|, 2)$. The likelihood of the observed data, the $\{\delta_{ij}\}$, is then

$$L = \prod_{i<j} f(\delta_{ij}) \propto \prod_{i<j} \exp\left\{-\frac{1}{4}\left[(\delta_{ij} - |\mu_i - \mu_j|)\right]^2\right\}$$

so that the log-likelihood is

$$\log L \propto -\sum_{i<j}\left[(\delta_{ij} - |\mu_i - \mu_j|)\right]^2 = -S_2$$

Ramsay (1982), in particular, has demonstrated how such methods can be based on a solid probabilistic footing: 'Underlying much of the work reviewed here is the question of stat-istical rationale. Maximum likelihood estimation is usually better than no rationale at all in that one is forced to consider explicit models for the random components in the data . . . . The Bayesian approach has much to offer' (pp. 301–302).

Up to this point, we have described how to find values representing an existing set of objects. If the result is to be used as a measuring instrument, we also need to be able to find values for new objects. Gower and Hand (1996), in the more general multidimensional scaling case in which vectors of values represent each object, term this exercise *interpolation*. We shall assume that the new object is described in terms of its dissimilarities from the original n objects. That is, we are given the values $\delta_{i(n+1)}, i = 1, \ldots, n$.

One possibility would be to repeat the exercise of finding values for the objects, using the new object in addition to the original n objects, so that all the summations are over $i = 1, \ldots, n + 1$ instead of $i = 1, \ldots, n$. Unfortunately, this is likely to change the values for the previous n objects. It is as if the graduations on a ruler shifted around in a nonlinear manner depending on what new object was measured. A more sensible alternative is keep the n values fixed, and find a value for the new object so that the aggregated differences between the dissimilarities between this object and the original n objects are minimized. For example, using measure S_2 above, we would find the value s_{n+1} such that

$$S_2 = \sum_{i=1}^{n} \left(\delta_{i(n+1)} - |s_i - s_{n+1}| \right)^2$$

was minimized, where $\delta_{i(n+1)}$ is the dissimilarity between the ith object and the new $(n + 1)$th object. Such minimizations will typically have to be carried out numerically.

3.5 Latent variable approaches to indirect models

3.5.1 The basic model

The indirect measurement methods described in Section 3.3 are based on relatively simple underlying models. Indeed, it is possible (even likely) that in some cases the developers did not think of themselves as using a 'model' at all. An alternative approach, however, is explicitly to base the development on elaborate probabilistic models relating the constituent observations to the attribute whose value is being sought. These are *latent variable* models. The observed or *manifest* variables are those which are directly measured (although, as we noted at the beginning of Section 3.3, even these components may be the result of a more complicated process) and one seeks the value of an unobservable or latent variable which can explain the values taken by the manifest variables. For example, if the manifest variables are highly correlated, tending to take high or low values together, then perhaps this can be explained by the fact that they are all closely related to a single underlying variable, the latent variable. Then, given the values of the manifest variables for a particular object, our aim is to deduce the value of the latent variable which is most likely to have given rise to these manifest values.

Formally, then, we will need to find some way of combining the manifest variables, so that the combination yields an approximation to the value of the latent variable. We can turn this around: latent variables summarize, to a greater or lesser extent, the information in a set of manifest variables relating to a particular concept. From this alternative perspective it will be immediately apparent that such summaries can go awry in two ways.

Firstly, the summary might be biased, in the sense that it might not quite measure the concept we are really interested in. In fact, since we have no quantitative information about the concept we are really interested in apart from that contained in the manifest variables, we have to modify the description of what latent variables do. Rather than 'latent variables summarize, to a greater or lesser extent, the information in a set of manifest variables relating to a particular concept' we will have to take simply 'latent variables summarize, to a greater or lesser extent, the information in a set of manifest variables'. It will then be up to us to ensure that the manifest variables included in the model, and the way they are combined in estimating the latent variables, are such as to properly reflect the concept we are interested in. Secondly, the summary might not be very accurate, in the sense that it may have a large variance. Questions such as these, of how to define, explore and assess the accuracy of a measurement, are discussed in detail in Chapter 4.

The following description is based on the beautiful development in Bartholomew and Knott (1999). In what follows, we will denote the p manifest variables by $x_i, i = 1, \ldots, p$, with $\mathbf{x} = (x_1, \ldots, x_p)^T$ representing the vector of these, and we will denote the q latent variables by $y_j, j = 1, \ldots, q$, with $\mathbf{y} = (y_1, \ldots, y_q)^T$. The use of q, which may be greater than 1, indicates that we may feel that a single latent variable does not adequately capture the information in the p manifest variables, and we may want to describe the information in a space with two or more dimensions. In general, however, since we are trying to summarize the information in the p variables, q will be substantially less than p. In particular, our primary aim in this book takes $q = 1$.

We now have two sets of variables, $\mathbf{x}$ and $\mathbf{y}$, so these will have a joint distribution, $f(\mathbf{x}, \mathbf{y})$. For simplicity, in what follows, we will use the single symbol f to represent a probability function, with the argument, which will always be made explicit, indicating which function it is: thus $f(\mathbf{x}, \mathbf{y})$, $f(\mathbf{x})$ and $f(\mathbf{x}|\mathbf{y})$, for example, are three different functions.

Our aim is to estimate a value for the $\mathbf{y}$ vector when we are given a value for $\mathbf{x}$. More generally, we want to find the conditional distribution $f(\mathbf{y}|\mathbf{x})$, from which we can estimate a value for $\mathbf{y}$ – for example, by taking the modal value, or the expected value, or the median value. Now, by Bayes's theorem, we have

$$f(\mathbf{y}|\mathbf{x}) = \frac{f(\mathbf{x}|\mathbf{y})f(\mathbf{y})}{f(\mathbf{x})} \propto f(\mathbf{x}|\mathbf{y})f(\mathbf{y}) \tag{3.10}$$

The problem is that $\mathbf{y}$ is unobservable. This means that, of all the distributions in these expressions, the only one we observe is $f(\mathbf{x})$. Since

$$f(\mathbf{x}) = \int f(\mathbf{x}|\mathbf{y})f(\mathbf{y})dy \tag{3.11}$$

we need to decompose this function to yield estimates of $f(\mathbf{x}|\mathbf{y})f(\mathbf{y})$. Unfortunately, this *inverse problem* does not have a unique solution, so that we need to introduce some constraints. Put another way, we need to define a class of models within which we will seek a solution (the decomposition $f(\mathbf{y}|\mathbf{x}) \propto f(\mathbf{x}|\mathbf{y})f(\mathbf{y})$ is universally true, so that it does not constitute a model).

The constraint we introduce is that the relationships between the components of $\mathbf{x}$ are solely due to their mutual dependence on the latent variables $\mathbf{y}$. That is, amongst cases with the same vector $\mathbf{y}$, information about any of the x_i does not give us any information about any of the remaining x_i. This assumption is one of *conditional independence*, and can be expressed as $f(\mathbf{x}|\mathbf{y}) = \Pi_i f(x_i|\mathbf{y})$, so that our conditional distribution becomes

$$f(\mathbf{y}|\mathbf{x}) \propto f(\mathbf{y})\prod_i f(x_i|\mathbf{y}) \tag{3.12}$$

In practice, we will estimate the model from the data (as described below), so that the relationship $f(\mathbf{x}|\mathbf{y}) = \Pi_i f(x_i|\mathbf{y})$ will never be observed exactly (unless, that is, $q = p$). This means that there is room for doubt. In particular, we have to decide whether a single y variable yields a sufficiently good approximation, or that we need two y variables, or three, and so on.

Even adopting the form $f(\mathbf{x}|\mathbf{y}) = \Pi_i f(x_i|\mathbf{y})$ is not sufficient to give us a unique decomposition of $f(\mathbf{x}) = \int f(\mathbf{x}|\mathbf{y})f(\mathbf{y})d\mathbf{y} = \int \Pi f(x_i|\mathbf{y})f(\mathbf{y})d\mathbf{y}$. This is easily seen by changing the variable of integration, $\mathbf{y}$. Any change in $f(\mathbf{y})$ can be compensated for by a change in $f(\mathbf{x}|\mathbf{y}) = \Pi_i f(x_i|\mathbf{y})$, which leaves $f(\mathbf{x})$ unaltered. There is thus still an infinite number of pairs $f(\mathbf{x}|\mathbf{y})$, $f(\mathbf{y})$ which will yield a solution. Transformations between these different representations, yielding different predicted values for $\mathbf{y}$, would be the set of permissible transformations. Only statistical statements which are invariant to those transformations would be legitimate. We could explore the set of transformations and statements which are invariant to them, or we could attempt further to restrict the model, by applying some additional constraints. In *factor analysis*, a special case and perhaps the earliest example of such models, the latent variables are generally constrained to have unit variance and zero mean. This special case is discussed in more detail below.

We need to choose a family of distributions for the $f(x_i|\mathbf{y})$. To some extent, these will be determined by the nature of the x_i – whether they are continuous, categorical, binary, etc. The exponential family is a useful and fairly general family, which includes Bernoulli, binomial, Poisson, normal and gamma distributions as special cases. It has the form

$$f(x|\theta) = F(x)G(\theta)\exp[\theta u(x)]$$

Thus

$$f_i(x_i|\theta_i) = F_i(x_i)G_i(\theta_i)\exp[\theta_i u_i(x_i)], \qquad i = 1,\ldots,p \tag{3.13}$$

We will adopt this form, with the value of the parameter θ_i being determined by $\mathbf{y}$ from a linear model

$$\theta_i = \sum_{j=0}^{q} \beta_{ij} y_j \tag{3.14}$$

where we have introduced a constant $y_0 = 1$ for convenience.

Using (3.13) in (3.12), we find

$$f(\mathbf{y}|\mathbf{x}) \propto f(\mathbf{y})\prod_i f_i(x_i|\mathbf{y})$$

$$= f(\mathbf{y})\prod_i F_i(x_i)G_i(\theta_i)\exp[\theta_i u_i(x_i)]$$

$$\propto f(\mathbf{y})\exp\left[\sum_i \theta_i u_i(x_i)\right]\prod_i G_i(\theta_i)$$

$$= f(\mathbf{y})\exp\left[\sum_i \sum_j \beta_{ij} y_j u_i(x_i)\right]\prod_i G_i(\theta_i)$$

$$= f(\mathbf{y})\exp\left[\sum_j y_j X_j\right]\prod_i G_i(\theta_i)$$

where

$$X_j = \sum_i \beta_{ij} u_i(x_i) \tag{3.15}$$

We see from this that all the information in $\mathbf{x}$ about the conditional distribution $f(\mathbf{y}|\mathbf{x})$ is given in the sums X_j. $\mathbf{X} = (X_1, \ldots, X_q)^T$ is thus a sufficient statistic for $\mathbf{y}$. Since, moreover, the dimension of $\mathbf{X}$ is the same as that of $\mathbf{y}$, $\mathbf{X}$ is a minimal sufficient statistic for $\mathbf{y}$. More than this, however, we can now apply a remarkable theorem: if two individuals have respective $\mathbf{x}$ vectors $\mathbf{x}_1$ and $\mathbf{x}_2$, and respective $\mathbf{X}$ components X_{j1} and X_{j2}, then, for any monotonic non-decreasing function h,

$$E\big[h(y_i)\,|\,\mathbf{x}_1\big] \geq E\big[h(y_j)\,|\,\mathbf{x}_2\big] \tag{3.16}$$

if and only if $X_{j1} \geq X_{j2}$ (see, for example, Bartholomew and Knott, 1999, p. 34; or Bartholomew, 1996, p. 170). If we take the expected value of the posterior distribution, $E_{y_j}[f(y_j|\mathbf{x})]$, as our 'measure' of the jth latent variable for a case with manifest variable vector $\mathbf{x}$, then this will be monotonically related to the score X_j. The X_j, simple weighted sums of the $u_i(x_i)$, provide ordinal scale values of the jth latent variable. This is true regardless of the distribution $f(\mathbf{y})$. Thus, if we use the X_j as measures of the latent variable, we can make statistical statements provided these statements are invariant to ordinal transformations.

The outline above is all very well for determining the values of the latent variables for members of the sample on which the analysis was carried out. But what about new cases? How does one measure the value of the latent variable for a new subject? One possibility would be to redo the analysis, including the new subject(s) in an expanded sample. However, this could result in changes to the values for previous members of the sample. Another possibility is to regard the initial sample as spanning, in some sense, the population of interest (e.g. it might be a random sample from this population), and treating it as a *calibration sample*. New cases will then be assigned scores using the previously estimated values of β_{ij}. This is similar to the use of different minerals as a calibration sample when measuring hardness using the Mohs hardness scale (see Chapter 2), although in this case a 'continuous' score results, with new cases not merely being assigned to intervals between calibration sample elements. Also, of course, dispute could arise about the adequacy of the calibration sample, in a way it cannot with Mohs scale.

Before we move on to examine special cases, a brief discussion of the role of representational versus pragmatic choices in latent variable models is appropriate. If one really believes that the model for the relationships between the latent variable and the manifest variables is a good model (that the world 'behaves' according to the model), then this is a strong representational model. However, if one is merely regarding the model as a practically convenient approximation, then it is more appropriate to regard it as having a major a pragmatic component. For example, we could adopt the best-fitting single latent variable by which to define our measurement of that underlying concept, even if a two-latent-variable model provided a superior fit. If the model does reflect the 'truth', then the choice of manifest variables will not matter – they will not distort the fit of the model. On the other hand, if the model does not reflect the 'truth' – as may well be the case if an explicitly pragmatic approach is being adopted – then the choice of variables will influence the definition of the resulting measurement scale. Put another way, if the model is improperly specified, the choice of manifest variables will affect the weights used in the weighted sum. With a properly specified model different choices of manifest variables will, apart from measurement error issues, lead to measurement instruments which yield perfectly correlated results; however, with an improperly specified model different choices of manifest variables will yield results which may not be highly correlated.

3.5.2 Factor analysis

Perhaps the most important single special case of the above latent variable model is factor analysis. This is the oldest and best developed such model. The $\mathbf{x}$ variables are regarded as coming from a multivariate normal distribution

$$f(\mathbf{x}) = N_q(\boldsymbol{\mu}, \boldsymbol{\Sigma}) \tag{3.17}$$

and this distribution is decomposed as $f(\mathbf{x}) = \int f(\mathbf{x}|\mathbf{y})f(\mathbf{y})d\mathbf{y}$ by taking

$$f(\mathbf{x}|\mathbf{y}) = N_p(\boldsymbol{\mu} + \boldsymbol{\Lambda}\mathbf{y}, \boldsymbol{\Psi}) \tag{3.18}$$

with

$$f(\mathbf{y}) = N_q(\mathbf{0}, \mathbf{I}) \tag{3.19}$$

$\boldsymbol{\Lambda}$ here is a matrix of coefficients relating the manifest and latent variables, so that (3.18) specifies that the manifest and latent variables are related by a regression model. The only difference between this and standard regression is that the 'independent' variables of the regression, the $\mathbf{y}$ variables in our notation, are not actually observed. The latent variables in this model are called *factors*, and the elements of $\boldsymbol{\Lambda}$ are called *factor loadings*. $\boldsymbol{\Psi}$ is a diagonal covariance matrix, with one value on the diagonal corresponding to each manifest variable.

Using (3.18) and (3.19) in (3.11), we obtain

$$f(\mathbf{x}) = N_p(\boldsymbol{\mu}, \boldsymbol{\Lambda}\boldsymbol{\Lambda}^T + \boldsymbol{\Psi}) \tag{3.20}$$

so we see that this is a model of the required form, in which

$$\boldsymbol{\Sigma} = \boldsymbol{\Lambda}\boldsymbol{\Lambda}^T + \boldsymbol{\Psi} \tag{3.21}$$

That is, we are approximating the true covariance matrix of the manifest $\mathbf{x}$ variables by a covariance matrix of the form $\boldsymbol{\Lambda}\boldsymbol{\Lambda}^T + \boldsymbol{\Psi}$, with $q < p$. In particular, if we believe that the common aspect of the manifest variables can be represented in terms of a single underlying variable, which we are trying to measure indirectly via this model, then $q = 1$. Expression (3.21) shows that the variance of the kth manifest variable can be decomposed as

$$V(x_k) = \sum_{j=1}^{q} \lambda_{kj}^2 + \psi_k \qquad k = 1, \ldots, p \tag{3.22}$$

The first term here arises as sum of contributions, each of which relates to one of the common factors, so this term is called the *communality*. The remaining term, the ψ_{ki}, is specific to the kth manifest variable. The ratio

$$\sum_{j=1}^{q} \lambda_{kj}^2 \left/ \left(\sum_{j=1}^{q} \lambda_{kj}^2 + \psi_k \right) \right.$$

tells us the proportion of the variance of the kth manifest variable which can be attributed to the latent factors. This can thus be used as a measure of the effectiveness of the model for the kth manifest variable. In the case of a single latent factor, the ratio reduces to $\lambda_k^2/(\lambda_k^2 + \psi_k)$.

The posterior conditional distribution of **y** given **x** is then

$$f(\mathbf{y}|\mathbf{x}) = N_q(\boldsymbol{\Lambda}^T(\boldsymbol{\Lambda}\boldsymbol{\Lambda}^T + \boldsymbol{\Psi})^{-1}(\mathbf{x} - \boldsymbol{\mu}), (\boldsymbol{\Lambda}^T\boldsymbol{\Psi}^{-1}\boldsymbol{\Lambda} + \mathbf{I})^{-1}) \tag{3.23}$$

and some summary statistic from the distribution, such as its mean, can be used as the measured value for **y**, with, for example, its standard deviation being used as a measure of precision of the measured value. In fact, in this normal case we have

$$E(\mathbf{y}|\mathbf{x}) = \boldsymbol{\Lambda}^T(\boldsymbol{\Lambda}\boldsymbol{\Lambda}^T + \boldsymbol{\Psi})^{-1}(\mathbf{x} - \boldsymbol{\mu})$$

$$= (\boldsymbol{\Lambda} + \boldsymbol{\Psi}\boldsymbol{\Lambda}^{T-1})^{-1}(\mathbf{x} - \boldsymbol{\mu})$$

$$= (\boldsymbol{\Lambda}^T\boldsymbol{\Psi}^{-1}\boldsymbol{\Lambda} + \mathbf{I})^{-1}\boldsymbol{\Lambda}^T\boldsymbol{\Psi}^{-1}(\mathbf{x} - \boldsymbol{\mu})$$

Now, for the normal case in which the conditional distribution of x_i is $N(\mu_i, \sigma_i^2)$ with known σ_i^2, the X_j in (3.15) takes the form $\sum_{i=1}^{p}\beta_{ij}x_i/\sigma_i$, or, in matrix terms $\mathbf{X} = \boldsymbol{\Lambda}^T\boldsymbol{\Psi}^{-1}\mathbf{x}$, with $\boldsymbol{\Lambda} = \{\beta_{ij}\}$ and $\boldsymbol{\Psi}^T = (\sigma_1^2, \ldots, \sigma_p^2)$. This leads to

$$E(\mathbf{y}|\mathbf{x}) = (\boldsymbol{\Lambda}^T\boldsymbol{\Psi}^{-1}\boldsymbol{\Lambda} + \mathbf{I})^{-1}(\mathbf{X} - E(\mathbf{X}))$$

It follows from this that if $\boldsymbol{\Lambda}^T\boldsymbol{\Psi}^{-1}\boldsymbol{\Lambda}$ is diagonal then the **y** are independent and are linearly related to **X**. We are concerned with the special case of a single **y** variable, and can therefore just use the corresponding univariate X as our measure of the latent variable.

Although we are not especially concerned with the case of $q > 1$, it is worth noting an extra complication this introduces, compared to the $q = 1$ case: when $q > 1$, even with the constraint that $f(\mathbf{y}) = N_q(\mathbf{0}, \mathbf{I})$, the model is not unique. For example, suppose we transform **y** by an orthogonal transformation to $\mathbf{y}^* = \mathbf{M}\mathbf{y}$. Since $\mathbf{M}^T\mathbf{M} = \mathbf{I}$, we see that

$$f(\mathbf{x}|\mathbf{y}^*) = N_p(\boldsymbol{\mu} + \boldsymbol{\Lambda}\mathbf{M}^T\mathbf{y}^*, \boldsymbol{\Psi}) \tag{3.24}$$

and

$$f(\mathbf{y}^*) = N_q(\mathbf{0}, \mathbf{I}) \tag{3.25}$$

The pair (3.9) and (3.10) yielded equation (3.11),

$$f(\mathbf{x}) = N_p(\boldsymbol{\mu}, \boldsymbol{\Lambda}\boldsymbol{\Lambda}^T + \boldsymbol{\Psi})$$

for $f(\mathbf{x})$, and the pair (3.15) and (3.16) yield

$$f(\mathbf{x}) = N_p(\boldsymbol{\mu}, \boldsymbol{\Lambda}\mathbf{M}^T\mathbf{M}\boldsymbol{\Lambda}^T + \boldsymbol{\Psi}) = N_p(\boldsymbol{\mu}, \boldsymbol{\Lambda}\boldsymbol{\Lambda}^T + \boldsymbol{\Psi})$$

because of the orthogonality of **M**, so that the two pairs cannot be distinguished. In geometric terms this represents a rotation in the space spanned by the latent variables in **y**. When $q = 1$, which is the main concern of this book, **y** is univariate so that rotation is not possible so that this problem does not arise (we have already constrained the scale of **y**). There are various strategies and conventions for removing the indeterminacy by choosing a particular rotation. For example, since the latent variables will be interpreted in terms of their relationship with the manifest variables via the loading matrix $\boldsymbol{\Lambda}$, a natural strategy is to find a rotation in which the terms of this matrix are as near either 0 or 1 as possible. Further details of such methods are given in books on factor analysis, such as Everitt (1984) and Basilevsky (1994).

A special case of this model arises when $q = p$ since then $\boldsymbol{\Sigma}$, being symmetric, can be written as

$$\boldsymbol{\Sigma} = \mathbf{B}\mathbf{D}\mathbf{B}^T$$

where $\mathbf{B}$ is an orthogonal matrix with columns the eigenvectors of $\boldsymbol{\Sigma}$, and $\mathbf{D}$ is diagonal with elements the corresponding eigenvalues. This can be rewritten as

$$\boldsymbol{\Sigma} = \mathbf{B}\mathbf{D}^{1/2}\mathbf{D}^{1/2}\mathbf{B}^T = (\mathbf{B}\mathbf{D}^{1/2})(\mathbf{B}\mathbf{D}^{1/2})^T + \mathbf{0}$$

This has the same form as (3.21), with $\boldsymbol{\Lambda} = \mathbf{B}\mathbf{D}^{1/2}$ and $\boldsymbol{\Psi} = \mathbf{0}$. It follows from (3.18) that the conditional distribution of $\mathbf{x}$ given $\mathbf{y}$ collapses to a single value, $\mathbf{x} = \boldsymbol{\mu} + \mathbf{B}\mathbf{D}^{1/2}\mathbf{y}$. That is, $\mathbf{x}$ is a linear function of $\mathbf{y}$ (and $\mathbf{y}$ is a linear function of $\mathbf{x}$), so that this represents an exact transformation of the $\mathbf{x}$-space, rather than a simplifying approximation in terms of a few latent variables. This transformation is known as the *principal component* transformation, and the rescaled $\mathbf{y}$ variables, $\mathbf{z} = \mathbf{D}^{1/2}\mathbf{y}$, are the principal components. If only some (r, say) of the eigenvalues are large, and the remainder are small, then a principal components analysis will yield results similar to a factor analysis based on r latent factors.

The model given in expressions (3.18) and (3.19) can be alternatively written as

$$\mathbf{x} = \boldsymbol{\mu} + \boldsymbol{\Lambda}\mathbf{y} + \mathbf{e} \tag{3.26}$$

$$f(\mathbf{y}) = N_q(\mathbf{0}, \mathbf{I})$$

$$f(\mathbf{e}) = N_p(\mathbf{0}, \boldsymbol{\Psi})$$

with $\mathbf{y}$ and $\mathbf{e}$ independent. If the $\mathbf{x}$ variables are standardized to have zero mean, then (3.26) simplifies to $\mathbf{x} = \boldsymbol{\Lambda}\mathbf{y} + \mathbf{e}$. In the case of a single underlying factor, the score on the kth manifest variable for the ith subject is then $x_{ki} = \lambda_k y_i + e_{ki}$. The key thing here is the separation into a term for the test (the λ_k) and a term for the subject (the y_i). Our particular interest and focus in this book is, of course, the latter – the 'ability' of the subject – but in other contexts other aspects are of interest.

3.5.3 Poisson models

The factor analysis model described above assumes that the manifest variables are normally distributed, and extracts the underlying latent variable from them. This is not an adequate description of all manifest variables. Sometimes the manifest variables are counts, with conditional distributions which may be adequately modelled by Poisson distributions which depend on the latent variable. That is, we model the conditional distribution of the ith manifest variable as

$$f_i(x_i|\mathbf{y}) = \frac{e^{-\mu_i(\mathbf{y})}\mu_i(\mathbf{y})^{x_i}}{x_i!}$$

with

$$\mu_i(\mathbf{y}) = \exp\left[\beta_{i0} + \sum_{j=1}^{q} \beta_{ij} y_j\right]$$

so that, in the notation of Section 3.5.1, $\theta_i(\mathbf{y}) = \log \mu_i(\mathbf{y})$, The X_j then become

$$X_j = \sum_{i=1}^{p} \beta_{ij} x_i$$

In the case of a single latent variable with $\beta_{i1} = 1$, $\mu_i(\mathbf{y})$ simplifies to $\exp[\beta_{i0} + y]$, so that the parameter of the Poisson distribution is simply a product of two values, one ($\exp(y)$)

describing the object being measured (or the person taking the test), and the other $(\exp(\beta_{i0}))$ describing the particular measurement being applied (or the test being taken). This product formulation was originally developed by Rasch (and is described, for example, in Rasch, 1977). He began with a scenario in which each of several readers read each of several tests, and assumed a Poisson distribution for the number of errors. Then the conditional distribution of the number of errors made by the rth person in the ith test, a_{ri}, given that the total number of errors this person makes in the ith and jth tests is $a_{ri} + a_{rj} = a_{r+}$, is

$$p(a_{ri}|a_{r+}) = \frac{p(a_{ri} \cap a_{r+})}{p(a_{r+})} = \frac{p(a_{ri})p(a_{r+}|a_{ri})}{p(a_{r+})} = \frac{p(a_{ri})p(a_{rj})}{p(a_{r+})}$$

$$= \frac{\dfrac{e^{-\mu_{ri}} \mu_{ri}^{a_{ri}}}{a_{ri}!} \cdot \dfrac{e^{-\mu_{rj}} \mu_{rj}^{a_{rj}}}{a_{rj}!}}{\dfrac{e^{-\mu_{r+}} \mu_{r+}^{a_{r+}}}{a_{r+}!}} \tag{3.27}$$

where μ_{ri}, μ_{rj} are the parameters for the Poisson distributions of the number of errors made by the rth person in the ith and jth tests respectively, and μ_{r+} is the parameter for the Poisson distribution of the total number of errors made by the rth person in these two tests.

Now the Poisson distribution has the property that the sum of two Poisson-distributed random variables has a distribution which is also Poisson, with parameter the sum of the parameters of the component distributions. Thus

$$p(a_{r+}) = \frac{e^{-\mu_{r+}} \mu_{r+}^{a_{r+}}}{a_{r+}!} = \frac{e^{-(\mu_{ri}+\mu_{rj})}(\mu_{ri} + \mu_{rj})^{a_{r+}}}{a_{r+}!} \tag{3.28}$$

Now, adopting the decomposition of $\mu_i(y_r)$ given above, namely

$$\mu_i(y_r) = \exp[\beta_{i0} + y_r] = \exp(\beta_{i0})\exp(y_r)$$

which for notational simplicity we will write as $\mu_i(y_r) = \tilde{\beta}_i \tilde{y}_r$, (3.27) and (3.28) yield:

$$p(a_{ri}|a_{r+}) = \frac{\dfrac{e^{-\mu_{ri}} \mu_{ri}^{a_{ri}}}{a_{ri}!} \cdot \dfrac{e^{-\mu_{rj}} \mu_{rj}^{a_{rj}}}{a_{rj}!}}{\dfrac{e^{-(\mu_{ri}+\mu_{rj})}(\mu_{ri} + \mu_{rj})^{a_{r+}}}{a_{r+}!}}$$

$$= \frac{\dfrac{e^{-\tilde{\beta}_i \tilde{y}_r} \tilde{\beta}_i^{a_{ri}} \tilde{y}_r^{a_{ri}}}{a_{ri}!} \cdot \dfrac{e^{-\tilde{\beta}_j \tilde{y}_r} \tilde{\beta}_j^{a_{rj}} \tilde{y}_r^{a_{rj}}}{a_{rj}!}}{\dfrac{e^{-(\tilde{\beta}_i \tilde{y}_r + \tilde{\beta}_j \tilde{y}_r)}(\tilde{\beta}_i \tilde{y}_r + \tilde{\beta}_j \tilde{y}_r)^{a_{r+}}}{a_{r+}!}}$$

$$= \binom{a_{r+}}{a_{ri}} \left(\frac{\tilde{\beta}_i}{\tilde{\beta}_i + \tilde{\beta}_j}\right)^{a_{ri}} \left(\frac{\tilde{\beta}_j}{\tilde{\beta}_i + \tilde{\beta}_j}\right)^{a_{rj}}$$

that is, a binomial distribution which depends only on the test parameters, and not on the person parameters. This crucial fact, that the conditional distribution separates the test parameters from the person parameters, was spotted by Rasch (aided by the Nobel prize winning economist Ragnar Frisch – see Rasch, 1977). It means that one can estimate the characteristics of the tests people are taking without the results being contaminated by which people happen to have taken the tests. A similar derivation also allows one to deduce parameters for the people, uncontaminated by which tests they happen to have taken. For this reason, Rasch called the phenomenon *specific objectivity*. Here is an extract from the foreword, by Benjamin D. Wright, to Rasch (1980), a book first published in 1960:

> *Probabilistic Models for Some Intelligence and Attainment Tests* is the most important work on psychometrics since Thurstone's articles of 1925–1929 and the 1929 monograph by Thurstone and Chave. The psychometric research done by Rasch between 1951 and 1959, which he explains and illustrates in this book, marks the point at which psychometrics moved from being purely descriptive to become a science of objective measurement. The psychometric grail of psychological measures that transcend the questions or items producing them was glimpsed by Thurstone in the 1920s when he set down his requirements for valid measuring.
>
> > A measuring instrument must not be seriously affected in its measuring function by the object of measurement. To the extent that its measuring function is so affected, the validity of the instrument is impaired or limited. If a yardstick measured differently because of the fact that it was a rug, a picture, or a piece of paper that was being measured, then to that extent the trustworthiness of that yardstick as a measuring device would be impaired. Within the range of objects for which the measuring instrument is intended, its function must be independent of the object of measurement.
> >
> > Thurstone (1928, p. 547; 1959, p. 228)
>
> Objective measurement, that is, measurement that transcends the measuring instrument, not only requires measuring instruments which can function independently of the objects measured, but also a response model for calibrating their functioning, which can separate instrument and object effects. But, because Thurstone and his followers did not parameterize or condition for persons individually in their response models, they never attained objectivity. Even today there are psychometricians who try to achieve objective measurement with methods built on samples of normally distributed random individuals and response models that deny parameter separation.

The central role of the binomial distribution in this is also used directly in the models discussed in the next subsection.

3.5.4 Binary responses

Systems in which the manifest variables are Bernoulli are particularly important, and have been explored in several areas. They arise in education, where the manifest variables are test *items*, scored right or wrong. They arise in the social sciences, in such areas as attitude and opinion testing, where the responses may be agree/disagree. When the latent variable is continuous, such models are often called *latent trait* models. Although they are a special case of the latent variable model outlined above, one of their initial developments was in the form of *item response models*, sometimes called *Rasch models* in honour of Georg Rasch, referred to above.

In this case, the conditional distribution of the manifest variables has the form

$$f_i(x_i|\mathbf{y}) = \pi_i(\mathbf{y})^{x_i}(1 - \pi_i(\mathbf{y}))^{1-x_i}$$

where $\pi_i(\mathbf{y})$, the probability that the ith manifest variable will score a 1, is modelled as

$$\theta_i(\mathbf{y}) = \text{logit}\,\pi_i(\mathbf{y}) = \sum_{j=0}^{q} \beta_{ij}y_j$$

as above. This leads to

$$\pi_i(\mathbf{y}) = \frac{\exp\left(\displaystyle\sum_{j=0}^{q} \beta_{ij}y_j\right)}{\left\{1 + \exp\left(\displaystyle\sum_{j=0}^{q} \beta_{ij}y_j\right)\right\}}$$

Most of the literature on item response theory does not develop things from this general perspective in which the model is merely a special case of that outlined in Section 3.5.1. Indeed, as is so often the case (generalized linear models being the perfect example), many special cases have first to be explored before their common nature is recognized, and item response theory was one of the special cases of the latent variable model developed in considerable detail. As a result, it has developed its own terminology and special cases (and followers).

Item response theory takes the manifest variables, the items, as having different degrees of difficulty, and the subjects taking the test as having different degrees of ability. The difficulty and ability are taken to be measured on the same dimension, and the aim is to position the items and subjects on this dimension. Given a subject's ability and an item's difficulty, the model gives the probability that a subject will answer the test item correctly. The work on item response theory has generally approached things from the fixed effects perspective, seeking to estimate the ability score directly rather than indirectly from its conditional distribution.

The simplest form of item response model is the *one-parameter model*, originally developed by Rasch himself. We initially outline this in a simple notation. Denoting the ith item's difficulty by β_i and the rth subject's ability by y_r, the probability that the subject will get the item correct (score 1) is

$$\pi_i(y_r) = \frac{e^{(y_r - \beta_i)}}{1 + e^{(y_r - \beta_i)}} = \frac{1}{1 + e^{-(y_r - \beta_i)}} \tag{3.29}$$

For a given item, such a function of y_r is called the *item response function* or *item characteristic curve*. These functions are parallel, but location-shifted, with those with the greater item difficulty parameters lying further to the right (for a given value of ability, those more to the right will have a smaller proportion of subjects giving correct answers).

The simple model given in equation (3.29) is all very well, but it is rather limiting. In particular, if two items both have difficulty β then a subject with ability y_r will have the same probability of getting the item right. We can relax this by introducing the notion of *item discrimination*. In terms of model (3.29), we introduce a new, item-specific, parameter a_i so that

$$\pi_i(y_r) = \frac{e^{a_i(y_r - \beta_i)}}{1 + e^{a_i(y_r - \beta_i)}} = \frac{1}{1 + e^{-a_i(y_r - \beta_i)}} \tag{3.30}$$

Now, the probability that a subject with ability parameter y_r gets an item with difficulty parameter β_i correct depends not only on the difference between these values, $y_r - \beta_i$, but also on the item discrimination parameter, a_i. For items with a high value of a_i, a small difference $y_r - \beta_i$ will have a big impact on the probability of getting the item correct. Parameter a_i thus tells us how well the item separates out subjects with different abilities (i.e. how well the item 'discriminates' between subjects).

This may be more clearly expressed if we work on a log-odds scale, since then

$$LO(y_r) = \log\frac{\pi_i(y_r)}{1 - \pi_i(y_r)} = \log\left[\frac{\dfrac{e^{a_i(y_r - \beta_i)}}{1 + e^{a_i(y_r - \beta_i)}}}{\dfrac{1}{1 + e^{a_i(y_r - \beta_i)}}}\right] = a_i(y_r - \beta_i)$$

so that the slope of the log-odds function is given by a_i.

In terms of the general model of Section 3.5.1, $\beta_{i0} = -a_i\beta_i$ and $\beta_{i1} = a_i$. β_{i1} is thus the discriminability of the item. The Bernoulli distribution belongs to the exponential family and has $u_i(x_i) = x_i$, so that, from (3.15),

$$X_j = \sum_i \beta_{ij} x_i$$

That is, the sufficient statistic for **y** is simply the sum of the β_{ij} weights for those variables on which the subject has scored positively.

In fact, it has been found useful to extend the model of (3.30) yet further, in a way which steps out of the elegance of the latent variable approach (an illustration of the life of its own that the item response model has developed). Subjects may get the right answer purely by chance (e.g. simply by guessing in an examination). The models above assume that getting an item correct is a sign of greater ability, so somehow we must allow for items which are correct by accident. An appropriate extension is given by

$$\pi_i(y_r) = c_i + \frac{1 - c_i}{1 + e^{-a_i(y_r - \beta_i)}} \tag{3.31}$$

This introduces a third parameter, c_i, for the ith item. Suppose that someone has no ability, and simply guesses. This is equivalent to an ability parameter $y = -\infty$. In models (3.29) and (3.30) this gives $\pi_i(y) = 0$. In model (3.31), however, it gives $\pi_i(y) = c_i$. c_i thus tells us what fraction of items with difficulty β_i we would expect a subject who simply guessed to get right. The model in (3.31) is, for obvious reasons, sometimes called the *three-parameter model*.

A difficulty arises with model (3.31), namely that the model is not identified: alternative values for the parameters will yield identical values for $\pi_i(y_r)$. In particular, if we apply a linear transformation to β_i, yielding $\beta_i' = \gamma\beta_i + \delta$, the same linear transformation to y_r, yielding $y_r' = \gamma y_r + \delta$, and the rescaling transformation $1/\gamma$ to a_i, yielding $a_i' = a_i/\gamma$, then $\pi_i(y_r)$ is unchanged. To overcome this problem it is necessary to impose some constraints on the parameters. A common constraint is to arrange that the distribution of the abilities y_r has mean zero and standard deviation unity in a given population of subjects. On the other hand, this has the disadvantage that different populations of subjects, each artificially standardized in this way, cannot be compared. This can be overcome by aligning the difficulty and discrimination scores of the items, by applying the linear transformation which produces the same mean and standard deviation for the ability scores in the two samples. Thus, suppose that one sample has mean subject score $\bar{x}_1$, with standard deviation s_1, and the other has mean subject score $\bar{x}_2$, with standard deviation s_2. Then the item scores a_i and β_i in the first sample are transformed to

$$a_i^* = a_i \frac{s_1}{s_2} \quad \text{and} \quad \beta_i^* = (\beta_i - \bar{x}_1)\frac{s_2}{s_1} + \bar{x}_2$$

This can also be applied if a single sample of subjects is exposed to both sets of items.

If the same items have been used with two different populations, then one can find the transformation of the item scores from one group such that the overall differences between the subject ability scores in the two populations are minimized – for example, by minimizing the sum of squared differences between the subject scores in one group and the transformed subject scores in the other group (cf. the use of the least-squares criterion in multidimensional scaling).

The models described above are static, in the sense that all the subjects answer the same items, and the set of items is fixed. If the tests are administered by computer, however, some very exciting possibilities arise. In particular, one can then choose the items to be answered by each subject on the basis of that subject's performance so far. This allows one to match the item's difficulty to the subject's (estimated) ability, to maximize the information obtained about each subject, so that each subject's ability parameter is estimated most accurately. It also means that a less able subject is not discouraged by facing many questions they have no idea how to answer, and a very able subject is not bored by having to wade through many elementary questions.

The subject score which emerges from the above is y. Sometimes alternative transformations of y have merits. For example, for the ith item the probability

$$\pi_i(y) = c_i + \frac{1 - c_i}{1 + e^{-a_i(y - \beta_i)}}$$

is sometimes a useful transformation, taking the score to the unit interval and having a meaningful interpretation as a probability. Furthermore, we can sum these transformed scores over all of the items, to yield the expected score that the subject would have obtained had they answered all the items (sometimes this is called the *item pool score*).

In the above, we have used the logistic function to model the relationship between ability, item difficulty, and the probability of scoring a 1 on any particular item – that is, on any particular manifest variable. Other functions could be used. The Guttman scale, described above, is the particular case in which a step function is used. The cumulative normal distribution is another obvious choice, and one which was used in similar contexts prior to the widespread adoption of the logistic function. As it happens, the two functions are very similar, although the logistic function has more attractive mathematical properties. There is a strong link between the model based on the cumulative normal distribution and the normal theory factor analysis outlined above. This binary response model is obtained from a model in which the manifest variables have conditional normal distributions, as outlined above, and in which these are split at some threshold to yield binary variables. Bartholomew and Knott (1999) call this an *underlying variable model*. In many circumstances it is natural to conceptualize the binary variables as having arisen by partitioning an underlying continuous variable (e.g. yes/no responses in an opinion survey), whereas in others it is not (e.g. right/wrong answers in simple arithmetic questions). Hand *et al.* (1998) explore a similar issue in defining classes in discriminant analysis. Bartholomew and Knott (1999, p. 87) demonstrated that the two approaches, that based on assuming a continuous underlying variable which is partitioned, and that based directly on a response function, are equivalent.

3.6 More complicated models

The measurement structure in Section 3.5 is based on the notion that there are observable indicator variables which can be measured, and that these are related to the (unobservable) attribute in such a way that any dependence between the indicator variables is a consequence of their mutual dependence on the underlying attribute. This sort of structure is perfectly reasonable if the indicator variables are regarded as *consequences* of the underlying attribute, or if they are regarded as *aspects* of the underlying attribute, but it seems less reasonable for indicators which might be *causes* of the underlying attribute. In health-related quality of life measurement, for example, while both pain and psychological bullying at work might lead to poor quality of life, surely there is no reason to expect them to be a consequence of poor quality of life. There is thus no reason to expect them to be independent if one controls for quality of life – the dependence or lack of it will be an artefact of the data one happens to have collected. This idea has been explored by Hand and Crowder (2003) in the context of measuring customer creditworthiness in retail banking. The set of indicator variables is partitioned into two types: those describing primary characteristics of the individual customer which we take as possible influences on creditworthiness and which we shall denote by $\mathbf{z}$, and those describing how that customer has behaved in the past with other banking products, which we shall denote by $\mathbf{x}$. So, for example, $\mathbf{z}$ will consist of variables such as time at present address, time with employer, age, income and occupation, while $\mathbf{x}$ will consist of variables such as frequency of having missed credit card repayment dates, number of adverse County Court Judgements, frequency of previous mortgage churn, and so on. Our aim is to extract a single feature from these variables which can act, in some sense, as a measure of the customer's creditworthiness, and which, as above, we will denote by y. Now, while it is known, for example, that both age and time at current address are (probabilistically) related to creditworthiness, there seems little reason to model these as independent conditional on a given level of creditworthiness. The correlation which may well exist between these two variables is likely to owe more to the passage of time than to anything else. Indeed, even if time at address was thought to be influenced by creditworthiness, it is difficult to see how age could be! In contrast, one may well consider it reasonable to model aspects of previous credit behaviour as conditionally independent, given the value of the creditworthiness attribute, in the manner of the models outlined in preceding sections. Overall, then, we have two sets of variables, an 'input' set and an 'output' set, and the underlying attribute is regarded as an intermediary variable via which the influence of the input variables on the output variables is channelled.

This model is similar to a *multiple indicator multiple cause* or *MIMIC* model (e.g. Bollen, 1989), although the development of these has generally focused on the multivariate normal case.

The joint distribution of the $\mathbf{x}$, y and $\mathbf{z}$ variables is

$$f(\mathbf{x}, y, \mathbf{z}) = f(\mathbf{z}) f(y \mid \mathbf{z}) \prod_{i=1}^{p} f(x_i \mid y)$$

so that the posterior distribution of y is

$$f(y \mid \mathbf{x}, \mathbf{z}) \propto f(y \mid \mathbf{z}) \prod_{i=1}^{p} f(x_i \mid y)$$

Hand and Crowder describe model fitting and prediction methods for this structure.

3.7 Quantifying categorical scales

Categorical scales are those which take a small number of possible values. Such scales may be ordinal or nominal, and sometimes it is convenient to attempt to assign numerical values to the categories. Clearly, such values should satisfy any constraints one knows the categories to satisfy. In the case of ordinal scales, a constraint is that the order must be preserved, but, as I remarked in Chapter 2, Abelson and Tukey (1959) have pointed out that even in situations where one cannot confidently assign numerical values to different levels of an attribute, there is often some information beyond the mere order of the objects. Typically, however, such constraints, even if they imply more than mere ordinality, are insufficient to completely determine the numerical values which should be assigned. In this common situation, one needs to adopt some additional, external and pragmatic, criterion which is to be optimized by the choice of numerical scale values.

Abelson and Tukey (1959, 1963) explored a maximin criterion based on the following idea. Let $(X_1, X_2, \ldots, X_n)$ be a chosen sequence of values to be assigned to the n categories of a variable. Suppose that $(Y_1, Y_2, \ldots, Y_n)$ is the unknown 'ideal' sequence (i.e. the 'values one would assign if one had full scale knowledge'). Then the correlation coefficient tells us how closely the chosen values match the ideal values. Unfortunately, of course, since we do not know the ideal values, we cannot maximize this correlation. Instead, Abelson and Tukey (1959) suggest choosing that set $(X_1, X_2, \ldots, X_n)$ which maximizes the minimum possible correlation over possible sets $(Y_1, Y_2, \ldots, Y_n)$. 'This results in a guarantee that r^2 cannot be less than a certain value ... so long as Nature obeys the inequalities.'

The Abelson and Tukey strategy is clearly univariate, concerning itself with a single variable. In many, perhaps most, situations, however, our ultimate interest lies in *relationships between variables*. Scaling methods have been developed which hinge on this. In general, in parallel with the Abelson and Tukey strategy, we will seek quantifications of the variables which optimize some aspect of their relationship. A very important special case of this arises when the different variables are thought to be measuring the same underlying attribute. In such a case, a sensible approach would be to choose the quantifications of the variables so that their *correlation* is maximized. If only two categorical variables are involved, then this completely defines the problem (assuming standard product moment correlation is being used). However, if more than two variables are involved, some decision is needed on how to aggregate the pairwise correlations. A popular approach is to sum the squared correlations.

This can be achieved in various ways. In fact, the history of data-analytic tools to achieve this objective is long and convoluted, with the same idea being developed several times from entirely different perspectives. For example, Richardson and Kuder (1933) suggested the first of the methods outlined below over 70 years ago. Tenenhaus and Young (1985) integrate these various strands in an elegant synthesis, and the following description is based on their work. Greenacre (1984) gives a comprehensive overview, especially of the bivariate case.

Suppose that each of n individuals is described by p categorical variables $\mathbf{x}_j, j = 1, \ldots, p$, with the jth variable having k_j categories. The occupied categories of the jth variable can be represented in an $n \times k_j$ binary indicator matrix $\mathbf{X}_j$, taking the value 1 if the ith individual lies in category l of the jth variable and 0 otherwise. The matrices can then be concatenated to form an overall $n \times k$ indicator matrix $\mathbf{X}$, with $k = \sum_{j=1}^{p} k_j$, and $\mathbf{X} = (\mathbf{X}_1, \ldots, \mathbf{X}_p)$. Our aim is to find numerical quantifications for the categories of the variables and for the individuals. Denote the quantification of lth category of variable j by ϕ_{jl}, so that vector

$$\boldsymbol{\phi}_j = (\phi_{j1}, \dots, \phi_{jk_j})^T$$

contains the quantifications of the jth variable, and vector

$$\boldsymbol{\phi}^T = (\boldsymbol{\phi}_1^T, \dots, \boldsymbol{\phi}_p^T)$$

contains all of the quantifications. Similarly, define a vector $\boldsymbol{\psi}^T = (\psi_1, \dots, \psi_n)$ to represent the quantifications of the n individuals.

Now, given the quantification of the variables, a reasonable quantification of the individuals would be as a sum, or more conveniently, an average, of their category scores:

$$\boldsymbol{\psi} = \frac{1}{p}\mathbf{X}\boldsymbol{\phi} \tag{3.32}$$

Similarly, given a quantification of the individuals, a reasonable quantification of the category of a variable would be the average of the scores of the individuals who fall in that category of that variable:

$$\boldsymbol{\phi} = \mathbf{D}^{-1}\mathbf{X}^T\boldsymbol{\psi} \tag{3.33}$$

where $\mathbf{D}$ is the $k \times k$ diagonal matrix containing the n_{jl}, the numbers of individuals lying in category l of variable j:

$$\mathbf{D} = \mathrm{diag}(n_{11}, n_{12}, \dots, n_{1k_1}, n_{21}, \dots, n_{2k_2}, n_{31}, \dots, n_{pk_p})$$

Unfortunately, equations (3.32) and (3.33) cannot hold simultaneously, simply because (3.32) means that the ψ_i lie within the range of the ϕ_{jl} (because ψ_i is defined as an *average* of the ϕ_{jl}) while (3.33) means that the ϕ_{jl} lie within the range of the ψ_i. Some kind of normalization is needed to enable (3.32) and (3.33) to be satisfied simultaneously. Making such an adjustment leads to

$$\boldsymbol{\psi} = \frac{\rho}{p}\mathbf{X}\boldsymbol{\phi} \quad \text{and} \quad \boldsymbol{\phi} = \rho\mathbf{D}^{-1}\mathbf{X}'\boldsymbol{\psi} \tag{3.34}$$

From (3.34), we see that

$$\boldsymbol{\psi} = \frac{\rho^2}{p}\mathbf{X}\mathbf{D}^{-1}\mathbf{X}'\boldsymbol{\psi} \quad \text{and} \quad \boldsymbol{\phi} = \frac{\rho^2}{p}\mathbf{D}^{-1}\mathbf{X}^T\mathbf{X}\boldsymbol{\phi} \tag{3.35}$$

so that the quantifications are given by the eigenvectors of the matrices $\mathbf{X}\mathbf{D}^{-1}\mathbf{X}^T/p$ and $\mathbf{D}^{-1}\mathbf{X}'\mathbf{X}/p$. For uniqueness we need to normalize $\boldsymbol{\Psi}$ and $\boldsymbol{\phi}$, so we set

$$\frac{\boldsymbol{\phi}^T\mathbf{D}\boldsymbol{\phi}}{np} = 1 \quad \text{and} \quad \frac{\boldsymbol{\psi}^T\boldsymbol{\psi}}{n} = 1 \tag{3.36}$$

It is easily seen that the matrices $\mathbf{X}\mathbf{D}^{-1}\mathbf{X}^T/p$ and $\mathbf{D}^{-1}\mathbf{X}^T\mathbf{X}/p$ have trivial eigenvectors consisting of vectors of 1s, associated with eigenvalue 1. The other eigenvectors satisfy the centring constraints

$$\mathbf{U}_k\mathbf{D}\boldsymbol{\phi} = 0 \quad \text{and} \quad \mathbf{U}_n\boldsymbol{\psi} = 0 \tag{3.37}$$

with $\mathbf{U}_r$ a vector of r 1s. Of course, in general there are multiple eigenvectors satisfying (3.35). Study of these leads on to biplot displays for simultaneously representing individuals and variables (Gower and Hand, 1996). In this book, however, we are especially concerned with the single eigenvector associated with the largest (non-trivial) eigenvalue. It is not difficult to show that this eigenvalue is less than 1.

This approach to deriving a scaling for the categories of the variables, and simultaneously for the individuals, is called *dual scaling*. It has been developed by, in particular, Nishisato (1980).

The dual scaling approach has an elegant simplicity, as well as a natural intuitive appeal. However, an alternative approach based on more conventional statistical ideas can be derived from analysis of variance (Guttman, 1941). If we define the individuals' scores as in (3.32), $\psi = X\phi/p$, and adopt the constraints in (3.36) and (3.37), $\phi^T D\phi/np = 1$ and $U_k D\phi = 0$ respectively, then we can define a ratio of between- to within-individual sums of squared scores. Here, 'within' means between the scores given by the p different variables for each individual and 'between' means between the overall scores ψ_i for each individual. Thus the 'within' scores for the ith individual are the values of ϕ_{jl} that they actually take. In matrix terms, this ratio is

$$\eta^2 = \frac{\phi^T X^T X\phi}{p\phi^T D\phi} \tag{3.38}$$

Our aim, then, is to find the quantification ϕ which maximizes this ratio, and this is found to be the first non-trivial eigenvector of the matrix $D^{-1}X^T X/p$, the same result as above. The value of η^2 is the same as the largest non-trivial ρ^2 in eigenequations (3.35).

A third approach is closely related to principal components analysis, and has been extensively developed by Benzécri (1992) and others in the form of correspondence analysis. The constraints we adopted above, that $\phi^T D\phi/np = 1$ and $U_k D\phi = 0$, mean that maximizing (3.38) is equivalent to maximizing the variance of the vector of individuals' scores $\psi = X\phi/p$. This is equivalent to solving the eigenequation

$$\theta = \frac{\nu^2}{p} D^{-1/2} X^T X D^{-1/2}\theta \tag{3.39}$$

leading to eigenvectors θ related to the eigenvectors ϕ of $D^{-1}X^T X/p$ by

$$\phi = (np)^{1/2} D^{-1/2}\theta$$

Finally, yet another approach was explored by McKeon (1966) and others. They show that the sum of squared correlations between the scaling of the individuals, ψ, and the scaled values obtained from the quantifications of the variables, $X_j\phi_j$, is maximized by choosing ϕ as above. This corresponds to a generalization of canonical correlations analysis. In the case of just two categorical variables, the special case with which we opened this section, the method is equivalent to a canonical correlations analysis between two sets of binary variables, one set being indicator variables for the categories of one of the categorical variables and the other set being indicator variables for the categories of the other categorical variable. Put another way, the method seeks to find the quantifications of the two categorical variables such that the correlation between the two variables is maximized.

Yet another group of researchers who discovered these ideas is the Gifi group in the Netherlands (Gifi, 1990). They emphasized the tool as yielding the quantifications which maximize the homogeneity between the variables, as measured by their correlation matrix, and hence called the method *homogeneity analysis*. To illustrate the perspective taken by this group, we will change our notation. So, let $(v_1, \ldots, v_p)$ represent the values of p vectors of continuous variables, each of length k, the number of objects measured. These vectors will be completely homogeneous if $\mathbf{v}_1 = \ldots = \mathbf{v}_p$. The extent to which a set of vectors is homogeneous can then be measured by the variation between those vectors. In particular, for example, we could measure the total deviation between the $\mathbf{v}_i$ and $\bar{\mathbf{v}} = \Sigma \mathbf{v}_i/p$, the centroid of the $\mathbf{v}_i$. If this 'total deviation' is measured using the standard Euclidean norm, and if we simply average this deviation over the $\mathbf{v}_i$, then a measure of heterogeneity is

$$\sigma^2 = \frac{1}{kp} \sum_i \sum_j (v_{ij} - \bar{v}_j)^2$$

If all of the variables are first standardized (so that the mean of the k observations on $\mathbf{v}_i$ is 0, and the standard deviation of the k observations on $\mathbf{v}_i$ is 1, for each i) then $\sigma^2 = 1 - \bar{r}$, where $\bar{r}$ is the average of the correlation coefficients between the $\mathbf{v}_i$: the greater the mean correlation, the lower the heterogeneity and the greater the homogeneity between the variables. In the limit of all of the variables yielding identical results, $\bar{r}$ is 1 and σ^2 is zero – perfect homogeneity.

More generally, we can extend the definition of homogeneity by requiring that the vectors be similar under a given class of transformations. For example, we might permit linear rescalings, so that $\mathbf{v}_i \mapsto a_i \mathbf{v}_i$ for constant a_i, or a more general class of transformations. In the linear rescaling case, the heterogeneity measure σ^2 then becomes

$$\sigma^2 = \frac{1}{kp} \sum_i \sum_j (a_i v_{ij} - \bar{v}_j)^2$$

where now $\bar{\mathbf{v}} = \Sigma a_i \mathbf{v}_i/p$. To find the optimal rescaling transformation – that which minimizes heterogeneity – we need to choose suitable a_i. This leads to a cyclic *alternating least-squares algorithm*, in which, from a given starting point, one cycles through the two steps:

(i) Minimize σ^2 by setting $\bar{\mathbf{v}} = \Sigma a_i \mathbf{v}_i/p$.
(ii) Minimize σ^2 by choice of the a_i. This, of course, is simply a regression problem – with a_i the regression coefficient of $\bar{\mathbf{v}}$ on $\mathbf{v}_i$.

By the monotone convergence theorem it is easy to see that this process must converge, but it has trivial solutions $a_i = 0$, so that some normalization is required, such as fixing $\Sigma_j v_{ij}^2$ for all i. Normalizing the variance of the elements of $\bar{\mathbf{v}}$ to 1 results in $\bar{\mathbf{v}}$ being the first principal component of the $\mathbf{v}_i$.

Linear rescalings are a fairly limited class of transformation, and one might consider more flexible approaches. Gifi (1990) then considers the particular class of nonlinear transformations in which each variable is discretized into a small number of cells, and then each cell is permitted to take an arbitrary value (and variables which are already discretized are, of course, taken in the cells they already have). This leads to $v_{ij} \mapsto \mathbf{b}_{ij}^T \boldsymbol{\mu}_i$, where $\mathbf{b}_{ij}^T$ is an indicator vector, taking the value 0 everywhere but in one position, and where $\boldsymbol{\mu}_i$ is a vector of k parameter values (cell values) for the ith variable), so that

$$\sigma^2 = \frac{1}{kp} \sum_i \sum_j (\mathbf{b}_{ij}^T \boldsymbol{\mu}_i - \bar{v}_j)^2$$

Minimizing this will lead to optimal scores for each cell of each variable.

Gifi (1990) extends these ideas in various ways, for example by restricting the values assigned to the cells depending on aspects such as ordinality. The idea of transforming raw variables to optimize some criterion can also be extended to continuous variables, with various nonlinear transformations being permitted.

3.8 Item analysis

Indirect measurement involves combining values from many manifest variables so as to yield a measured value of some attribute which cannot be directly observed. In many indirect procedures, the overall measurement is simply the sum or a weighted sum of many observable scores. The more of these observable scores, manifest variables, or items which are included in such a procedure, then, by virtue of basic statistical laws, the more consistent will be the overall score, reducing the effect of any random influences. On the other hand, the more items which are included, the more difficulty one might find in obtaining valid responses – respondents will tire as the number of items increases, and might even refuse to take part if there are too many. Often there is considerable latitude about which items might be used. *Item analysis* is the process of investigating and choosing which items to include.

Of course, before one can begin to investigate which items to include, one must generate a set of items. If one has a well-developed theory about the underlying attribute to be measured, then one will create items which tap into that theory. Of course, if the theory is faulty, or is later rejected in favour of an alternative, then the scale may lose its merits. Furthermore, the significance and implication of items chosen in this way is often readily apparent to someone taking the test. This leads to complications of the kind mentioned in Section 4.2.1: it is not difficult to see what a question such as 'Do you ever feel down?' is aiming at.

In some situations there is a gold standard. For example, one might be aiming to predict future job performance. In this case one might have information which has been collected from an earlier cohort of candidates, so that data on both their job performance and item scores is available. This can then be used to construct an instrument for measuring/predicting likely future job performance of new candidates. This sort of strategy is used for constructing screening instruments: a data set containing values of both potential predictor variables and an indicator for the presence/absence of the disease in question is collated, and a predictive model is built which can serve to predict the likely class of new respondents solely from the values of their predictor variables.

In such cases, when there is a known outcome, criterion variable, or gold standard, standard statistical item selection methods can be used. The aim of these methods is to identify a set of items which, together, provide good predictive accuracy for the gold standard. They include forward and backward stepwise methods, as used in linear and logistic regression. Any such analysis should take care to avoid overfitting: the phenomenon in which a statistical model fits the available data very well, but does not generalize well to new data. Such problems arise when the models are highly parameterized (or equivalently, as in this case, have many potential subsets of variables from which to choose) so that they produce a good match to the idiosyncrasies of the available data, rather than to the underlying distributions.

When there is a gold standard, and statistical item selection methods are used, one can, at least in principle, generate a large set of items without regard to their relationship to the criterion to be measured, relying on the data analysis to choose an effective subset. This can lead to individual items which do not bear an obvious relationship to the criterion to be measured. The underlying theory here is rather different from that of most of the indirect measurement procedures described in this chapter. These methods are based on the notion that one has some theory of the general form which exists ('I believe manifest variables x, y and z are related to latent variable w, by such and such a model, so the value of w can be deduced from the value of x, y, and z') or is desired ('I want manifest variables x, y and z to be related to latent variable w, by such and such a model, so the value of w can be deduced from the value of x, y, and z'). The factor analysis approach is an example of the former of these. In contrast, the present case is based on a theory of the form 'For previous subjects I have observed variables x, y, z and w, so I can build a statistical model which will enable me to predict w from x, y and z for future subjects'. The model in the former case is a theoretical or iconic model. The model in the latter case is an empirical model. This distinction is discussed in Chapter 1.

When there is a very large number of items from which to choose, it is usual to thin these down by an initial stage in which the items are examined individually, rather than in combination. In cases where a criterion variable is available, a common strategy is to measure the relationship between individual items and the criterion value in a data set which includes both. In cases when no criterion variable is available, a common strategy is to choose variables according to the strength of the relationship between them and the overall score (see the discussion of Likert scales above). As we mentioned in the context of Guttman models above, this strategy can also be viewed as a test of the representational model.

3.9 Multiple-choice questions

In many indirect measurement procedures, especially in psychology and education, the individual test items are multiple-choice questions, with only one correct answer and a small number of incorrect answers. The tests are less distracting to those taking them if the number of possible answers is the same for each question. To illustrate, suppose that there are n questions and that each question has a total of c possible answers, with only one answer being correct for each question. Then the score distribution of someone who does not know the answers to any of the questions, and answers at random, will follow a binomial $B(n, 1/c)$ distribution. This has mean n/c.

This is an ideal, and in fact assumes not only that the respondent does not know the correct answer to any of the questions, but also that they do not have any knowledge about which of the incorrect possible answers are incorrect. If they do know, for any question, that a particular $c - 2$ of the wrong answers are wrong, then their chance of picking the correct answer is in fact $1/2$. If this is true for all of the questions, their mean score will be $n/2$. (If the pass mark of an examination was 40%, this could have serious implications. On the other hand, it does show that the candidates know something, even if they do not know the correct answers.)

If an item is such that nobody gives the correct answer, then this item is of no value for measurement. In an educational testing context, we might describe the item as too difficult for its purpose. Likewise, items which are too easy will be answered by everyone, and hence will not contribute to the measurement process. (This is not to say that they will not serve some other purpose: some easy questions near the start of a test may help students to relax into the examination.)

Some other types of measurement procedure superficially resemble multiple-choice questionnaires, but are different. We have already seen this above, and an example is the General Health Questionnaire mentioned in Section 3.3.2 and described in more detail in Section 6.5.2. Such instruments have multiple questions, each of which has several possible answers, but there is no question of 'right' or 'wrong' for these items. Rather, each possible answer represents a position on an underlying scale of severity (e.g. better than usual, same as usual, worse than usual, much worse than usual), so that their sum forms a Likert scale. The items in such a test will be chosen partly on the grounds that they span the psychological phenomenon of interest, and partly on the grounds that their combination has good predictive power for the problem in question.

Perhaps the most common criticism of multiple-choice tests is that they tend to test details and do not adequately test the larger analytic and synthetic skills: they give limited opportunity to test someone's ability to work out a strategy for tackling a problem, and for putting together the various components needed to solve that problem (see, e.g., Jones and Applebaum, 1989), or, to put it more bluntly, they do not tap creative thinking. Although natural, the evidence in favour of these criticisms seems weak. In particular, the results of the multiple-choice tests are highly correlated with other forms of test.

3.10 Conclusion

This chapter has described some of the main approaches to constructing measurement scales. As has been illustrated, the procedures involve a balance between the representational and pragmatic aspects, with the position of the fulcrum of that balance depending on the attribute being measured and the aims of the measurement. Direct physical measurement is almost entirely representational, whereas many of the behavioural measures have a large pragmatic component, simultaneously defining and measuring the attribute.

Even more extreme pragmatic measurements than those described above are sometimes used. For example, an airline might rank type of aircraft on a preference scale, according to an aggregate of loudness of engines, speed, fuel efficiency, number of passengers, and so on, and a quality-of-life scale may combine aspects of completely distinct features of life. The point is that, even in situations where a concept has multiple dimensions, it is often necessary to summarize these into a single measure, so that operational decisions can be made, such as funding allocations. Although one may appear to be adding apples and oranges, one needs to combine these different aspects in order to make decisions.

4

Accuracy of measurement

... nearly all the grandest discoveries of science have been but the rewards
of accurate measurement and patient long-continued labour in the minute
sifting of numerical results.
Lord Kelvin, in his Presidential address to the British Association for the
Advancement of Science, 1871, quoted in Skinner (1991)

4.1 Introduction

Precise observations and measurement had become important in science following the efflorescence of scientific ideas and understanding which occurred in the seventeenth century (see, for example, Jardine, 1999). Later, around the second half of the eighteenth century, precision in physical measurements gained practical importance, with the industrial revolution. To make reliable machines, one needs components which fit together well. Moreover, if a broken part of a machine is to be replaced by an equivalent part to repair the machine, then accuracy in the construction of the replacement part is paramount. The history of the development of an accurate marine chronometer, described in Sobel (1995), provides a nice illustration of the need for accuracy in constructing machines, but the physical sciences in general show the synergistic leapfrogging act which relates measurement accuracy and scientific understanding.

Of course, increases in the accuracy of measurement of physical attributes depend heavily on the skills and motivations of those experts and craftsmen making the measurements and the measuring instruments. By the nineteenth century, these craftsmen were living in a culture which increasingly valued numerical accuracy, not only perceiving moral virtues in it but also seeing it as a matter of national pride. Indeed, as was noted in Chapter 1, it was social and economic needs rather than the physical sciences which were the original drivers of the need for measurement accuracy. To know what taxes could be raised or to know how large an army he could muster, a ruler needed to know how many people he had in his realm. To know how much to pay for a shipment of grain, a merchant needed to know how many sacks there were, and how much each sack weighed. To know when he would get back to port, a sailor needed to know his current position. The growth of states, and the increase in size of states and nations, led to an increasing need for measurement accuracy.

Without such accuracy, effective government is not possible. Measurement accuracy, then, is the common thread which ties social and economic needs on the one hand with scientific and engineering needs on the other. M. Norton Wise (1995b, p. 358) puts this exquisitely: 'Precision instruments are packaged trust. And the trust they carry is a social accomplishment.' He is talking about physical measuring instruments, but the sentiment applies more widely: accuracy of measurement is packaged trust. It is only by trusting the accuracy of our measurements that we can make effective governmental decisions and build reliable machines.

Turning to science, accuracy of measurement allows regularities and patterns to be spotted. Kepler's laws could only be derived from Tycho Brahe's observations because the latter were sufficiently accurate for the lawlike relationship to be discerned. Had their measurement error been much greater, no patterns would have been evident. Or perhaps, to put it another way, had the measurement error been much greater, then replicated measurements and advanced statistical techniques would have been needed to detect the lawlike relationships, and these would not exist until the later part of the twentieth century. A nice illustration of the importance of accuracy in length measurement is given by the fact that Isaac Newton used the Paris foot when deriving the force of gravity:

> In his early twenties Newton had attacked the problem of gravitation with the best available estimates of the Earth's dimensions, which were, as we know now, about 15 percent short. He could not fit the falling toward Earth of the orbiting Moon into the same equations as the falling toward earth of apples and such, until revised and increased measurements of Earth segments (in Paris feet) were made.
>
> (Klein, 1974, p. 72)

Accuracy of measurement also permits anomalies, departures from the predicted theory, to be spotted: it is only when you can measure consistently to one part in a million that you can spot a deviation of 0.001 as very substantial. If your theory predicts no such deviation, then your theory needs re-examination. Perhaps the discrepancy between the astronomical predictions of Newtonian mechanics and that observed in the advance of the perihelion of Mercury provides an example, even though in this case the theoretical prediction came first.

Accuracy of measurement is not achieved without effort. Indeed, greatest accuracy, at the frontiers of science, requires greatest effort, and probably the most expensive or complicated of measurement instruments and procedures. If science is to be universal, in the sense that anyone who studied the same material would be led to the same conclusions, then, it might be argued, science cannot rely on instrumentation or procedures which are available to or can be used by only the few. Indeed, there have been numerous examples, throughout the history of science, where the original results could not be replicated by other researchers. Then questions arise as to the truth of the conjecture underlying structure (cold fusion and some recent non-replicable results in nanotechnology spring to mind). Joseph Priestley (1733–1804) made this very argument when confronted by Antoine Lavoisier's (1743–1794) experiments on the composition of water. Golinsky (1995) quotes Priestley (1796) as saying, on the synthesis of water, that the experiment required 'so difficult and expensive an apparatus, and so many precautions on the use of it, that the frequent repetition of the experiment cannot be expected; and in these circumstances the practised experimenter cannot help suspecting the certainty of the conclusion'. Replication, he is saying, is fundamental to science, and if the conclusion depends on accuracy which is so difficult that few if any can replicate it then perhaps the conclusion is suspect. It follows from this that science

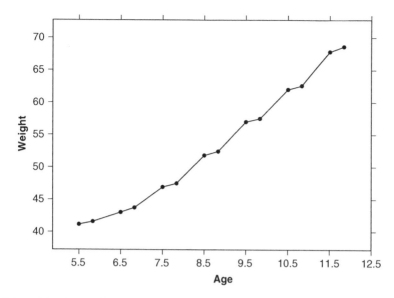

Figure 4.1 Weight changes for 10,000 children in a 1930s trial to study the effect of milk on growth (Leighton and McKinlay, 1930).

is most able to advance when instrumentation technologies have themselves advanced to the stage that the measurements are easy to make. Indeed, we are currently seeing the same effect in a different context with the ready availability and ease of use of computers.

To improve measurement accuracy it is necessary to understand the measurement error process. When humans are involved in recording or transcribing figures, particular types of errors can be made. Brunskill (1990) illustrates errors in the coding of birth weights: 14 oz recorded as 14 lb, birth weights of one pound (1 lb) being read as 11 lb, and misplaced decimal points – for example, 510 g recorded as 5100 g. It is worth noting that all of these errors yield overreporting of birth weights, which might lead to a mistaken impression of an excess of overweight infants at low gestational ages (David, 1980; Neligan, 1965). Another example occurs in reported road traffic accidents, where serious accidents are reported rigorously and accurately, while less serious ones may not be reported at all. This leads to a distortion in the apparent distribution of accidents.

Furthermore, recordings or measurements taken by humans are often subject to subconscious bias. Figure 4.1 shows the mean weights for the control group from a classic study of 10,000 children carried out in the 1930s to explore the effect of free school milk (Leighton and McKinlay, 1930). Two points are given for each age class, the two points resulting from measurements in February and June, respectively. The stepwise nature of the curve is unexpected, and is possibly explained by differences in the weights of clothes worn by the children on each occasion. In fact, these data were also subject to other possible sources of distortion, including the fact that those doling out the milk had some freedom of choice, and it seems likely that they may have tended to give milk to the children who would apparently benefit more from it.

A well-known phenomenon when humans record data from analogue instruments is *digit preference*, whereby people have a natural subconscious tendency to round values to nearby convenient values, such as integers. An example is given in Figure 4.2, which shows a

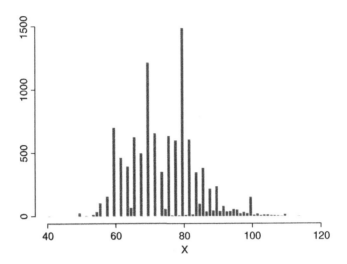

Figure 4.2 Diastolic blood pressure for 10,772 men.

histogram of diastolic blood pressure for 10,772 men (I am grateful to Dr Fergus Daly for these data). In fact there are several curious things about this display. Digit preference is manifest in the large peaks at 60, 70, 80 and 100 mmHg. It is unlikely that blood pressure naturally peaks at these values, and a more likely explanation is distortion in the data recording process. Other examples of digit preference are given in Dellaportas *et al.* (1996), Wilhelm (2000), Roberts and Brewer (2001) and Wright and Bray (2003). Of course, peaks in data at integer values are not necessarily attributable to digit preference, and Hand *et al.* (2000) give several illustrations of data sets where such peaks do arise naturally.

Another curious feature of Figure 4.2 is the dearth of odd-numbered values for low measured blood pressure. In fact, further investigation unearthed the facts that the instrument's graduations were rounded to even numbers, ending in 0, 2, 4, 6 or 8, and that, if an anomalously high value of blood pressure was recorded, those making the recording were instructed to repeat it and average the results. The few low odd values of blood pressure which do appear are probably again due to digit preference – they all end in 5.

We do not intend to give the impression that humans are the only source of measurement error problems. Computers presumably bear at least some of the responsibility for sending a 104-year-old a letter asking her to attend primary school. One of the author's graduate students, studying meteorological records, spotted that occasional apparent excessively high winds were an artefact of the measurement instrument automatically periodically resetting itself. On the other hand, blame does not always lie at the computer's door. Berry and Linoff (2000) quote a company executive as saying: 'The data is clean because it is automatically generated – no human ever touches it'. Berry and Linoff then describe how 20% of transactions in the data set described files that had apparently arrived before they were sent, and went on to say that 'not only did people never touch the data, but they didn't set the clocks on the computers either'.

Missing values are a particularly gross form of measurement error, and they are extremely common in some areas – perhaps especially areas involving human answers to questions. Sometimes entire records are missing, so that overall distributions are distorted (see, for example, Copas and Li, 1997). Often, however, just some of the fields in each record are

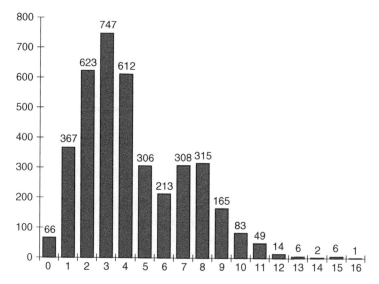

Figure 4.3 Histogram of number of missing values for each of 3884 applicants for personal loans.

missing. Figure 4.3 shows how many of 3884 applicants had different numbers of missing values in a set of 25 items from an application form for personal loans. Only 66 applicants provided complete information, and one applicant had 16 of the 25 values missing (history does not record whether this applicant was granted a loan). Only five of the variables had no missing values, and two had over 2000 missing values.

Missing fields in records pose an easy measurement error problem in the sense that one knows when a value is missing. (Or, at least one does if missing-value codes are appropriately chosen: if the variable is a person's age, then a missing-value code of 99, not an unknown choice to represent 'missing', would lead to mistaken conclusions.) On the other hand, it may not be so easy to know what to do about them. Various strategies have been developed, including imputation (e.g. Little and Rubin, 1987; Rubin, 1987). The strategy must depend on why the data are missing. Suppose that the value a variable X would take, if it were to be observed, is x, and define a random variable B which takes one value (say, 0) if X is observed and another value (say, 1) if X is not observed. Then we can distinguish between situations in which there is no relationship between X and B and those in which they are related. In the former case, things are relatively straightforward, and valid inference can usually be based on the incomplete data. In the latter case, however, the observed data will be a distorted sample from the entire distribution. For example, perhaps cases with larger values of X are less likely to have X recorded. Obviously, in this case, there is the potential for mistaken inferences to be made. Little and Rubin (1987) discuss such issues in detail.

Of course, none of this is new. Shryock (1961, p. 103) quotes Quetelet as complaining, in 1846, that medical data were usually 'incomplete, incomparable, suspected, heaped up pell-mell … and nearly always it is neglected to inquire whether the number of observations is sufficient to inspire confidence'.

In the introduction to Wise (1995a), the author points out that most historical studies of the accuracy of quantification have concentrated on how people achieved precision, but that there are other aspects. These are the reasons why accuracy is valued. The papers in Wise (1995a) explore such issues.

The ease of making mistakes with units, and the importance of getting the units of measurement right, is nicely illustrated by the correspondence in *The Times* following a brief article the preceding day. The article (19 December 2002, p. 4) read:

Britain's largest bat, the greater mouse-eared bat, which was officially declared extinct in the UK 12 years ago, has been rediscovered hibernating in an underground hole in West Sussex. They can weigh up to 30 kg and have ears as long as 3 cm.

The next day the letters column included the following (*The Times*, 20 December 2002, p. 21):

Sir, The discovery of a greater mouse-eared bat is cause for celebration, the more so as this species was thought to be extinct.

However, 30 kg, the weight you quote, is a lot of bat – more like several large turkeys, or my seven-year-old nephew in his Batman suit. Have you any idea of the collateral damage a single 30 kg bat could do if it recklessly flew into you one night? Or the type of diet required to sustain such a remarkable animal?

As I hear of no sudden decline in the cat population of West Sussex, I conclude that your report is incorrect.

MR TIM BLOOMFIELD

Sir, Startled to read that a bat weighing in at 30 kg was back, I hastened down to our local supermarket, hoping to arrange 30 bags of sugar in a bat-like formation so that I could properly appreciate this discovery.

I left the manager in no doubt that he should take more interest in our wildlife as he escorted me out of his shop. I think he was miffed that even his best turkeys came nowhere near 30 kg; even allowing for the added water content, the best he could offer was a mere 15 kg.

THE REVEREND BRIAN E. PHILLIPS

Sir, If the greater mouse-eared bat weighs some 66 lb, as a resident of West Sussex I feel I should stay indoors at night until it is declared extinct again.

MR BILL CAIRNS

Sir, Why worry about possible terrorist attacks? Should we not be more concerned about the dangers of meeting with a 30 kg bat when it comes out of hibernation?

MRS JANIS MASON

Sir, Monty, our somewhat overweight yellow labrador, just about weighs 30 kg on a good day.

The thought of him equipped with leathery wings and airborne fair boggles the imagination.

MR GODFREY J. CURTIS

Sir, No wonder the mouse-eared bat is hiding in a hole in the ground.

It is probably waiting for the results of the airport inquiry to see if there will be a runway big enough to use for take-off.

MR DEREK CANNELL

Sir, My children are struggling with the concept of Santa Claus and flying reindeer.

To add 30 kg bats to the night skies is pushing their imagination a little too far.

MR ANDREW BLACKMORE

The newspaper then includes a footnote which says: 'A greater mouse-eared bat usually weighs about 30 g, not 30 kg.'

Sometimes, translating between units of measurement can lead to a misplaced sense of accuracy. In a letter to *The Times* on 6 January 2003 commenting on a report of ice blocks 10 m or '32.8 ft' high, Mr Dudley Smith remarks 'Would it not have been better to have expressed this as 32 feet nine and five eighths inches, for the benefit of those of us not conversant with decimal feet?'

The bulk of this chapter is concerned with two distinct aspects of accuracy: validity and reliability. The former indicates whether or not a measurement process yields systematically distorted results, while the latter indicates whether or not a measurement process yields randomly inaccurate results. The last section of the chapter briefly discusses how the measurement procedure itself might influence the results obtained, especially in quantum measurement.

4.2 Validity and reliability

Terminology in the area of measurement accuracy is not always used consistently. Sometimes this is because authors are being deliberately casual, because the precise interpretation does not matter. At other times the authors are not aware of subtle distinctions, and this can lead to problems when the various interpretations have different implications. In this chapter, I try to stick to the following usage:

- *Accuracy* is a general term, describing how closely a measurement reproduces the value of the attribute being measured. There are, of course, several assumptions in that definition: that there is a true value to be measured, for example.
- *Validity* describes how well the measured variable represents the attribute being measured, or how well it captures the concept which is the target of measurement.
- *Reliability* or *precision* describes the differences between multiple measurements of an attribute.

Validity and reliability thus both describe aspects of measurement accuracy. From a statistical perspective, validity may be regarded as similar to bias, in the sense that a biased measurement is somehow missing the fundamental target, whereas reliability may be regarded as similar to variance, in the sense that an unreliable measure varies between measurement occasions. Accuracy, overall, might be thought of as mean squared error. To be accurate, a measurement procedure must have low bias and low variance.

Validity is the extent to which a measurement procedure measures what we want it to measure. There is clearly an interplay here between the definition of the measurement procedure and the definition of the concept being measured: an alternative way of looking at validity is that it is an aspect of how we interpret the results of a measurement procedure. In this context, McDowell and Newell (1996, p. 30) remark that

immense amounts of time have been wasted over arcane speculation on what certain scales of psychological well-being are actually supposed to measure. This can best be avoided by closely linking the validation process to a conceptual expression of the aims of the measurement and also linking that concept with other, related concepts to indicate alternative possible interpretations of scores.

One might even go so far as to say that the measurements or measurement procedures themselves are not really the subject of the validation at all. Instead it is the use made of the measurement which may or may not be valid. An IQ test will not be valid as a measure of social anxiety, for example. Issues of validity are generally of greater concern in areas such as psychology and the social sciences, where the measurement procedures typically have a stronger pragmatic component.

Turning to reliability or precision, this is a desirable characteristic in a measurement system. Indeed, both 'precision' and 'reliability' are generally perceived as positive attributes in a wider social context. Reliability is a measure of the consistency of the result of a measurement across repetitions. If repeated measurement of the same attribute of the same object, assumed not to have changed between measurements, yields dramatically different results, the measurement procedure, and by inheritance the measurement itself, is said to be unreliable. We see here one of the practical differences between the physical and social or behavioural sciences. It is possible to imagine repeating a physical measurement (say, of length or weight) under effectively identical conditions, but it is not so clear what repeating a psychological or clinical measurement under identical conditions means. If the measurement consists of a questionnaire, then there is the risk of the subject learning (perhaps subconsciously) what the procedure is. Even physical clinical measurements are subject to this phenomenon: blood pressure is often high on a first reading, and becomes lower on repeated readings as the patient begins to relax (but beware of regression to the mean). Of course, often the variation between measurements in the physical sciences is required to be very small compared with those in the psychological and social sciences. An engineering tolerance of a very small percentage of the length of a component may be needed when building a machine. Further discussion of accuracy of physical measurements is given in Chapter 7.

Perhaps we should note parenthetically here that although the precision of a measurement is reflected by the number of digits needed to describe it, the converse does not hold. The fact that a numerical result is reported to many digits does not necessarily reflect the accuracy with which that result was obtained. We have all seen examples of this, emerging from modern statistical software packages: for example, a random sample of ten values, each represented by two digits, with the mean presented by the software to six decimal places of accuracy. This is at best lazy and at worst misleading.

Of course, modern computers are not the only source of inflated accuracy. Priestley had doubts about the stated accuracy of Lavoisier's measurements in his experiments on the composition of water, referred to above, and descriptions of the size of the French population in the seventeenth and eighteenth centuries typically gave an 'exact' value, even if this had been obtained by some estimation process (Chapter 8; see also Rusnock, 1995). Perhaps this is an appropriate place to comment that all measurements of continua are inaccurate, in that they will be recorded to a finite number of decimal places. Only counting of discrete entities can ever hope to be completely accurate. The problem of completely accurate measurements of continua really requiring an infinite number of decimal places caused the problems the Pythagoreans encountered with irrational numbers, described in Chapter 1.

The distinction between reliability and validity is nicely put by Campbell and Fiske (1959), who say: 'Reliability is the agreement between two efforts to measure the same trait through maximally similar methods. Validity is represented in the agreement between two methods to measure the same trait through maximally different methods.'

It might seem that validity is fundamental, in that an invalid measurement procedure is measuring something other than what is needed, while reliability is merely desirable.

However, sometimes lack of reliability can outweigh an emphasis on validity. In dieting, for example, the aim is generally to lose body fat to improve physical appearance, but body weight is used as the measure. Procedures for measuring body weight are highly reliable, and the relationship between weight and appearance is good enough, even if weight and appearance are not really the same thing.

Since the social and behavioural sciences make so much use of indirect measurement, deducing the value of an attribute of interest from its relationships to variables which can be directly measured, validity is of crucial concern in these areas: it is all too easy to be confused about these related variables and about their relationships. In the physical sciences, validity has been less of an issue. In both areas, however, reliability is important, and has been the focus of much attention. A key difference is that often in the physical sciences it is possible to estimate the reliability fairly easily, whereas in the social and behavioural sciences this is not so. These points are explored in more detail below.

4.2.1 Validity

Systematic errors – or invalid measurements – can arise in many ways. For example, subjective health measurements may yield results distorted from the 'truth' one is seeking by virtue of the very nature of the illness, from the way individual respondents tend to respond to questions (e.g. some respondents have a greater tendency to give socially acceptable responses), or from the wording of the questions.

The effect of the wording of questions on the way people respond has been the focus of much research. Sometimes a question-wording influence is hardly surprising, as in one example the author came across, which asked 'Do you feel that anyone should be forced to pay a union boss for permission to earn a living? Answer YES or NO.' More subtle such problems arise from:

- questions being insufficiently precise. A question such as 'Are your working conditions good?' is less likely to yield a useful response than questions which focus on particular aspects of the working conditions. The question 'How many cigarettes do you smoke each day?' is likely to elicit a much more useful response than 'Do you smoke a lot?'
- language that the respondents find difficult to understand. For example, in surveys aimed at eliciting medical or financial details, one should not use technical terms. A more subtle point here is the use of words which may be quite familiar to a questionnaire designer, but may not be familiar to many of the respondents (acquaint, reside, pecuniary, ...).
- ambiguous questions. A classic is 'Have you stopped beating your wife yet?', but such problems also arise with questions which conceal multiple parts: 'Do you like ice-cream, chocolate, and cake?'
- double negatives: 'Would you not say no to improved working conditions and lower pay?'
- questions which permit of only one reasonable response. 'Would you like a better public transport system?' is unlikely to yield many negative answers.
- certain types of topics which invite a social pressure to respond in certain ways. Women tend to underestimate the amount they drink and young men may overestimate it, people may not honestly answer questions about their willingness to vote for an unpopular political party, and so on.
- retrospective questions based on memory of past incidents. Accuracy of recall can depend on how important were the incidents to the respondent, as well as how long ago they occurred.

In general, it is important to be very clear about exactly what one wants to measure. We may well want to use different treatments for someone whose pain is primarily organic and for someone whose pain is largely psychological, even if the reported level of pain is the same. To influence the pain we need to be able to decompose the mechanism through which the pain is experienced. Leavitt's Back Pain Classification Scale, mentioned in Chapter 6, is an example of a scale which was developed to distinguish between pain of an organic origin and pain related to emotional disorders.

In fact, there are two schools of thought about the question of subjectivity in psychological and social measurements of individuals, and careful thought is needed about whether the aim is to measure some property external to the mind of the subject, or to measure that property as perceived by the subject. (Probability, of course, with its subjective and objective interpretations, provides another nice illustration of the difference.) If the first is the aim, perhaps with the choice of treatment depending on it, we need to remove any subjective bias. Many strategies have been developed for this, such as asking questions which conceal the true objective, having someone else do the rating, or concealing the questions amongst many others. Sometimes, as with the General Health Questionnaire and the Health Opinion Survey, the name of the scale is deliberately misleading. Of course, ethical issues arise whenever one misleads respondents. The second aim is appropriate if the subjective effects are to be regarded as contributing to the measured value. For example, a tendency to exaggerate in order to attract attention, the extent of social support network, and the desire to appear socially acceptable will all influence a subjective report of health status – but if the perceived health status is what one wants to measure then perhaps it would be a mistake to seek to eliminate these effects. The apparent bias arising from these sorts of causes has led to many criticisms of subjective health scales, not always justified in view of the real aims. In statistical terms, it is analogous to the bias in parameter estimation induced into linear regression by errors in the predictor variables. This matters if the objective is to estimate the regression parameter, but it does not matter if the objective is to formulate a rule for predicting the values of the response variable from new predictor values which are subject to the same error mechanism.

Several different types of validity have been defined. *Content validity* indicates how well the items used in a test reflect the items which could have been used. Questions to consider here are whether all the items used are relevant to the underlying concept being measured and whether all relevant aspects are included. For example, in a test of mathematical ability, one would not include items relating to word usage. And at the other extreme, one would not restrict the mathematics test items to items relating just to algebra if the ability being measured was also taken to include arithmetic and geometry. Clearly investigating the content validity of scales for measuring some attributes is easier than for scales measuring others: the universe covered by the notion of mathematical ability is presumably better defined than that covered by, for example, social awareness. Perhaps it is better to include too many items than too few, to ensure that the universe is covered. Indeed, the possibility of exploring content validity only arises when multiple items are used: 'with a single measure of each variable one can remain blissfully unaware of the possibility of measurement error' (McIver and Carmines, 1981).

Often expert reviews of the items are carried out, but statistical methods such as factor analysis are also used. This can indicate whether there is a single common underlying factor which explains the relationships between items. Factor analysis will also give an indication of whether the items divide into natural groupings and can hence shed light on construct validity, described below. Of course, if the measurement procedure is predominantly pragmatic, then the factor analysis will simply reflect the sets of items used in the construction. There has been some misunderstanding of the role and use of factor analysis in this regard – outlined in a quality of life measurement context in Fayers and Hand (1997, 2002), Fayers *et al.* (1998) and Fayers and Machin (2000).

The term *face validity* also sometimes arises in this context, and is used to describe how well a measurement procedure actually measures what it is supposed to measure.

Criterion validity indicates whether the test correlates highly with other measures of the same thing. Sometimes this goes even further, and there is an accepted gold standard (an ideal 'true' measure) against which the proposed measure can be assessed. For example, this can happen if the measure is being used to predict some future outcome (when the term *predictive validity* is sometimes used), or if the new measurement procedure is being developed as a substitute for a slow or expensive existing procedure (e.g. replacing an invasive surgical assessment procedure by a questionnaire). The terms *correlational validity*, indicating the correlation between the new measure and the criterion, and *concurrent validity*, indicating that the criterion is concurrent (not a future event) with the measure, are also sometimes used. The special case of predicting outcome classes is of particular importance.

Criterion validity is all very well, but it relies on the existence of other, external measures. Often there is no such thing, though there may be alternative proposed measures of the concept in question. *Construct validity*, on the other hand, involves the internal structure of the measure, and also its expected relationships with other, external measures. Construct validity thus refers to the theoretical construction of the test: it very clearly mixes the measurement procedure with the concept definition. Sometimes the term *convergent validity* is used to describe the expected correlation between a measure and similar measures of the same concept (of course, this reduces to criterion validity if one such measure is a gold standard, and then many of the conceptual difficulties evaporate). Often multiple alternative measures exist and multivariate techniques can be used to compare them. In cases where the theory suggests that two measures should not be correlated (e.g. in quality of life measurement, because they measure distinct aspects) the term *divergent validity* is sometimes used. Of course, this gives more limited information about whether the measurement procedure under investigation is actually measuring what one wants it to measure – there is an infinite number of things which are *different* from whatever one is interested in. Even for those other measures with which the new measure is expected to show some correlation, it is doubtful that the theory will indicate the strength of those correlations. In any case, due to lack of reliability in the various measures, all correlations will be attenuated somewhat, so that theoretical correlations, even if one could state them, would be reduced. The use of correlation coefficients is discussed in more detail below.

Unfortunately, construct validity can appear to be lacking because it genuinely is lacking, or for various other reasons (Cronbach and Meehl, 1955). It is also possible that the theory relating the various measures is wrong, that the tests of the hypothesized relationships are inadequate in some way, or that there is something incorrect in the definition and construction of the other variable(s) involved in the relationship.

We remarked above that the concept of validity is related to the concept of bias. Bias, of course, has both a well-defined technical meaning in measurement and also a less precisely defined everyday usage. (It also has well-defined technical meanings in other contexts, such as a steady voltage or magnetic field applied in an electrical system, or the threshold parameter in the hidden nodes of neural networks.) The bias of a measurement is the difference between the expected value of that measurement, $E(X)$, over repetitions of the measurement procedure under identical circumstances, and the true value, τ:

$$\text{bias} = E(X) - \tau$$

According to my dictionary, the everyday meaning of bias is 'a predisposition or prejudice'. Statistical bias thus refers to systematic errors, with no cultural, moral, or ethical implication, whereas everyday bias consists largely of such cultural, moral, or ethical implications. Statistical bias represents a systematic mismatch between the true and measured values, just as (lack of) validity does.

It will come as no surprise, in the light of the above discussion, to learn that validating an instrument can cause the evolution of that instrument as understanding of its scope and limitations develops in use. In this vein, perhaps we should emphasize the importance of being clear, before attempting to measure validity of whatever kind, about what would constitute support for the validity of the scale. Without a clear framework, it is easy to regard any result as supportive: a positive correlation is a positive correlation, even if it is very weak. All of this is a characteristic of science as a whole, of course. Perhaps the best one can hope for is a carefully argued process of attempting to establish validity, so that other potential users can agree that what has been done is solid. A process of 'due diligence', perhaps, or 'protocol' in the sense of Bird *et al.* (2003) (see Section 8.2 below).

4.2.2 Reliability

There are two fundamental issues in the measurement of reliability. The first is how to define and measure the reliability of a single measuring instrument or procedure. That is, how would the observed values of this instrument vary if it was applied repeatedly to the same object under identical conditions? This tells us about the measurement error intrinsic to the procedure. The other fundamental issue is how to define and measure the consistency of two (or, indeed, more) instruments (or procedures or raters) – that is, how closely the measurement results of the two instruments agree when they are applied to the same object under identical conditions. We discuss the first of these issues in this section, and the second in Section 4.3.

Measurement error leading to unreliability can have many different causes. Bragg (1974, p. 4) gives seven reasons why a micrometer might give different values when applied repeatedly to the same object. These are changes in the operator (these could be different people using the micrometer, or could be changes within a single operator, such as tiredness), changes in the room temperature (micrometers are generally made of metal, which expands with increasing temperature), dirt on the object being measured, tension on the spindle, errors in resetting the micrometer to zero (compare weighing scales which show a couple of pounds before you even get on them), simple human misreading, and interpolation errors (e.g. the digit preference mentioned above).

The reliability of a measurement procedure (Wittenborn, 1972; Dunn, 1989; McDowell and Newell, 1996) is the extent to which the procedure yields reproducible and consistent results if the measurement procedure is repeated under effectively identical conditions. In the physical sciences it is often possible actually to repeat a measurement under effectively identical conditions. One can, for example, measure the weight of an object ten times. On the other hand, even in the physical sciences, measurement procedures are sometimes so complex and expensive that repetition is infeasible (e.g. in some particle physics experiments). The social and psychological sciences are even more difficult in this regard, and it is very often impossible to repeat a measurement under effectively identical conditions. A particular problem in these cases is that the system being measured often reacts to and adapts itself to the measurement procedure. For example, if the measurement is an indirect one obtained by combining the responses to a series of questions in some way, then effects such as fatigue and memory can influence the results. Repeated measurements would show a gradual change over time, so that the conditions would not be effectively identical, and the later measurements might be regarded as not properly reflecting the magnitude of the attribute being measured. Indeed, things can be even worse: in situations where the thing being measured represents some target, then a social system may deliberately adapt itself to improve the measurement results (see Chapter 8 for more discussion of this issue). We might thus define the reliability of a measurement procedure as the extent to which the procedure *would yield* reproducible and consistent results *if* the measurement procedure *were repeated* under effectively identical conditions. The essence of the idea is that if the results were to vary dramatically from measurement instance to measurement instance under effectively identical conditions then one would have limited confidence in the measurement obtained at the one time one did actually use the measurement procedure. The reliability of a measurement procedure is thus an indication of the *measurement error*, and *reliability coefficients* are used to measure this.

Classical test theory in psychology models the *j*th measurement on the *i*th subject as

$$X_{ij} = \tau_i + e_{ij} \tag{4.1}$$

where τ_i is the 'true value' of the subject and e_{ij} is the error of measurement. As mentioned in Chapter 1, the notion of a true value is a slippery one. In order to make it concrete, in classical test theory the true value is taken to be the expectation of X_{ij} over the population of possible measurements for the *i*th subject. It follows from this that $E_j(e_{ij}) = 0$, the expectation of the measurement error for the *i*th subject is zero, and also that $E_{ij}(e_{ij}) = 0$, the expectation of the mean values over all subjects is zero. If the measurement error is biased (e.g. always reads high), then the bias will appear as part of the τ_i, so to this extent this definition of 'true value' will deviate from any underlying or Platonic truth. It is also usually assumed that values of the errors measured on different occasions are independent both within and between subjects, and that the measurement errors and the true values are also independent. As a result of these assumptions, we have

$$V(X_{ij}) = V(\tau_i) + V(e_{ij}) \tag{4.2}$$

The reliability of a measurement procedure is then defined as

$$R = \frac{V(\tau_i)}{V(X_{ij})} \tag{4.3}$$

the ratio of the true score variance to the observed score variance. This measure is also called the *intraclass correlation coefficient*. Using (4.2), it may also be written as

$$R = \frac{V(\tau_i)}{V(X_{ij})} = 1 - \frac{V(e_{ij})}{V(X_{ij})} = 1 - \frac{V(e_{ij})}{V(\tau_i) + V(e_{ij})} \tag{4.4}$$

We see that the reliability coefficient takes values between 0 and 1, taking larger values the smaller is the variance of the measurement error relative to the variance of the true scores in the population.

The last sentence contains an important point. R is not an invariant of the measurement procedure, but rather depends on both the measurement procedure and the population to which it is applied. This is clear by virtue of the appearance of $V(\tau_i)$ in (4.3) and (4.4), since this will depend on the population to which the measurement procedure is applied. In classical test theory it is the measurement error component $V(e_{ij})$ which is taken to be an invariant of the measurement procedure. That is, $V(e_{ij})$ will be the same whatever population it is applied to. However, this does mean that interpreting $V(e_{ij})$ requires understanding the scale: whether a value of $V(e_{ij})$ is large or not will depend on the scale. In contrast, the reliability coefficient always takes values between 0 and 1, and so is an absolute measure in that sense. But if the same measurement procedure is applied to two populations, the first of which has a larger true score variance than the second, then the measurement procedure will have a higher reliability coefficient when applied to the first population. This, of course, can have implications when designing studies aimed at particular, homogeneous populations, since their smaller true score variance will mean the measurement procedure has lower reliability. We note parenthetically (and with tongue firmly in cheek) that there is an obvious message here for someone developing a test and who has a vested interest in showing how reliable it is: assess it on as heterogeneous a population as possible.

In fact, it is easy to relate the reliability coefficients when the measurement procedure is applied to two different populations, via the observed score variances using the property of constant measurement error variance. From (4.4), the variance of measurement error for the measurement procedure applied to population A is $V(e_{ijA}) = (1 - R_A)V(X_{ijA})$ and the variance of measurement error for the measurement procedure applied to population B is $V(e_{ijB}) = (1 - R_B)V(X_{ijB})$. Since the left-hand sides of these two expressions are assumed to be equal, we have

$$R_B = 1 - \frac{(1 - R_A)V(X_{ijA})}{V(X_{ijB})} \tag{4.5}$$

If measurements are repeated 'under effectively identical conditions' for a subject, then the τ_i values will be the same, but the e_{ij} values will be independently randomly drawn from the measurement error distribution. Now the covariance between the two observed measurements is

$$C(X_{ij}, X_{ik}) = C(\tau_i + e_{ij}, \tau_i + e_{ik}) = V(\tau_i)$$

by virtue of the various independence conditions. Also, $V(X_{ij}) = V(X_{ik})$ for all j and k. It follows that the correlation between the two observed measurements is

$$\rho(X_{ij}, X_{ik}) = \frac{C(X_{ij}, X_{ik})}{\sqrt{V(X_{ij})V(X_{ik})}} = \frac{V(\tau_i)}{V(X_{ij})} = R \tag{4.6}$$

This means that we can estimate the reliability of the measurement procedure – and hence the variance of the measurement error – by taking pairs of measurements 'under effectively identical conditions' and calculating the correlation between them. Of course, this assumes that the score on one measurement is not influenced by the fact that another measurement has been taken. As we remarked above, this is often a risky assumption in areas such as the behavioural sciences. It is important to note here that this derivation has depended on the assumption that the true scores, the τ_i, are the same within each pair of repeated measurements. This assumption cannot be assumed to hold when we discuss inter-rater reliability in the next section, with the consequence that correlation is an inappropriate measure of inter-rater reliability.

It is also worth noting that for measurements X_{ij} and their associated true values τ_i,

$$\rho(X_{ij}, \tau_i) = \frac{C(X_{ij}, \tau_i)}{\sqrt{V(X_{ij})V(\tau_i)}} = \frac{V(\tau_i)}{\sqrt{V(X_{ij})V(\tau_i)}} = \sqrt{\frac{V(\tau_i)}{V(X_{ij})}} = \sqrt{R} \qquad (4.7)$$

so that $R = \rho(X_{ij}, \tau_i)^2$ – the reliability coefficient is the square of the correlation between the observed measure and the true value.

Since our aim in taking measurements is to estimate the true score, uncontaminated by measurement error, we prefer measurement procedures which yield large values of R. In general, we can, of course, improve accuracy by taking multiple measurements (which have the same true score) and using their mean. Since the measurement errors are independent and have mean zero, the variance of the mean of m measurements will be $V(X_{ij})/m$. This translates into an improved reliability coefficient as follows. Here R_m is the reliability coefficient of a measurement defined as the mean of m simple measurements. From (4.1), we have

$$\frac{1}{m}\sum_{j=1}^{m} X_{ij} = \tau_i + \frac{1}{m}\sum_{j=1}^{m} e_{ij} \qquad (4.8)$$

so that, because of the independence assumptions,

$$R_m = \frac{V(\tau_i)}{V\left(\tau_i + (1/m)\sum e_{ij}\right)} = \frac{V(\tau_j)}{V(\tau_i) + (1/m^2)V\left(\sum e_{ij}\right)}$$

$$= \frac{V(\tau_i)}{V(\tau_i) + (1/m)V(e_{ij})} = \frac{mV(\tau_i)}{mV(\tau_i) + V(e_{ij})}$$

Now, using (4.2),

$$R_m = \frac{mV(\tau_i)}{mV(\tau_i) + V(X_{ij}) - V(\tau_i)} = \frac{mV(\tau_i)}{(m-1)V(\tau_i) + V(X_{ij})}$$

Finally, dividing top and bottom by $V(X_{ij})$ yields

$$R_m = \frac{mR}{(m-1)R + 1} \qquad (4.9)$$

This expression is called the *Spearman–Brown prophesy formula*. Rewriting it as

$$R_m = \frac{1}{1 + (1-R)/mR}$$

shows that arbitrarily high reliability (i.e. as near unity as you like) can be achieved by taking m sufficiently large. Of course, this is all predicated on the assumption that it is possible to take m measurements which have the same true score and calculate their average. In particular, if adding extra measurements means we include some which are less well correlated with the others, then the reliability of the overall test may decrease.

As discussed elsewhere in this volume, the social and psychological sciences are characterized by having many indirect measurement procedures, in which measurements are defined by combining component items. If these items are combined by summation, then we have a situation analogous to that of repeating a single measurement procedure and taking the mean (or total) of the separate results. Of course, issues of whether the tests can be thought of as equivalent and as repetitions of a measurement taken 'under effectively identical conditions' arise. Suppose, for example, that the jth item yields a score X_{ij} on the ith subject, and that this can be decomposed into two components: $X_{ij} = \tau_i + e_{ij}$, as in (4.1) above. If the reliability of individual items is R, then the reliability of a test formed by adding together m such items will be R_m, as above. Generally, $R_m > R$. This is an argument for having as many items as possible in such tests.

We noted, in (4.6), that if we apply the measurement procedure twice, then we are able to estimate its reliability. If a test consists of multiple items, each measuring the same thing, then we can apply this idea to estimate the reliability of the item measurements. When applied in this way, to calculate the reliability of a test formed from many items, R_m is often called *Cronbach's alpha* (after Cronbach, 1951). It is also then often expressed in a different form. From the definition of reliability coefficient we have

$$R_m = \frac{V\left(\sum_{j=1}^{m} \tau_i\right)}{V\left(\sum_{j=1}^{m} X_{ij}\right)} \tag{4.10}$$

where the variances are over the subjects, $i = 1, 2, \ldots, n$, in the sample. Since τ_i is constant for the ith subject,

$$R_m = \frac{m^2 V(\tau_i)}{V\left(\sum_{j=1}^{m} X_{ij}\right)} \tag{4.11}$$

Now

$$V\left(\sum_{j=1}^{m} X_{ij}\right) - \sum_{j=1}^{m} V(X_{ij}) = V\left(\sum (\tau_i + e_{ij})\right) - \sum V(\tau_i + e_{ij})$$

$$= m^2 V(\tau_i) + V\left(\sum e_{ij}\right) - mV(\tau_i) - mV(e_{ij})$$

$$= m^2 V(\tau_i) + mV(e_{ij}) - mV(\tau_i) - mV(e_{ij})$$

$$= m(m-1)V(\tau_i) \tag{4.12}$$

The second line here assumes that $V(e_{ij})$ is the same for all j, but this assumption can be relaxed. Substituting for $V(\tau_i)$ from (4.12) into (4.11) yields

$$R_m = \frac{m^2}{m(m-1)} \frac{V\left(\sum X_{ij}\right) - \sum V(X_{ij})}{V\left(\sum X_{ij}\right)} = \frac{m}{m-1}\left(1 - \frac{\sum V(X_{ij})}{V\left(\sum X_{ij}\right)}\right) \tag{4.13}$$

Here $V(X_{ij})$ is the variance of the jth test, calculated over the i subjects, and $V(\Sigma X_{ij})$ is the variance of the total test score. That is, the total score is calculated for each subject, and the variance of this total is calculated over the subjects. Cronbach's alpha is usually given in the form (4.13).

This derivation is based on the assumption that the items are measuring the same thing – have the same true score. If they do not, then Cronbach's alpha, calculated as in (4.13), provides a lower bound for the reliability of the procedure; that is, it is a conservative estimate of the reliability of the measurement procedure (see, for example, Novick and Lewis, 1967).

In the special case of all the items being binary, $V(X_{ij}) = P_j(1 - P_j)$, where P_j is the probability that the jth item will have a score of 1, and (4.13) becomes

$$R_m = \frac{m}{m-1}\left(1 - \frac{\sum P_j(1 - P_j)}{V\left(\sum X_{ij}\right)}\right) \tag{4.14}$$

This form is sometimes known as the Kuder–Richardson formula 20 or KR20 (after Kuder and Richardson, 1937).

Cronbach's alpha is based on assumptions about the internal consistency of the measurement procedure – that is, on assumptions about the independence between the different items composing the overall test. An alternative approach, which does not make these assumptions, is to apply the procedure on two separate occasions and compare the results. This yields the *test–retest reliability estimate*. The assumption here, noted above, is that memory and fatigue effects are negligible. This assumption may usually be reasonable in the physical sciences, but will need to be examined carefully in the social and behavioural sciences. Indeed, fatigue can affect both the object being measured (measuring the flexibility of a piece of metal will eventually result in its breaking; giving a person test after test will eventually result in exhaustion) and the person doing the measurement (measuring a rock on weighing scales eventually becomes tiring; administering a psychological assessment requires concentration, which will lapse after a time).

An even more extreme version of the memory effect was referred to above: this is when the thing being measured is influenced by the mere fact of being measured. This is a familiar concept in quantum theory, but also applies, at a more mundane conceptual level, in (for example) the behavioural sciences. Measuring someone's attitude to contravention of minor social mores may sensitize them to the issues, so that repeating the measurement at a later date may elicit a different response. This is very likely to occur when the attribute is being measured with the aim of changing it: dieters do not weigh themselves just for interest, but as a criterion against which change can be measured and as a stimulus for change.

A further complication is that the underlying attribute may naturally change over time (depression, attitudes to capital punishment, and mortality rates of hospitals may well change over quite short spaces of time). Indeed, detecting and quantifying such changes is often the aim of the analysis, so they are to be expected. Teasing apart measurement differences due to underlying changes in the condition of the person or thing being measured and lack of reliability in the measurement procedure is a difficult problem. One might try to juggle the time between repetitions of the measurements, so that it is not so large that significant change in the condition of the object being measured is expected to occur (and this, itself, requires assumptions about the rate of change) and yet not so small that there will be significant memory or fatigue effects. Alternatively, another possible solution hinges on postulating a model for the underlying changes. At the extreme we have the simplest model, which assumes that there are no underlying changes, so that any differences in the measurement results are solely attributable to lack of reliability. More complicated models, which involve more parameters, will need more data; that is, more than two measurements will be necessary. An example of such a model is that due to Wiley and Wiley (1970), which assumes that the true score at measurement occasion $t + 1$ is linearly related to the true score at measurement occasion t, for all t, but with the coefficients in the linear relationship allowed to change over time.

In the test–retest method, the same measurement procedure is applied to the same objects (normally these are human subjects) on two separate occasions. In the *alternative forms* approach, two different versions of the test are applied to the same people. The different versions are assumed to be parallel: that is, it is assumed that they have the same true scores for each object being measured, and the same measurement error variances. The items comprising the test are typically chosen randomly by splitting a larger pool. Another version simply notionally splits the entire set of items in a test into two groups, and calculates scores for each of these groups, thus permitting an estimate of reliability using (4.6). Of course, this gives an estimate of the reliability of the two 'half-tests' separately. To deduce the *split-halves estimate of reliability* of the overall test from this, we use the Spearman–Brown formula (4.9). Now R will be the estimated reliability of each of the two 'half-tests', and $m = 2$, so that the overall reliability of the test is $2R/(1 + R)$.

The classical test theory model assumes that the error variance, $V(e_{ij})$ is an invariant of the test, and does not depend on the true score distribution of the population to which the test was applied. In particular, this means that a test's error variance will be the same whatever the ability of the subject to whom it is applied. In item response theory models (see Chapter 3), however, the error variance is a function of ability, so that in this case more elaborate estimation procedures are needed.

4.3 Inter-rater reliability

Reliability of measuring instruments is clearly important, but consistency across different measuring instruments and across different people making the measurements is also crucial. As we remarked in Chapter 1, measurements travel: it is the consistency of measurements from time to time and from place to place which has made science so effective. If I weigh a sack of potatoes, and find it weighs 2 kg, and you also weigh it, using some other weighing machine, standard instrumentation (e.g. weighing scales) means that you will also obtain 2 kg, and we can talk meaningfully about the quantity of potatoes in the sack. On the other hand, if we pushed our measurement accuracy further, we will probably find differences between our measured values. If we had both obtained the weights in grams, you may have recorded 2011 g and I may have recorded 1987 g. This difference would probably not have troubled us, but had we been concerned with measuring a quantity of gold the difference may well have mattered. There are various influences which might have led to the differences between our measurements. The ambient temperature, the local gravitational force of the Earth, and so on, may all influence the results. However, there may also be fundamental differences between the machines we used, or perhaps between the way you and I conduct the measurement procedure (do I, for example, always put the machine on an inclined table?). These sorts of effects become even more important in the social and behavioural sciences, where personal and subjective influences are likely to be larger. If you and I both rate the papers submitted to a journal in terms of quality, it seems quite likely that our ratings will not agree perfectly.

The study of consistency between different measuring instruments or different measurers or even the same rater at different times is known as *inter-rater reliability* assessment or *inter-rater agreement* (see, for example, Fleiss, 1981, Chapter 13); the different instruments/people/occasions being the 'raters'. We see immediately that a key difference between inter-rater reliability and the reliability explored in the preceding subsection is that no notions of 'true scores' are needed: we just want to know to what extent the different raters agree with each other. This last sentence gives the key characteristic of measures of inter-rater reliability. They indicate the agreement or similarity, or, conversely, the difference or dissimilarity or even distance between raters. Measures of inter-rater reliability will indicate how different are the scores of different raters. As we shall see, some subtleties intrude – such as how to allow for agreement by pure chance. However, we begin by describing some suitable measures of difference between raters.

With numerical scales, the similarity between the ratings of an object by two raters is given by the difference between their scores. If you and I each weigh a bag of sweets, and you obtain 102 g and I obtain 97 g, then the difference is simply 5 g. If we both weigh a sequence of bags as they emerge from the packing machine, we might expect to obtain a distribution of difference values. Some summary of this distribution will provide a reasonable measure of how consistent we are in our measurements. For example, we might find that the mean value of this difference is 1.5 g, suggesting that you tend to produce higher measurement results than I do. Or we might find that the standard deviation of the distribution is 3 g, suggesting that our two measures can differ fairly substantially.

At this juncture we should point out that the use of the correlation coefficient to measure inter-rater reliability is generally incorrect. This is unlike the situation discussed in the preceding section. There we assumed that each object had a true score, and that the measurements fluctuated randomly about this, with the mean of the random fluctuation being 0 – or, to put it another way, that the distribution of repeated measurements on a particular object had a mean value equal to the true score. This assumption (along with various independence assumptions) led us to the result in (4.6) that the correlation between two measurements on a series of objects was equal to the reliability coefficient. In the present situation, however, we are not prepared to make the assumption that the distribution of repeated measurements on a particular object has the same mean value for both instruments. Indeed, one of the things we might hope to find is whether one of the measuring instruments tends to give higher values than the other; that is, that the distribution of measurements produced by one of the machines has a mean which is higher than the mean of the measurements produced by the other.

In general, the Pearson correlation coefficient measures how closely the results of two measurement procedures conform to a *linear* relationship. This means, for example, that two measures, one of which is consistently twice as large as the other, will show a correlation coefficient of 1, and that two measures, one of which is consistently 1 unit larger than the other will show a correlation coefficient of 1. The correlation coefficient does not tell us about the distribution of differences $(X - Y)$, that is, about how well the data conform to the model $X = Y$, but about the extent to which the data conform to the model $X = a + bY$, where a and b are fixed (but unknown) numbers. If two measurement procedures are supposed to give results on the *same* scale (and not simply two linearly related scales), as they are in our sweet-weighing example, then the correlation coefficient is an inappropriate measure, and agreement should be measured by summarizing the pairwise differences between scores as above. Further discussion of these points is given in Bland and Altman (1986) and Altman (1999).

Binary scales are, in a sense, at the opposite end of the continuum from numerical scales. Binary scales permit only two categories of response, which we can label 1 and 0, and which might represent pass/fail, accept/reject, yes/no, male/female, and so on. A simple difference can also be used here (again, taking the absolute value of the difference) so that agreement of two raters, whether both agree on a score of 0 or both agree on a score of 1, contributes 0 to the overall measure of difference between the raters and disagreement contributes 1 to the overall score. We will generally want to scale this by the total number of items assessed by the two raters, to yield a proportion, a measure between 0 and 1. For example, if two examiners rate a series of 20 students on a pass/fail basis, scoring a pass as 1 and a fail as 0, then this approach will yield the proportion, out of 20, on which the two examiners disagree. Alternatively, we might take the complement of this: the proportion of students, out of the 20, on which the examiners agree. This is called the *simple matching coefficient*, and is the most basic of similarity measures for binary data. Other measures for binary data weight the different kinds of agreement differently. For example, the *Jaccard coefficient* is applicable in situations where one of the grades (0, say) is far more frequent than the other, so that a $(0,0)$ agreement between two raters is relatively uninformative. This coefficient then ignores the cases in which both raters give a 0, simply expressing those cases where both give a 1 rating as a proportion of the number where at least one rater gave a 1. The *Dice coefficient* takes this even further: if $(0,0)$ matches are irrelevant, then $(0,1)$ and $(1,0)$ matches should lie between $(0,0)$ and $(1,1)$ matches. This coefficient thus weights $(1,1)$ matches by a factor of 2 when computing the proportion. That is, if $n(1,1)$ is the number of $(1,1)$ matches, $n(0,1)$ is the number of $(0,1)$ matches, and $n(1,0)$ is the number of $(1,0)$ matches, the Dice coefficient is the ratio

$$\frac{2n(1,1)}{2n(1,1) + n(0,1) + n(1,0)}$$

Binary data have the advantage that, with only two categories of response, there is only a single difference in value between the response categories. When the scoring system yields categorical data with more than two response categories there are multiple differences. For example, if examiners grade the students using A, B, and C grades then there are two differences (A–B and B–C). In this case, the categories are ordered, so the third difference, A–C, can be derived from the first two. In some situations only a nominal scale exists, so that the third difference is of independent interest. The most straightforward way to handle all such categorical data is to assign values to the differences between ratings. These can be simply via a ranking of the ratings (e.g. assign A, B and C numerical values of 1, 2 and 3, respectively) so that the A–B difference is 1, the B–C difference is 1, and the A–C difference is 2. Or, if more information is available, one might assign more informative numerical values, either to the individual grades or directly to the differences between grades (perhaps the difference between an A grade and a B grade is thought to be much more significant than that between a B and a C grade). In fact, as was discussed in Chapter 3, it is possible to adopt even more elaborate procedures for assigning numerical values. Thus, one can assign those numerical weights to the grades so as to optimize some measure of agreement between the raters.

When we introduced item response theory in Chapter 3, we described an elaboration of the model which controlled for the possibility of (random) guessing. We can do something similar when estimating a reliability coefficient. The need for such a strategy is illustrated by the following simple situation, in which two raters each rate 1000 students on a binary pass/fail scale. The results of this imaginary exercise are given in Table 4.1. This shows that rater X

Table 4.1 Ratings of 1000 students by two raters, X and Y, when the raters' scores are unrelated.

		Rater Y	
		0	1
Rater X	0	10	90
	1	90	810

scores $100 \,(=10 + 90)$ of the students 0, and $900 \,(=90 + 810)$ of them 1. Rater Y also scores 100 of the students 0 and 900 of them 1, but the two raters do not assign identical scores to all the students. In fact only $820 \,(=10 + 810)$ receive the same scores from the two raters.

A simple matching coefficient applied to these data would suggest an inter-rater reliability of 0.82 which, at first glance, looks pretty good. However, closer examination of Table 4.1 shows that the ratings given by the two raters are, in fact, independent. That is, given the marginals of the table (each rater assigning 90% to class 1), then the expected proportion of cases in which they agree is 0.82 if the assignments of the two raters have no relationship to each other at all. Matching coefficient values below 0.82 will only be attained if *worse than chance* assignments are made (i.e. there is a tendency for one rater to assign students to class 0 when the other assigns them to class 1).

Clearly the chance level of agreement, assuming no relationship whatsoever between the raters, is a good baseline measure for the reliability coefficient: we are really interested in how much larger than this is the value. Also, it will be convenient if we can construct a measure of agreement which ranges between 0 and 1; 0 when the agreement is equal to that which would be expected by chance, and 1 when the agreement is complete (i.e. the two raters agree on each student). Let O_A be the observed number of students on whom the raters agree, E_A be the number on which they are expected to agree by chance, if there is no consistency at all about their ratings, n the total number of students assessed, and let $O_D = n - O_A$ and $E_D = n - E_A$ be the observed and expected number of students on whom they disagree. Then a suitable measure would be

$$\kappa = \frac{O_A - E_A}{n - E_A} \qquad (4.15)$$

This takes the value 0 when $O_A = E_A$ and the value 1 when $O_A = n$, as required, and is known as *kappa*, or *Cohen's kappa*. *Kappa* takes negative values if worse than chance agreement is attained. Another common way of writing expression (4.15) as

$$\kappa = 1 - \frac{O_D}{E_D} \qquad (4.16)$$

Landis and Koch (1977) suggest some guidelines for how values of kappa between 0 and 1 should be interpreted (see Table 4.2).

The measure defined in (4.15) and (4.16) can be generalized by replacing O_A, E_A, O_D and E_D (and n, of course) by measures of agreement and disagreement other than simple counts. It can also easily be extended to categorical variables which have more than two categories, and to ordered categorical variables. In the latter case, the observed and expected values will be the category counts weighted by numerical values corresponding to the category codes, yielding the *weighted kappa*.

Table 4.2 Suggested interpretations of kappa values (Landis and Koch, 1977).

Kappa	Agreement strength
0.00	Poor
0.01 to 0.20	Slight
0.21 to 0.40	Fair
0.41 to 0.60	Moderate
0.61 to 0.80	Substantial
0.81 to 1.00	Almost perfect

Table 4.3 Two tables with the same κ but different proportions on which the raters disagree.

		Rater 1	
		0	**1**
(a) Rater 2			
	0	10	90
	1	90	810
(b) Rater 2			
	0	40	160
	1	160	640

Of course, it is rare that a single-number summary of a table is ideal for any purpose, and the kappa statistic also has its weaknesses. A strength of kappa is that it is deliberately designed to allow for chance agreement, but this has the inevitable consequence that identical values of kappa may correspond to situations where the proportion of cases on which the two raters agree is different. Table 4.3 gives a very simple illustration of this. In both Table 4.3(a) and Table 4.3(b) the two raters are behaving independently, so that in each table the observed numbers of disagreements equal the number of disagreements expected by chance and in each case $\kappa = 0$. However, in Table 4.3(a) the proportion of disagreements is 0.18, while in Table 4.3(b) it is 0.32. This is a consequence of the two tables having different marginals.

4.4 Other aspects of reliability

The classical test theory model for the observed value of a measurement is that it is a sum of a 'true' value and an error. In fact, of course, as will have been apparent from the preceding pages, there is typically more than one source of error. 'Errors', or at least differences in recorded values, may arise from a variety of causes: differences between measuring instruments, differences between those taking the measurements, differences between the circumstances under which the measurements are taken, inaccuracy with which the measurements can be read, and so on. We could thus decompose the error component into multiple sub-components. These different sources of error occur at different levels: inaccuracy of reading the measurements is possible even if the same instrument is being used all the time. It would thus be appropriate to model the recorded value using a multi-level approach (Goldstein, 1995). The term *generalizability theory* is sometimes used to describe multi-level modelling in the context of measurement error and test reliability. It will also be clear from this that some subtlety in designing reliability studies is needed in order to tease apart the various sources of error and to estimate their effects most accurately. Dunn (1989), especially Chapter 4, discusses these design issues.

4.5 Squeezing the pips

There are various measurement situations in which the very act of carrying out a measurement influences the thing being measured. For example, in the physical sciences measuring something necessarily involves interacting with that something (even if the interaction is simply the bouncing of a photon off it). This very interaction will change the state of the system being measured. At a macro level, this does not matter. Bouncing billions of photons off a car in police radar speed trap will not noticeably affect the car's speed. Squeezing an object between the ends of a calliper will not dramatically affect the size of the object (unless one squeezes very hard). But at a submicroscopic level things are different: if the thing being measured has a mass of the same order as the particle doing the measuring, then the effect will be dramatic. It will mean that the object of the measurement may have quite different values before and after the measurement.

Similar effects also arise in the social and behavioural sciences, where the effect of the results of measurement becoming known can be to modify the value in the future. We referred to memory and learning effects above. In economic contexts, *Goodhart's law* may be summarized as 'An economic time series that is targeted becomes distorted and unusable'. This law is discussed in Chrystal and Mizen (2003). Another example is the UK's 'Research Assessment Exercise' (RAE), described in Chapter 8, which involves attempting to assess the quality of research output by groups on the basis of a formula. Inevitably, once the details of the formula become known, universities attempt to adapt both the nature of their research output and the way in which their submissions are constructed to match the formula as well as they can. At the time of writing, it is becoming very clear that the shape of UK universities is changing in response to the results of the RAE: departments which did not achieve a high rating are being closed down (regardless of the quality of their teaching or the importance of their discipline to the national economy). There have also been reports of some universities transferring 'research inactive' members of staff out of a department, or only submitting that small proportion of the department who are research active, so giving a grossly distorted picture of the overall research quality of the university. This is clearly related to the topic of grade inflation discussed in Chapter 5, and, indeed, the proportion of submissions gaining the top RAE grades of 5 and 5* has increased over the years. Since, for example, the US has not reported an apparent dramatic increase in the quality of UK research over the period, it seems likely that the effect is largely illusory and does not reflect an improvement in the quality of research output.

Such a feedback phenomenon can also manifest itself in other ways. For example, the results of measuring someone's propensity to do well in certain types of jobs might influence their career choice. It could be that they would have done very well in the careers the measurement result influenced them against. This also manifests itself in areas such as credit scoring in retail banking. Individuals are scored on a 'creditworthiness' scale which seeks to predict their probability of defaulting on (say) a loan. Those predicted to be most likely to default are then turned down. True default outcome for these customers thus remains unknown, so complicating the process of evaluating the predictive measurement.

Sometimes it is possible to observe by stealth. In psychology, behavioural characteristics are normally studied by direct observation, with the observed aware of the observer. As a consequence, there is a very real danger that they may alter their behaviour. The *Hawthorne effect* describes how the fact that someone is apparently interested in a subject influences that subject's behaviour, leading to more positive responses. (I must insert a parenthetical remark here. The studies leading to the Hawthorne effect are described in Roethlisberger *et al.* (1939). They were carried out between 1927 and 1932 at the Hawthorne Works of the Western Electric Company in Chicago. However, Roethlisberger *et al.* (1939) revealed that the study had several major deficiencies, so that the results certainly do not support the conclusion implicit in the Hawthorne effect. For example, two of the five relay assemblers studied were withdrawn halfway through the experiment, apparently in response to declining output due to a relaxed management style, and replaced by two very highly motivated assemblers, who were keen to maximize the output of the group so as to earn more money.)

Sometimes it is possible to observe by concealed means (e.g. watching children at play via concealed cameras), but obviously subtle ethical issues are involved here (recall the mention of the General Health Questionnaire above). Sometimes clever probabilistic methods can be employed, so that overall statistical summaries of the response distributions can be obtained without the subjects revealing their own value (e.g. the randomized response method described in Section 5.2).

Quantum mechanics has a further complication. This is that there are certain pairs of attributes both of which it is not possible to measure precisely *even in principle*. No matter how clever you are, you cannot do it: it is not simply a matter of subtlety of the measuring procedure. These pairs include position/momentum and energy/time. Call the uncertainty of measurement of position Δx and the uncertainty of measurement of momentum Δp; then Heisenberg's uncertainty principle tells us that $\Delta x \Delta p \geqslant h/4\pi$, where h is Planck's constant. In fact this result follows from the Schwarz inequality, which, in another form, will be familiar to statisticians. The idea behind such non-compatible variables can be seen from the following analogy.

Suppose that a physicist, studying the relationship between the length and translucency of crystals of a particular substance, models the relationship by a linear form

$$t = \alpha + \beta l + \varepsilon$$

where t is the translucency, l is the length, and ε is a random error term assumed to be normally distributed, $N(0, \sigma^2)$. The physicist is particularly interested in two aspects of the process: the average length of crystal produced in one day's growth and the slope, β, of the relationship between length and translucency. He therefore builds a machine which automatically grows five crystals for a day, measures their lengths and translucencies, fits a regression model, and reports the mean length $\hat{l}$ and the regression parameter estimate $\hat{\beta}$.

Running this machine ten times gives him ten pairs of values $(\hat{l}, \hat{\beta})$. Studying these values, he sees that the ten observed $\hat{l}$ values have quite substantial differences between them, having a fairly large standard deviation, although the standard deviation of the observed $\hat{\beta}$ values is quite small. On close examination, he discovers that the large standard deviation in the $\hat{l}$ values is attributable to impurities in the original solution, so he repeats the ten experiments with a more carefully purified solution. This time he observes that the standard deviation of the $\hat{l}$ values is satisfactorily small, but unfortunately the standard deviation in the observed $\hat{\beta}$ values is now much greater.

Further work shows that no matter how he controls the purity of the solution he is unable to make the standard deviation of the mean crystal length and the standard deviation of the slope parameter both small. If one is very small, the other becomes large no matter what he does.

Fortunately, the physicist's wife is a statistician. When he presents this puzzling result to her (which he has now spent the better part of a year trying to resolve by refining his measuring apparatus) she points out that it cannot be done. In the physicist's simple linear regression, the variance of the estimated slope of the regression line is given by

$$V(\hat{\beta}) = \frac{\sigma^2}{V(l)} = \frac{\sigma^2}{5 \times V(\hat{l})}$$

From this, taking square roots, we have

$$\Delta\hat{\beta} \cdot \Delta\hat{l} = \frac{\sigma}{\sqrt{5}} \tag{4.17}$$

where ΔZ signifies the standard deviation of the random variable Z. That is, the accuracy with which the physicist can measure the estimated slope of the regression line and the accuracy with which he can measure the mean crystal length are inversely related. If he contrives to obtain a better estimate of the mean of the $\hat{l}$ distribution by decreasing $\Delta\hat{l}$, then the standard deviation of the estimate $\hat{\beta}$ inevitably increases. Note that either of the two measures can be measured with as much accuracy as one likes. The physicist can make $\Delta\hat{l}$ as near to zero as he likes or he can make $\Delta\hat{\beta}$ as near to zero as he likes. But he cannot make them both arbitrarily near zero: if one is small the other is large.

Of course, in the macro world of this example, one always has the possibility of increasing the size of each sample – the physicist can build a machine which grows more than five crystals a day. By making n big enough, the product $\Delta\hat{\beta} \cdot \Delta\hat{l} = \sigma/\sqrt{n}$ can be made as small as one likes, so that each of $\Delta\hat{\beta}$ and $\Delta\hat{l}$ can be made as small as one likes. However, the quantum world is different, and there is more to the quantum uncertainty principle. In quantum mechanics, an object is described by a *wave function*, which completely characterizes the state it is in. When an attribute is measured, this wave function 'collapses' to one of a set of particular wave functions called *eigenfunctions*, each describing a possible state of the object, and each associated with a particular member of a set of possible attribute values. The precise eigenfunction and attribute value which results is randomly chosen from a specified set, and is determined by a distribution given by properties of the original wave function, as explained below. Only when an object is in a state characterized by one of the eigenfunctions is it meaningful to describe the attribute as 'having a value'. This means that 'measurement' in quantum mechanics has a very different meaning from that in classical physics. In classical physics the state of an object and the results one would obtain by measuring its attributes are alternative descriptions of the same thing. In quantum mechanics, measuring an attribute of an object throws the object into a particular state, simultaneously producing a measured value. In classical mechanics the attribute's value exists prior to the measurement operation, but may be disturbed (to an unknown extent) by this operation. In quantum mechanics, the attribute's value does not exist prior to the measurement operation (unless the prior state happened to correspond to one of the eigenfunctions), but is then thrown to a known value (at random) by the measurement procedure.

Some pairs of attributes are *compatible*, in that they have identical sets of eigenfunctions. Suppose that A and B are two compatible attributes. Then measuring A, B, and A

again in quick succession (by this we mean so rapidly that there is no time for the system to change between measurements) will yield the same value at the two measurements of A. The first A measurement collapses the wave function into an eigenfunction of the A attribute. Since this is also an eigenfunction of the B attribute, the B measurement leaves it unchanged (the probability distribution is concentrated on this eigenfunction). That is, the intervening measurement of B will not affect the eigenfunction chosen by the first measurement of A. Hence, when the second A measurement is made we obtain the same result as before. In such cases, we say that A and B are simultaneously measurable.

However, other pairs of attributes are incompatible. That is, they have different sets of eigenfunctions. When we repeat the A, B, A successive measurements on incompatible attributes, the first A measurement collapses the wave function to some eigenvalue of A. However, now this is not an eigenfunction for the B attribute, so when B is measured, the system collapses to one of the B eigenfunctions, different from the original A eigenfunction. That is, the system will change its state. This means that, when the second A measurement is made, the eigenfunction that the system collapses to will be chosen from a distribution which is not concentrated on the result of the first measurement, but is spread over a set of possible eigenfunctions. It may yield a different value.

We can put these ideas more formally. An *operator* is a mapping which takes one function to another. So, an operator O will map a wave function $\psi(x)$ to some other wave function $O\psi(x) = \phi(x)$. If an operator applied to a particular wave function has the particularly simple effect of multiplying that wave function by a scalar, $O\psi(x) = c\psi(x)$, then the function $\psi(x)$ is an eigenfunction of that operator, and c is the corresponding eigenvalue.

Measurement operations yielding physically observable numbers are represented mathematically by operators, in that they map the wave function of an object to some other wave function. In fact, to any physical observable P there corresponds a linear *Hermitian* operator with a complete orthonormal basis of eigenvectors, which has a corresponding set of real eigenvalues. A Hermitian operator O is one for which the inner product of two wave functions ψ_1 and $O\psi_2$ is equal to the inner product of the two wave functions $O\psi_1$ and ψ_2. That is $(\psi_1, O\psi_2) = (O\psi_1, \psi_2)$, where $(\psi_1, \psi_2) = \int_{-\infty}^{\infty} \psi_1^*(x)\psi_2(x)dx$, with $\psi_1^*(x)$ being the complex conjugate of $\psi_1(x)$. We will use this property below.

When P is measured, the wavefunction collapses to one of the eigenfunctions of the operator corresponding to the measurement operation, with the result of the measurement being the corresponding eigenvalue. Thus the eigenvalues are the only possible results of such measurement. In particular, if an observable operator has a basis of eigenfunctions $\{\vartheta_i(x)\}$, with corresponding eigenvalues $\{t_i\}$, and if the system is in state $\psi(x)$ immediately before a measurement, then when the measurement is made the system has probability $|(\vartheta_i, \psi)|^2$ of collapsing to the ith eigenfunction, with corresponding measured value t_i. Note that it follows from this that $(\psi, \psi) = 1$.

Two operators, O_1 and O_2 are said to *commute* if the result of applying first O_1 and then O_2 is the same as the result of applying them in the reverse order: $O_1 O_2 \psi(x) = O_2 O_1 \psi(x)$, for all vectors ψ. Not all pairs of operators commute. In particular, as an example, consider the two operators $O_1 = x$ and $O_2 = d/dx$ (meaning that O_1 maps $\psi(x)$ to functions of the form $x\psi(x)$, $O_1\psi(x) = x\psi(x)$, and $O_2\psi(x) = d\psi(x)/dx$). Then

$$O_1 O_2 \psi(x) = O_1 \frac{d\psi(x)}{dx} = x\frac{d\psi(x)}{dx}$$

but

$$O_2O_1\psi(x) = O_2[x\psi(x)] = \frac{d}{dx}[x\psi(x)] = x\frac{d\psi(x)}{dx} + \psi(x)$$

Operators which commute are compatible in the sense used above. Operators which do not commute are incompatible. Many incompatible pairs of operators have the property that

$$O_1O_2\psi(x) - O_2O_1\psi(x) = c\psi(x) \tag{4.18}$$

which we can write for short as $O_1O_2 - O_2O_1 = c$. Since observable operators, that is, those corresponding to measurements, will yield values which arise randomly according to probabilities derived from the wave function, the results of such a measurement will be uncertain. The resulting values will thus have a standard deviation. Denote the standard deviation of measurements arising from O_1 by ΔO_1 and the standard deviation of measurements arising from O_2 by ΔO_2. Then, as we show below,

$$\Delta O_1 \Delta O_2 \geqslant \frac{|c|}{2} \tag{4.19}$$

This means that, if we arrange to take our measurements of the attribute corresponding to operator O_1 more and more accurately, then the inaccuracies in our measurements of the attribute corresponding to operator O_2 *will inevitably increase*. We can put this another way, which brings home how different these ideas are from the usual macroscopic measurement concepts. Equation (4.19) means that if the first attribute has a unique value, then it is meaningless to talk about the value of the second attribute: if the distribution of possible values of the first attribute has standard deviation 0, then equation (4.19) implies that the distribution of possible values of the second attribute must be infinite.

The proof of equation (4.19) essentially follows from the Cauchy–Schwarz inequality. Statisticians will be familiar with this inequality in another guise: the fact that the correlation between two variables must have an absolute value no greater than 1.

To prove equation (4.19), define two operators $\hat{O}_1 = O_1 - E(O_1)$ and $\hat{O}_2 = O_2 - E(O_2)$. Here, $E(O_i)$, $i = 1,2$, is the average of the values which would be obtained if the measurement was taken many times under identical conditions (it is simply the value averaged over the distribution function obtained from the wave function). The fact that both $\hat{O}_1$ and $\hat{O}_2$ are Hermitian follows from the fact that O_1 and O_2 are Hermitian and linear. Also, since the $E(O_i)$ are scalars, and again using the linearity, it is easy to show that

$$\hat{O}_1\hat{O}_2 - \hat{O}_2\hat{O}_1 = O_1O_2 - O_2O_1 \tag{4.20}$$

Now consider the inner product $(\psi,[O_1O_2 - O_2O_1]\psi)$. From (4.20) we have

$$(\psi,[O_1O_2 - O_2O_1]\psi) = (\psi,[\hat{O}_1\hat{O}_2 - \hat{O}_2\hat{O}_1]\psi)$$

$$= (\psi,\hat{O}_1\hat{O}_2\psi) - (\psi,\hat{O}_2\hat{O}_1\psi)$$

From the definition of Hermiticity, this is equal to

$$(\hat{O}_1\psi,\hat{O}_2\psi) - (\hat{O}_2\psi,\hat{O}_1\psi) = (\hat{O}_1\psi,\hat{O}_2\psi) - (\hat{O}_1\psi,\hat{O}_2\psi)^*$$

with the right-hand side following from the definition of the inner product. From this it follows immediately that

$$(\psi,[O_1 O_2 - O_2 O_1]\psi) = 2i\text{Im}(\hat{O}_1\psi, \hat{O}_2\psi) \tag{4.21}$$

where $\text{Im}(x)$ denotes the imaginary part of x. Taking the modulus of both sides of (4.21) yields

$$|(\psi,[O_1 O_2 - O_2 O_1]\psi)| = 2|\text{Im}(\hat{O}_1\psi, \hat{O}_2\psi)|$$

Now, since the modulus of the imaginary part of a complex number cannot be greater than the modulus of the entire complex number, we obtain

$$|(\psi,[O_1 O_2 - O_2 O_1]\psi)| \leq 2|(\hat{O}_1\psi, \hat{O}_2\psi)| \tag{4.22}$$

The Cauchy–Schwarz inequality applied to the right-hand side tells us that

$$|(\hat{O}_1\psi, \hat{O}_2\psi)| \leq \sqrt{(\hat{O}_1\psi, \hat{O}_1\psi)} \cdot \sqrt{(\hat{O}_2\psi, \hat{O}_2\psi)}$$

which, with (4.22) gives

$$|(\psi,[O_1 O_2 - O_2 O_1]\psi)| \leq 2\sqrt{(\hat{O}_1\psi, \hat{O}_1\psi)} \cdot \sqrt{(\hat{O}_2\psi, \hat{O}_2\psi)} \tag{4.23}$$

Now $(\hat{O}_1\psi, \hat{O}_1\psi) = (\psi, \hat{O}_1^2\psi)$ by the Hermiticity of $\hat{O}_1$. From the definition of $\hat{O}_1$ this is equal to

$$\left(\psi,[O_1 - E(O_1)]^2\psi\right) = \left(\psi,[O_1^2 - 2O_1 E(O_1) + E(O_1)^2]\psi\right)$$

and by the linearity of the operators this becomes

$$
\begin{aligned}
(\psi, O_1^2\psi) &- 2E(O_1)(\psi, O_1\psi) + E(O_1)^2(\psi, \psi)\\
&= E(O_1^2) - 2E(O_1) \cdot E(O_1) + E(O_1)^2 \cdot 1\\
&= E(O_1^2) - E(O_1)^2\\
&= (\Delta O_1)^2
\end{aligned}
$$

where the second line follows from the othonormality of ψ and the use of the probability function (ψ, ψ). That is, we have established that $(\hat{O}_1\psi, \hat{O}_1\psi) = (\Delta O_1)^2$, with a similar result applying for $\hat{O}_2$. Using these results in (4.23) gives

$$|(\psi,[O_1 O_2 - O_2 O_1]\psi)| \leq 2\Delta O_1 \cdot \Delta O_2$$

which, with (4.18), is equivalent to (4.19). Equation (4.19) is similar to equation (4.17), for our crystal grower – with the crucial difference that there is no scope for 'increasing the sample size'.

Measurement in psychology

Beauty is measured in units called helens. The standard unit is the millihelen,
which is the amount of beauty required to launch a single ship.

Anon

5.1 Introduction

Measurement in psychology has often been controversial. Indeed, the tension generated by attempts to measure psychological concepts has been the driving force for much of the research on the meaning and methods of measurement, described in Chapters 2 and 3. Here are some quotations which will give the flavour of the debates:

It is not uncommon for psychologists and other social scientists to investigate a phenomenon at great length without knowing what they're talking about. So it is with *attitude*. While 20,209 articles and books are listed under the rubric 'attitude' in the *Psychological Abstracts* from 1970 to 1979, there is little agreement about the definition of *attitude* and hence what aspects of attitude are worth measuring.

(Dawes and Smith, 1985, p. 509)

Why do not psychologists accept the natural and obvious conclusion that subjective measurements of loudness in numerical terms (like those of length ...) are mutually inconsistent and cannot be the basis of measurement?

(Norman R. Campbell; see Ferguson *et al.*, 1940)

To the extent that measurement in [the human sciences] is muted or impossible, researchers and practitioners will experience difficulty predicting, transcending their subjective impressions, learning or making inferences from others' observations, considering and dealing with error, promoting justice, accumulating knowledge, and making fine distinctions. In short, their work will be severely hampered.

(Bradley and Schaefer, 1998, p. 90)

One of the fundamental reasons for the controversy is the sheer complexity of the behavioural and social sciences. This is sometimes not appreciated by researchers in the natural sciences, who contrast the achievements of the two domains without taking the differing complexities into account. Eugene Wigner, who won the 1963 Nobel Prize for physics for his contribution to the theory of atomic nuclei elementary particles, wrote: 'if there were no phenomena which are independent of all but a manageably small set of conditions, physics would be impossible' (Wigner, 1960, p. 4). In psychology, quite the opposite is true: essentially all variable are related and the trick is to tease apart this complex tangle. Whereas, in physics *all* electrons are *identical*, in psychology *no* two people are identical.

Early motivation for formal mental testing methods came from a variety of sources. One was the dawn of the scientific discipline of psychology. The essence of science is that it deals with evidence; that is, with facts. The best kinds of facts are the results of an objective measurement procedure. This motivated Galton, Jastrow, J. Cattell and others to develop formal measurement tools to provide the facts from which a scientific psychology could be constructed. But science is a disinterested exploration of the universe. More immediate administrative and decision-making needs also provided impetus for the development of psychological testing procedures. This contrast is nicely illustrated by Murphy and Davidshofer (2001), who open the preface of their book with the sentence 'Tests are used to make decisions' and who later say 'we do not cover topics that, while important, have little relevance for decision making. Thus we do not discuss at length topics such as attitude measurement or the assessment of values'. Testing for decision making has a long history. Government selection procedures for civil servants have long used this process. The Chinese used written tests for this purpose three millennia ago, and the US government adopted such tests in the latter half of the nineteenth century. Binet's original work on 'intelligence testing' also fell into this decision-making mould. It was motivated by the wish to identify children who were unsuitable for standard schools in France. Later on, during the First World War, the US military developed tests to screen out men who were unsuitable for the army, and in 1926 the American educational establishment devised the Scholastic Aptitude Test to assist in decisions on university admissions.

Tests used for different purposes may have to satisfy different criteria. In particular, for scientific purposes, accuracy is paramount, whereas for screening purposes speed and ease of application are also important (of course, these are appalling generalizations, and I hope the reader will forgive me). Moreover, tests used for the practical purpose of prediction can (at least in principle) be evaluated against the gold standard of the attribute they are predicting, whereas those used for more fundamental scientific measurement may not have a gold standard. Perhaps more important than this, however, is the fact that tests aimed at predicting performance can be strongly pragmatic, with little underlying representational theory. Such tests can be highly effective, without there being any understanding of why they should be so, and without any model relating the various components. One might then wonder if such tests merit the description 'psychological', as in the 'science of the mind': certainly, they may be used to predict behaviour, but no theory need be used in their construction. In an extreme case, one could simply take measurements on a vast number of potential predictors, and then use a search method to choose those to use in a regression model which was maximally predictive. We can also stress the pragmatic nature of such tests and *define* the thing being measured in terms of the instrument we are using to measure it. The classic example of this is to define intelligence as that which is measured by an intelligence test. This is all very well, of course, but one must tread carefully.

As we have already discussed, the notion that one may map from an empirical phenomenon to a numerical representation, which can then be manipulated arithmetically to draw conclusions about the empirical system, has not always been obvious. The gradual dawning of this realization parallels the growth of civilization. In some cases, such as physical concepts like weight and height, we have now been making such mappings for thousands of years, so that they appear to us as part of the way the world simply is. In others, however, the notion has only recently taken hold, or is perhaps still battling to take hold. In psychology, the idea that one may represent phenomena and attributes quantitatively is about a century old. As Heidbreder (1933, p. 83) wrote: 'To Fechner's contemporaries, the remarkable feature of the psychophysical methods was the fact that they were quantitative. To measure mental processes was considered a startling innovation; to experiment with them in a manner that gave quantitative data marked the dawn of a new day.' The early work was based on measuring physiological or psychophysiological processes, such as reaction time. This is natural, since they have a closer link to more direct physical concepts (see the similar development of medical measurement described in Section 6.2). Gradually, however, those developing the tests recognized that such tests missed important aspects of the attributes they wished to measure. Indeed, when the tests were being used to predict a future attribute, it would seem to make sense that the tests should measure similar attributes. Thus Binet developed instruments which did not simply use measures of elementary psychophysiological processes but tested more elaborate cognitive processing and memory tasks (e.g. Binet and Henri, 1895). One development by Binet is particularly worth commenting on. He noted that there was a relationship between the difficulty of a task and the average age at which children could master it. This allowed him to give the age at which a normal child would achieve any given test score. Inverting this allowed him to express a child's score in terms of the age at which a normal child would also achieve that score. Children's scores could then be given in terms of their 'mental age'. Such a readily comprehensible interpretation was a great advantage.

One of the primary difficulties in areas such as psychological measurement is that the empirical system being modelled is often not well defined. This means that there is considerable freedom in choosing the pragmatic constraints: one can choose to include different questions in a test and to combine those questions in different ways. If different tests are regarded as measuring the same thing, it is hardly surprising that different researchers will draw different conclusions. Hence statements such as that of Hogan and Nicholson (1988, p. 622) that 'the literature is replete with examples of researchers testing substantive hypotheses with homemade and unvalidated scales; when it is later discovered that the scales did not measure what they purported to measure, the entire line of research is called into question'. In general, if a measurement scale is invalid or unreliable in some way, then it cannot be used to provide evidence for the purposes of falsification: there may be a natural tendency to reject the poor data rather than the theory under test, but this does not represent support for the theory.

The distinction between traits and states underlies much of the debate in psychological measurement, especially in measuring aspects of personality. Traits refer to relatively static properties of individuals, which do not change according to circumstances. A person may be honest, reliable, trustworthy, inconsiderate, lazy, or careless, for example. States refer to the condition or situation in which a subject finds himself or herself. We can make an analogous distinction for physical objects. We might describe the mass of a given rock as one of its traits: it is a fixed property of the rock, not influenced by external events. In contrast, the speed at which a hurled rock is travelling would be a state: it is a property of the rock which is influenced by the external environment. To a large extent, the two approaches have generated distinct psychological scientific literatures. The controversy was stimulated by

Mischel (1968), who argued that the relative influence of the propensity to behave in a certain way, as described by someone's traits, was less important than the particulars of the circumstances in which they found themselves. Various authors responded to this challenge. In their review, West and Graziano (1989) produced evidence for long stability of behavioural propensities, although noting the sometimes substantial effect of external influences. Sometimes, of course, the role of both state and trait is recognized. The *State-Trait Anxiety Inventory* (Spielberger *et al.*, 1970) has separate scales for each of temporary, context-dependent (state) anxiety and longer-term, more stable (trait) anxiety, and the instructions given to the subjects on how to respond to the two sets of items differ.

The behaviour of a subject is a question of state, and this can be directly observed. This makes state measurement a more direct measurement process. In contrast, a subject's trait has to be deduced from indirect observations, so that it is more a latent variable. Clearly measurement issues require more careful thought and development in the latter case. In her review of 50 years of development in psychological measurement, Anastasi (1985, p. 134) says:

> Factor analysis is no longer regarded as a means of searching for *the* basic, fixed, universal units of behavior. Rather, it is recognized as a method for organizing empirical data into useful categories through an analysis of behavioral uniformities. Like the test scores and other observational data from which they are derived, factors are descriptive, not explanatory. They do not represent underlying causal entities.

Thus, not only are such constructs indirect (and probably quite complex), but they also have a major pragmatic component.

The trait–state distinction is related to the nomothetic–idiographic measurement distinction. *Nomothetic* measurements are those referring to groups of subjects: they are abstractions summarizing the common characteristics of all of the individuals in the population being studied. By aggregating observations from multiple individuals, the nomothetic approach takes advantage of basic probability laws to reduce the effect of measurement errors and irrelevant sources of differences between individuals, which do not contribute to the underlying attribute of interest. In contrast, *idiographic* measurements focus on single subjects, studying the attributes of each individual. There has been a heated debate between the two approaches, though the nomothetic approach is currently dominant. The nomothetic approach has been criticized on the grounds that interest in psychology generally lies in individuals rather than groups of individuals. This criticism is not entirely appropriate: random effects approaches permit models of each individual to be constructed from aggregate observations. There is no reason, in such models, for the model form for each individual to be constrained to take the same form as that of the group aggregate. In psychology, random effects are often given the curious and rather misleading name 'individual differences'. The idiographic approach (sometimes called the '$n = 1$' approach) has been criticized as not science: if science is attempting to extract underlying commonalities and uniformities from the turmoil and complexity of the real world, then studying individuals in isolation from others rather misses the point.

A key pragmatic decision made during the design of a psychological test instrument is the choice of norm group used to calibrate it. The choice of norm group must depend on the use to which the instrument is to be put. One would not calibrate an instrument to measure reading ability in the subnormal by using a group of literary critics. An instrument calibrated on

18-year-old males may not very accurately measure other ages or females. Often, calibrations on several different groups are given, so that the instrument has wider applicability. Note that the word 'calibration' here is used in a sense slightly different from, say, calibrating a weighing scale. In the latter case, 'calibration' might mean adjusting the baseline so that the machine is accurate. In the present case, 'calibration' is effectively producing a different scale for each distinct norm group. Weights produced by the two different weighing scale calibrations can be meaningfully compared, but one would not compare the reading ability of a subnormal patient and a literary critic on separately calibrated scales (though this is not to say that one could not adopt a calibration group which included subnormal patients and literary critics). Calibration, in the sense used here, is describing part of the model construction process – and all statistical models are only valid for data drawn from the distributions on which they were constructed. In a sense many physical measurements are calibrated on samples from distributions which are universal. Weight, for example, is calibrated to standard scales.

More generally, pragmatic decisions which must be made during instrument construction refer to the conditions under which the instrument is to be applied and the process by which a test administrator is to give the test (one-on-one, in a group of examinees, time restrictions, etc.) as well as to how to score the test. In particular, different administrators are likely to obtain slightly different results, even if all else is equivalent, but such variation can be reduced by training the administrators. Highly standardized conditions can reduce measurement error, but this may be at a cost in terms of how widely applicable the instrument is.

An important distinction in psychological measurement is between objective and subjective test scoring procedures. An example of the former is a simple count of the number of correct answers in a mathematics test. An example of the latter is the mark given to an essay. It goes without saying that the latter has much more scope for poor reliability. Training assessors again plays a key role here. Scoring apart, tests themselves can involve different degrees of subjectivity: we might seek an objective measure of an attribute, or a measure of how the subject perceives that attribute. Murphy and Davidshofer (2001), for example, give three general categories of test:

(i) Where the subject performs a given task. For example, sitting a school examination, solving a puzzle, taking an IQ test, or responding to stimuli presented on a computer.
(ii) Where the subject's performance is observed. This might be under natural conditions – for example, one might observe how people interact with others as part of an employee selection process. Or one might measure verbal skills, slide quality, and so on in a test of lecturing ability.
(iii) Where the subject describes their own feelings, beliefs, etc.

The extent to which the subjective component is an aspect of what is being measured differs between these three categories.

Many psychological measurements take the form of a set of items, often questions, which are then combined in a weighted sum to yield an overall score. The decision to use a weighted sum may be based on either empirical or pragmatic considerations. For example, if the decision derives from a factor analysis of variables all thought to represent the underlying concept being measured, then least-squares posterior analysis leads to a weighted sum for scoring a new individual, and here the measurement is representationally defined, in that it is based on a model of the empirical relationships between observed variables. In contrast, one might simply choose the weights so that the resulting scale reflects the component scores in the way one wishes – a pragmatic synthesis. Of course, even such a pragmatic combination has its representational aspects. For example, if the weights are restricted to be positive, the decision to use a weighted sum reflects the fact that an increase in the score of any component results in an increase in the overall score.

Much of the technical literature on measurement methods in psychology is concerned with the details of classical test theory models, item response theory models, and psychometric models, which we discussed in Chapter 3. Much of the remainder of the technical literature on measurement methods in psychology is concerned with measurement reliability and validity, discussed in Chapter 4. Since all of these topics have been dealt with elsewhere in this book, most of this chapter focuses on issues which are unique to psychological measurement. An exception is Section 5.8, which contrasts informal subjective judgemental procedures with more formal statistical or actuarial approaches. This contrast has been intensively studied in psychology, but has also been examined in other areas, and we refer briefly to those areas. I have also chosen to include educational testing in this chapter, though this is clearly a vast subject which more properly merits an entire chapter (or even a library) of its own.

5.2 Difficulties in psychological measurement

All measurement activities present challenges, but psychological measurement is perhaps especially challenging. Meier (1994) drives this home with his choice of title: *The Chronic Crisis in Psychological Measurement and Assessment*. One fundamental problem, clearly differentiating measurement in such areas from measurement in the physical sciences, is that the system being measured is a dynamic changeable system which can interact with the measurement process in conscious as well as subconscious ways. The result is a great scope for errors of various kinds. Errors can be induced at various levels, including errors arising from aspects of the way the measurement procedure is designed or conducted, and errors arising from the subjectivity of the respondent (this latter assumes that one is trying to measure some underlying attribute).

For example, Paul *et al.* (1986) list the following as potentially inducing measurement error, most of which can be seen as a result of poor design of the measurement procedure: carelessness, fatigue, boredom, information overload, emotional strain, attention shifts, equipment failure, variations in lighting and temperature, and external distractions. As far as test construction is concerned, it is well known that the wording of questions can change the tendency to respond in particular ways. For example, 'Do you think your boss is doing a good job?' may well not elicit the opposite response pattern from the question 'Do you think your boss is doing a bad job?'. Perhaps a more inclusive 'Do you think your boss is doing a good job or a bad job?' would be more appropriate (provided, of course, the possible answers were not simply 'yes' and 'no'!). Precision in the questions matters, of course. 'Do you think your boss manages his staff well?' may be the intended meaning, or perhaps 'Do you think your boss organizes the projects well?', and so on. Ambiguous questions may be interpreted in different ways. Difficult questions, or questions which do not permit an easy and quick answer, may induce the subject to pick a response at random. Questions which include double negatives are asking for trouble. Furthermore, the mere fact of knowing that one is being studied can change the way one responds (a 'reactive' or 'Hawthorne effect', but see Section 4.5; this is a possible partial explanation for the placebo effect in medicine). Some of these causes yield random errors, which, thanks to the law of large numbers, can be reduced by aggregating observations over multiple measurements. This is one motivation for including many items within each test (e.g. Epstein, 1979), and also for testing many subjects. (Another reason for testing many subjects, and one which is sometimes overlooked in discussions of this issue, is that, for psychological test results to be useful, one must be able to relate them to the scores of other subjects – a small calibration sample limits the value of a test.) Sometimes unhelpful subjects may simply reply at random. One strategy to detect such behaviour is to include a set of items to which everyone would normally respond in a predictable way. An unusual response pattern is indicative of ignoring the questions. Another strategy is for the test developer explicitly to model a random response pattern, so that one can be recognized when it appears.

Errors arising from the subjectivity of the respondent include:

- strength of motivation or interest. This may depend on the likely consequences of different answers.
- social desirability, in which respondents tend to give the answer they believe is expected or desired. Statistical methods for controlling for this have been developed, but they do depend on a model of its extent. More generally, of course, scales can include subscales deliberately designed to assess the extent to which a given respondent will tend to give a socially desirable response.
- cultural differences, which can affect the cognitive schema through which the subject processes information.
- acquiescence, which is the tendency to agree with a statement, regardless of its content. There seems to be some evidence that this effect is most pronounced when the testing aim of the item is less obvious.
- response set distortion, in which a subject tends to answer in the same way to all questions. For example, the subject may have a tendency to answer in the affirmative to all questions. Strategies for avoiding this include phrasing some items positively and some negatively. For example, we might ask 'Do you enjoy your everyday job?' but 'Do you dislike your journey to work?'.

Of course, often errors or distortions arise as a consequence of an interaction between poor test design and the subjective behaviour of the respondent. Indeed, one might argue that this is always the case: if an error can arise because a respondent is behaving in an undesirable way, the test has been poorly designed. Important examples of this arise in medical screening, job selection, bank loans, and insurance claims, where the respondents may be inclined to inflate their health, ability, financial status, or damage, respectively. Indeed, in any area where one is attempting to measure sensitive topics (e.g. income, sexual activity) distortion might be expected. Various strategies have been developed to detect this, including searching for unusual profiles of responses, and randomized response measurement procedures. In one such scheme, respondents are randomly assigned to the sensitive question or some other inoffensive question, with respective probabilities p and $1 - p$. If π, the proportion who will answer yes to the inoffensive question, is known, and if the overall proportion who answer yes is seen to be P, then the proportion who answered yes to the sensitive question is $[P - (1 - p)\pi]/p$. This can be calculated without knowing whether an individual respondent has answered the sensitive question of interest or the other question. The 'other question' could even be determined by a random choice, provided the proportion giving each response is known. Sometimes the apparent complexity of randomized response schemes may compromise their effectiveness.

Distortion arising from unwanted interaction between tester and testee is also an example of an interaction between poor test design and the behaviour of the respondent. For example, the sex or attitude of the tester may make a difference to the results. Ethical issues aside, sometimes the measuring instrument can be concealed or its purpose disguised. The early psychologist Sir Francis Galton described a strategy for overcoming one possible interaction between the tester and testee in one particular context by cleverly concealing his measurement instrument:

> Many mental processes admit of being roughly measured. For instance, the degree to which people are bored, by counting the number of their fidgets. I not infrequently tried this method at the meetings of the Royal Geographical Society, for even there dull memoirs are occasionally read … The use of a watch attracts attention, so I reckon time by the number of my breathings, of which there are 15 in a minute. They are not counted mentally, but are punctuated by pressing with 15 fingers successively. The counting is reserved for the fidgets. These observations should be confined to persons of middle age. Children are rarely still, while elderly philosophers will sometimes remain rigid for minutes altogether.
>
> (Galton, 1909, p. 278)

One of the key ways in which the person being measured can interact with a psychological measurement instrument which is used repeatedly over time is that they can learn and remember the procedure and the scale. Testing memory by asking someone to memorize a list of words is all very well, but problems are encountered if one wants to study how memory capacity changes over time. The same list of words – the same test – can hardly be presented the second time. Clearly some equivalent test is needed.

Because of the relative ease with which they can be obtained, self-reports are widely used in some areas of psychology. In interpreting the results of such measurements, care needs to be taken to be sure that one is measuring what one wants to measure. For example, in measuring health-related quality of life (Section 6.5.4), is it the patient's perception which is of interest, or is it some external and more objective measure which one is seeking? More generally, one may embed the self-reported response to a question in a larger model of the responding process, in which those factors likely to influence the response are also measured. The problem of whether a measurement is 'distorted' is a subtle one in psychology. Information relevant to such models can be obtained by asking additional questions. For example, by including questions to which one would expect a known answer one can probe tendency to lie or to answer randomly.

Perhaps because of these difficulties, the existence of which is readily apparent, a great deal of effort has been expended on ensuring that psychological and psychiatric measurements are accurate. In many cases this means that they are more accurate than (physical) measurements which one might superficially expect to be easier to take. Much the same applies in medicine – see, for example, the discussion of X-ray measurements in Section 6.5.5. Of course, we do not mean to give the impression that psychological measurements are always accurate: despite the emphasis on measurement issues, far too many researchers create *ad hoc* 'measurement instruments' for one-off use, with next to no evaluation. Obviously it depends on precisely what one includes as a 'measurement instrument', but this is illustrated by an advertisement in *APA Monitor* (American Psychological Association, 1992) which suggested that some 20,000 psychological, behavioural, and cognitive instruments are devised each year. The number of tests is also illustrated by the *Mental Measurements Yearbooks* (the Buros Institute, University of Nebraska) and in *Test Critiques* (e.g. Keyser and Sweetland, 1984) which publish reviews by experts, giving details of available tests, along with assessments of their quality.

Since self-reports are subject to many obvious and less than obvious sources of distortion, a huge effort has been expended on developing instruments in which the measurement is carried out by someone else. But, of course, even these are not immune to distortion. Murphy and Davidshofer (2001, p. 461) classify the errors that separate raters make into three types: halo errors, leniency errors and range restriction errors. *Halo errors* are those in which scores are higher (lower) than they should be by virtue of the rater having an inflated (deflated) prior view of the subject. Such errors may be detectable by anomalous profiles in which (for example) pairs of subscales produce high (low) values together when they do not normally do so. From a statistical perspective, halo errors arise from a common latent factor, and detection is by outlier detection methods. *Leniency errors* (or their complement, criticalness errors) arise when the rater has shifted the scoring scale, so that there is a tendency for subjects to score artificially high or low. Such errors may be detected by an inappropriate value of the mean score. *Range restriction errors* are similar to leniency errors, but reflect a shift in the dispersion of scores, so that, perhaps, a rater tends to score everyone fairly close to the mean. One sometimes finds such an effect in marking essays, where there is a tendency to strict the range of possible scores (e.g. to 20–80%) on the grounds that a 100% score in an essay is not possible – there can always be a better one. Such errors may be detectable by studying the standard deviation of the scores.

Errors can also arise because raters have not been calibrated to the same scale, so that they have an inbuilt tendency to produce different scores, or because they collect different information. The former source of error can be reduced by training the raters, and providing them with feedback on test cases so that they can adjust their scales. Differences arising from the collection of different information can be reduced by formalizing the data collection procedure. Indeed, there is evidence in medicine which suggests that the reported high effectiveness of computer diagnostic procedures is largely due to the necessary formalization of the data collection which precedes application of such methods. For example, such methods mean that a list of required data is (at least implicitly) available, and that there is less chance of some item of data being missing by oversight. Interviews are, of course, particularly subject to lack of calibration, and the structured interview has been introduced as a way to tackle this problem. These give instructions on such things as what information is to be requested, the wording of the questions, and how to probe to clarify an answer. Various authors have demonstrated that the adoption of such structure reduces variability and increases reliability – for example, Endicott and Spitzer (1978) in the context of psychiatry, and Wiesner and Cronshaw (1988) in the context of employment interviews.

Issues of validity have been explored in great depth in psychology. Because of this, it would have been entirely appropriate to have discussed them in this chapter. However, because validity is a fundamental aspect of measurement quality, I chose to discuss it in Chapter 4, alongside the discussion of measuring reliability.

One of the features of psychological, social, economic and educational measurement – measurement which relates to humans – which perhaps applies less to physical measurements is that it can become outdated (though as Chapter 7 shows, it also applies there). Within the European Union, tests based on calculations involving currencies other than the euro could produce substantially lower scores than they would have a few years ago.

5.3 Measuring intelligence

It is an observation of everyday life that something which we will label 'intelligence' exists, in that some people are more able or quicker to find solutions to novel problems than others. At an extreme, some people simply cannot solve some problems which most people can readily solve. Clearly there is some kind of continuum which distinguishes between people and on which, in principle at least, they can be placed. Of course, this is not to say that it is not a multidimensional continuum. The hypothesis that the continuum is unidimensional comes from the empirical observation that measures of apparently different aspects of cognitive behaviour are positively correlated (Spearman, 1904, 1927; Allinger, 1988). The observed scores on particular cognitive tests are then attributed partly to the underlying general intelligence level and partly to a component unique to the particular test (plus, of course, factors such as measurement error). Assuming that the unique components are unrelated, one can use statistical tools such as factor analysis to extract the common component – general intelligence. The term 'general mental ability' is sometimes adopted instead of intelligence, and the symbol g is often used to represent it.

This basic single general factor notion has been extended (Thurstone, 1935, 1938) by a model in which different groups of cognitive tests have their own underlying common factor, as well as contributions from the general factor. Thus, tests relating to verbal and language skills may have their own common factor, distinct from the common factor relating tests of numerical and symbol manipulations skills. This two-level model has been taken further, into a hierarchical or multi-level structure, with general intelligence at the top, then factors specific to various types of intelligence, which are then split further into particular abilities. See, for example, Vernon (1960), Carroll (1993) and McGrew and Flanagan (1998).

There are many more complicated models of intelligence. R.B. Cattell (1963) introduced notions of fluid and crystallized intelligence, the former representing reasoning ability and the latter representing one's acquired skills. Guilford (1967) replaced the hierarchical or *nested* approach by a crossing approach in which intelligence is defined in terms of the cross-classification of three basic dimensions. Gardner (1983, 1993; see also Goleman, 1995) has suggested that people can be intelligent in various different ways (originally he said seven), including, for example, interpersonal and social intelligence, something which does not figure in more conventional approaches. Sternberg (1985) has developed a more elaborate theory in which intelligence is split into distinct components for acquiring information, planning tasks, and carrying them out.

One of the challenges of psychological research is to design experiments which will allow one to reject one of these models in favour of another, especially given the natural variability inherent in such data. Indeed, this may not always be possible. To take a trivial example, consider just two correlated tests. This structure can be explained as two distinct variables, which happen to be correlated, or it can be explained in terms of a mutual latent factor which induces the correlation between the tests. Both are valid descriptions of the observed empirical data, and more is needed if one is to hope to reject either one. At this level they are merely alternative but equivalent ways of looking at the same empirical structure. In any case, in the context of intelligence, the evidence suggests that the general factor g provides the bulk of the predictive power for predicting performance on cognitive tasks, and that other lower-level factors add little.

Amongst the earliest tests of intelligence were those developed by Binet, mentioned above. These were aimed at children, and later evolved into the Stanford–Binet Intelligence Scale. Wechsler extended the ideas to adult testing, leading ultimately to the Wechsler Adult Intelligent Scale. We have described how the term 'mental age' became to be used. Another common term in intelligence testing is 'IQ'. The difficulty with using mental age by itself is that its meaning depends on the physical age of the child taking the test: a child scoring a mental age of x is above average if he is younger than x, and so on. This dependency can be tackled by standardizing the mental age by the physical age (with the result conventionally being expressed as a percentage). This is called the *intelligence quotient* or IQ: IQ = 100 × (mental age)/(physical age). This is superficially attractive, but its limitation soon becomes apparent when we begin to apply it to adults. Observation shows that although we continue to learn new things as we age, our cognitive abilities do not continue to improve: whereas it is obvious that a 10-year-old has more advanced cognitive powers than a 5-year-old, it is not clear that a 40-year-old has more advanced cognitive powers than a 35-year-old. This means that a plot of 'mental age' against physical age will flatten out beyond a certain point, so that the ratio of mental age to physical age will start to decrease. It means that if someone's mental age reached its maximum of 100 at, say, 20, then their IQ would decrease over time, becoming, for example, 25 when they were 80. Because of this disadvantage, nowadays IQ is not calculated as a ratio involving mental age. Instead the score is positioned on a scale which has been standardised to have a given mean and standard deviation (e.g. the Stanford–Binet has a mean of 100 and standard deviation of 16, and the Wechsler Adult Intelligence Scale has a mean of 100 and a standard deviation of 15).

Any test which purports to measure general intelligence, rather than merely one of the subtypes which have been postulated, must include within it components which assess the different types of cognitive processing. This may not be easy to achieve. Certainly, the notion that one can sample from the space of possible cognitive functions, in the same way as one can sample from the space of people who will be given the test when validating it, is difficult to defend. This is not an issue if one's model properly reflects the underlying empirical structure of the data generating process, but can lead to distortion and bias otherwise. Given the differences of opinion as to what is the proper empirical system which underlies intelligence, it would be a brave person who would argue that the particular items included in the test did not influence the result. This means that in practice IQ tests have a substantial pragmatic component: whether one likes it or not, IQ tests are simultaneously measuring and defining IQ. Whether this IQ is the same as intelligence is something the user must decide for themselves.

As will have become obvious, a vast amount of intellectual effort has gone into developing intelligence tests. Since they are used so widely, there is also considerable commercial commitment involved in developing and marketing such tests. 'Wechsler Adult Intelligence Scale' is a registered trademark of The Psychological Corporation.

The use of IQ tests has been controversial because average test scores differ between different racial, socioeconomic, and gender groups, with (on certain common tests) whites tending to have the highest averages. Explanations which have been proposed for this difference include, on the one hand, genetic factors leading to real differences between racial groups, and, on the other hand, environmental influences, in particular that the tests have been calibrated for a middle-class white culture, and so are invalid for other groups. It is now generally accepted that both genetic and environmental influences play a role, though there remains debate about their relative importance. It is also worth remembering that such controversies also generally relate to *averages* of distributions. Properties of such averages probably have little bearing on the score which would be obtained by any reader of this book, whichever ethnic, gender, or other group you belong to.

Awareness of the possible effect of cultural bias is long-standing, and many 'culture-reduced' tests have been developed. One of the most famous of these is Raven's Progressive Matrices test, in which the subject must choose one from a set of pictures to complete a rectangular array of pictures. However, even this test does show a difference between different groups, suggesting that the simplest cultural bias hypothesis (that language usage explains the difference) is not sufficient. The point is that, since all learning takes place within the context of a culture, it is probably hopeless to seek a culture-free test. One could seek to construct a test which yields equal mean scores for the different groups, but this is based on a moral decision that the groups should be treated equally, and is external to the nature of the attribute being measured. Similar problems arise in scoring applicants for bank loans and other financial products (Hand and Henley, 1997). In the USA, UK and other countries, legislation prevents the inclusion of items such as gender and age, on the grounds that these will unfairly discriminate against certain groups. One could argue that this very restriction is unfairly biasing the scoring instrument (we await the legal test cases) since it necessarily means that people from different risk groups are being lumped together. A better strategy, if the aim is to treat different groups equally, is to construct different scoring instruments for the different groups and then require that similar proportions from each group are offered the financial product. But this is based on the external requirement that the groups should be treated equally. The law tries to achieve this, but unfortunately fails because of correlations between the excluded variables and those which have been included.

The question of the influence of genetics and environment on intelligence is related to that of the plasticity of IQ. Flynn (1984) described large improvements in IQ scores in the US between 1932 and 1978, and others have reported similar increases (e.g. Humphreys, 1989, for the US; Lynn and Hampson, 1989, for the UK; Flynn, 1987, for 14 countries). There is evidence that the difference between cultural groups seems to be decreasing. Such rapid changes cannot be attributed to genetic evolution, and it is not obvious what the cause is. Neither is it obvious that people nowadays are brighter than they were a few generations ago. The issue of grade inflation in the context of educational assessment is discussed in Section 5.9.2.

The rest of this section briefly describes two important intelligence tests.

5.3.1 The Stanford–Binet Intelligence Scale

Recent versions of this test choose the items on the basis of a three-level model of intelligence. This has general intelligence at the top level, split into crystallized ability, fluid-analytical abilities, and short-term memory. At the bottom level, there are 15 subtests: vocabulary, comprehension, absurdities, verbal relations, quantitative, number series, equation building, pattern analysis, copying, matrices, paper folding and cutting, bead memory, memory for sentences, memory for digits, and memory for objects. As described above, the scales are graduated in terms of the mean performance of different age groups. The test is adaptive, in the sense that the initial questions are intended to be simple for the child's age, with successive ones becoming more difficult. Through this means a *basal level* and *ceiling level* are determined, the former being the highest level at which an examinee answers all questions correctly, and the latter the lowest level at which they answer all questions incorrectly. Each of the fifteen subtests yields its own scale and score, and these can be combined (via given tables) to yield an aggregate general intelligence score, with a mean of 100 and standard deviation of 16.

The current Stanford–Binet test is the culmination of three-quarters of a century of development, gradually building on and refining earlier versions, but maintaining the same general principles. The effort that has gone into this test, and the quality of the result, is indicated by the fact that the Stanford–Binet measure of general intelligence has a reliability of at least 0.95 for all ages. This scale is a perfect illustration of the fact, mentioned above, that the clear difficulty of measuring psychological attributes has led to considerable careful research effort, so that often these scales are more reliable than those used in other, apparently more 'transparent', areas of measurement.

5.3.2 The Wechsler Adult Intelligence Scale (WAIS)

As with other established intelligence tests, the WAIS has undergone various improvements and developments. The current version, the WAIS-III, has 14 separate subtests. Subsets of these are combined to yield a full-scale IQ score, a performance IQ score, and a verbal IQ score. The 14 subtests are: information, digit span, vocabulary, arithmetic, comprehension, similarities, letter–number sequencing, picture completion, picture arrangement, block design, object assembly, digit symbol coding, symbol search, and matrix reasoning. A subject need not attempt all items in a subscale: a fairly straightforward opening question is given, and points are awarded for all simpler items if this one is answered correctly, and if a subject fails some of the more difficult items, the most difficult are not presented. Tables are given for converting raw scale scores into deviation IQs on a scale with a mean of 100 and a standard deviation of 15.

As with the Stanford–Binet scale, the theoretical underpinning, and continuous development on large numbers of subjects, of the WAIS have resulted in a scale which has impressive properties: the full-scale IQ score has a reported reliability typically greater than 0.95 and the performance and verbal scores have reported reliabilities above 0.90.

5.4 Measuring personality

Personality, of course, is not a unique attribute: people differ in various ways. The ancients classified people as having sanguine, melancholic, phlegmatic or choleric personalities. The complex nature of personality means that it cannot be measured without first building a model for it, and describing its various attributes. A huge amount of empirical research has been carried out in this area, and this book is not the place to review this work. Although there remains considerable controversy, important aspects of personality are generally thought to include extroversion, emotional stability, agreeableness, conscientiousness, and openness to experience. One might, therefore build measuring instruments to assess each of these.

Early empirical work on personality assessment, as with intelligence assessment, received a fillip from the First World War, when instruments were developed to try to predict who might be particularly susceptible to nervous breakdown under fire. This principle was later used in the Minnesota Multiphasic Personality Inventory (MMPI) which was developed to assist in diagnosing mental disorders. Projective techniques (such as the Rorschach inkblot test, sentence completion tests, and word association tests) were also developed, which were thought to give insight into the subconscious. Incidentally, the Rorschach test lacks reliability and validity (whatever 'validity' means for such a test) and is of little use in assessing people, despite its continued widespread popularity. In general, projective tests are drifting away from measurement as considered in this book, and should perhaps be regarded as mere indicators even at their best.

5.5 Clinical assessment

The aims of clinical psychological assessment include diagnosis and prognosis: the evaluation of a patient's current state and of their likely future state, perhaps as a result of treatment intervention. The range of clinical psychological assessments is vast. At one extreme it draws upon psychoanalysis and at the other neuropsychological assessment. Psychoanalysis is very much an idiopathic approach, involving a one-to-one relationship between the analyst and a patient, in which the patient essentially provides an unconstrained description of what is going on in his or her mind. Treatment generally takes several years and there is considerable controversy about whether psychoanalytic treatment is effective. Taylor (1984) includes the marvellous statement that 'psychoanalysis is a subject which cannot be summarily described without becoming meaningless to non-psychoanalytic readers', inviting the comment that it may only be a slight exaggeration to say that this seems almost equivalent to 'only the believers can understand', so that we seem to be drifting away from science. Neuropsychological assessment, at the other extreme, is an interdisciplinary subject, mainly related to neurology and psychology, and touching on disciplines such as physiological psychology and psychiatry, the latter especially as regards organic disorders of the mind. The aim of neuropsychology is to provide understanding, description and assessment of psychological impairments resulting from brain disease or injury. It may also be concerned with evaluating what jobs and other tasks such impaired patients could achieve. Broadly speaking, there are two interlinked approaches to neuropsychological assessment. One is based on studying the effects of damage to particular parts of or organs within the brain. The other is more holistic, and tries to characterize the types of malfunctions that can occur in different mental systems, such as memory and affect. In between the extremes of psychoanalysis and neuropsychology lie many other psychological clinical assessment tools and strategies. The MMPI, referred to above, is one of the most widely used of these intermediate approaches. It consists of several hundred statements, to which the subject responds 'true', 'false', or 'cannot say' and yields scores on ten scales, including depression, paranoia, schizophrenia and social introversion. The result is thus not a measurement of a single attribute of a patient, but rather a profile of scores. One aspect of the MMPI which is of particular interest to us is the fact that it is based on an empirical predictive model rather than an iconic model embodying a theory of the relevant psychopathology. That is, it is based on finding a statistical model which will discriminate between different levels of the particular psychopathology (e.g. normals versus patients with paranoia), where more extensive clinical investigations have been used to identify the diagnosis of each patient. Since this book is concerned with the ideas, tools, and principles of taking measurements, which can then be used as the basis of a diagnostic strategy, rather than diagnosis itself, we will not discuss such intermediate methods further. The measurement of depression is briefly discussed in Section 6.5.2.

Neuropsychological tests include those aimed at particular functions (e.g. the Wechsler Memory Scale), those intended as screening tests (e.g. the 7-Minute Screen), and also those aimed at producing a profile of test scores which can be used to diagnose particular types of damage (e.g. the Halstead–Reitan Neuropsychological Test Battery). Interested readers will find extensive and elaborate discussions of neuropsychology in Lezak (1995) and Groth-Marnat (2000). One, perhaps unusual, aspect of neuropsychological measurements is that they are typically concerned with measuring how much *less* than normal is some function: damage generally does not improve the brain's performance (and even in those rare cases when improvement is superficially indicated, closer examination usually shows it to be associated with a broader picture of deterioration). 'Normal' here may be a norm group, with which the subject can be compared, or it can be the subject's performance prior to the injury or illness. The latter is a more useful comparison, if typically more difficult to make.

Typically neuropsychological tests have weak representational and strong pragmatic components. They may be related to the underlying empirical attribute only to the extent that it is known that, for example, performance is degraded by damage of some kind, and is more severely degraded the greater is the damage. But they are converted to useful measurement scales by pragmatic constraints on the measurement process and the way the results are reported. For example, tests of memory may involve learning and later recalling a string of digits or words. Theory tells us that someone with brain damage is likely to have impaired recall, so we may expect the extent of damage to be related to the length of string they are able to recall. This is clearly at best an ordinal relationship – so permitting only ordinally invariant statements about the underlying empirical system (memory) to be made. However, we can certainly make statements about the relationship between the recalled string length and other variables. We might, for example, be able to make such a statement about the correlation between string length and the time since onset of an illness. This might be useful for prognostic purposes. This has added a pragmatic aspect to the representational aspect of recalled string length. When string length is constrained by this additional pragmatic aspect, we can make statistically meaningful statements which include operations such as sums, averages, and differences, statements which would not be permissible under the purely representational constraints.

Spreen and Strauss (1991) pointed out that many tests of specific function have been developed for their own use by research teams, so that performance characteristics of such tests are not available.

5.6 Psychophysics

The earliest attempts to quantify psychological concepts appear to be those connected with exploring the relationship between the physical intensity of a stimulus and the psychological perception of that intensity. One can use this relationship to transform the reported subjective rating into a measure of the physical intensity. That is, the human can act as a measuring instrument. Of course, how accurate this relationship is will depend on a variety of factors. One of the factors is the nature of the stimulus concerned (light, weight, heat, pressure, etc.). Reflecting such differences, development of the *psychophysical law* relating psychological perceptions and physical magnitudes has undergone a series of generalizations. One of the key concepts involved in this development is that of the *just noticeable difference*, often denoted *jnd*. For a stimulus a, this is the smallest increase δa for which $a + \delta a$

can be distinguished from a. However, this is not a precise definition. We might find that, in repeated exposures to a and $a + \delta a$, a subject correctly identifies the latter a certain proportion of the time, and define this as a noticeable difference if this proportion is greater than t, but then there arises the question of choosing t. Often a range of values of t is used. In terms of the jnd, an early recognition was that the ability to distinguish between the sizes of stimuli depended on their absolute magnitude. For example, we are able to judge more easily that 0.3 kg is heavier than 0.1 kg than that 20.2 kg is heavier than 20.0 kg. This fact led Ernst Weber (1795–1878) to suggest that the jnd was proportional to the absolute level of the physical magnitude (Weber's law; see Weber, 1846). To put it another way, he suggested that the accuracy of assessing the magnitude is inversely related to the magnitude of the stimulus. Using this, and making the additional assumption that jnds at different values of the underlying physical magnitude corresponded to equal units on the sensation scale, led Gustav Fechner (1860) to propose that the relationship between the psychological scale and the physical one was logarithmic. His argument went as follows. Suppose that the jnd is ca, with c a small constant (e.g. 0.01). Then, letting $\psi(a)$ denote the psychological (sensation) scale value corresponding to physical magnitude a, a change of ca in the physical magnitude will induce a change of approximately $d\psi/da \times ca$ in the perceived magnitude. For this to be constant for all values of a we require

$$\psi(a) = k \log a + c \tag{5.1}$$

The familiar decibel scale for loudness (Section 7.2.9) has this form. Let a_0 be the intensity of a reference sound. Then, using logarithms to base 10, and taking $k = 10$ and $c = -10\log_{10} a_0$, we have that the perceived loudness of a sound of intensity a is

$$L(a) = 10\log_{10} a - 10\log_{10} a_0 = 10\log_{10}(a/a_0)$$

Fechner's proposal seemed adequate for almost a century, although situations showing that it was not perfect gradually became apparent (even so, how many other scientific laws in psychology have lasted as long?). Luce and Edwards (1958) also attacked the derivation of (5.1) on theoretical grounds. They pointed out that Weber's law and the constancy that jnds of stimulus correspond to equal differences on the sensation scale are statements about *differences* ($\Delta\psi/\Delta a \times ca$ is constant), not about *differentials* ($d\psi/da \times ca$ is constant). Differentials arise in the limit as $\Delta\psi$ and Δa tend to zero. Luce and Edwards then show that the two assumptions, (i) that jnds at different values of the underlying physical magnitude correspond to equal units on the sensation scale, and (ii) that one can replace the statement involving differences by a matching statement involving differentials, are contradictory. In particular, for all but a few special cases for a proposed relationship between a and the jnd at a, we obtain values for the perceived jnd which cannot be constant. (Weber's law is one of the few exceptions.)

Luce and Edwards (1958) went on to explore ways of deriving psychophysical functions relating jnds on the stimulus scale to assumed equal differences in response on the sensation scale. The conceptually simplest approach is to cumulate jnds gradually. This is perfectly feasible if only a few are to be cumulated – for example, if the range of the stimulus is covered by a few jnds, or at most a few tens of jnds. In many cases, however, the range is covered by more than this: 'we doubt very much if the jnds for pitch will ever be added this way, since there are several thousand of them.' (Luce and Edwards, 1958, p. 234). Luce and Edwards also presented more advanced mathematical methods, but concluded (p. 231) 'we fear that no working psychophysicist will find in our mathematics a tool for determining a summated jnd scale any better or more efficient than the simple graphic pro-cedure of adding jnds up one at a time'. They also explored the relationship between scales constructed by this method and those produced by other methods, such as fractionation or direct magnitude estimation, pointing out that the different methods unfortunately lead to different scales.

Finally, I cannot resist a personal comment on the Luce and Edwards (1958) paper. This paper is extremely elegant. It shows, in a wonderfully clear and accessible manner how Fechner made a fundamental error, and also shows a correct solution to the problem. It is a beautiful paper. On these grounds, it is well worth reading. However, it is also immensely frustrating. It is one of those papers which, when you get to the end, you wonder if it was worth the effort. As the authors say (p. 237): 'This error does not seem to have any significant impact upon the controversy over the relation between cumulated jnd scales and scales based on fractionation and direct estimation data because most psychophysicists have, in fact, ignored the recommended (incorrect) procedure and have stubbornly summated jnd's in the obvious and correct way.' These issues are also discussed in Luce (1993).

Around the middle of the twentieth century, in parallel with the Luce and Edwards work, Stevens (1957, 1960, 1961, 1962) suggested instead that a power law

$$\psi(a) = \alpha a^\beta \tag{5.2}$$

was a more appropriate reflection of reality and adduced experimental evidence in support of this. For example:

> The power function turns out to be a good approximation to the relation between stimulus and response on at least 14 different prothetic continua. Values of the expo-nent β have been found to range upward from about 0.3 for loudness. Typical values for some of the other continua are: brightness 0.33 to 0.5 (depending on area of target, state of visual adaptation, etc.), apparent length 1.1, duration 1.15, numerousness 1.3, heaviness 1.45, velocity 1.77. In an experiment recently completed, we have found that the apparent intensity of an electric shock, produced by a current (60 cps) through the fingers, grows as a power function of the current, with an exponent of the order of 4.
>
> (Stevens, 1959, pp. 36–37)

We should note, however, that the value of the exponent does depend on the experimental conditions and circumstances, such as the range of stimulus values.

The figure of 1.1 for lengths (or short lengths, at least), which we might interpret as 'near unity', is of particular importance because of the common use of linear scales to represent subjective phenomena. For example, in *visual analogue scales* (Chapter 3), the subject may be asked to position themselves on a line (of length, say, 10 cm) labelled with adjectives indicating opposite extremes at each end. An exponent of unity suggests that the intensity of feeling (or whatever the scale is measuring) can legitimately be approximated by the position on the line. (The data on subjective ratings of the lengths of pieces of string given in Dunn (1989, p. 6) supports a value near 1 for this exponent.)

Various procedures have been developed for elucidating the relationships between phys-ical and perceived stimulus intensity. A basic procedure is to familiarize a subject with cat-egorized values of the stimulus intensity, measuring new stimuli by simply having the subject assign the stimulus to a category. Extending this, one can ask subjects to assign a

Table 5.1 Scoville scores of some hot foods.

Dry ground paprika	0–150
Tabasco sauce	2000
Fresh green *jalapeño* peppers	1500–4500
Dry ground *jalapeño* peppers	up to 30,000
The hottest known pepper, the *habañero*	100,000–350,000
Pure capsaicin	*c*. 16,000,000

series of stimuli to a set of k equally spaced ordered categories (see the discussion of Thurstone scales in Chapter 3). It is interesting to note that, regardless of the precise details of the physical/perceptual intensity relationship, the relationship which emerges from this exercise is generally a smooth function. A further extension (*magnitude estimation*) presents a 'standard stimulus' to the subject and requires them to give the ratio of the magnitude of new stimuli to this standard. A variant of this (*ratio estimation*) presents stimuli in pairs and requests the perceived ratio of the pair. All of these above are passive procedures, in that the subject simply has to describe, in some form or other, the sensations they are experiencing. An active procedure requires the subject to generate a stimulus of a required value. For example, one might attach a tone generator to a dial controlling volume, and request the subject to produce a stimulus of half the magnitude of a given stimulus. McDowell and Newell (1996, p. 17) remark that 'people can make numerical estimates of subjective phenomena in a remarkably consistent manner, even when the comparisons are abstract, indeed, more abstract than those involved in subjective health measurements'. Falmagne (1985) provides a *tour de force* of a discussion of psychophysical theory.

Sometimes, in specialized areas of psychophysics (though they may not always have been recognized as this), strongly representational measurement procedures have been adopted to define a measurement unit for an attribute via a proxy attribute. Thus, for example, spicy foods such as chilli peppers are said to taste 'hot', and a measure of the degree of this 'hotness' has been defined in chemical terms. Such foods contain a chemical called 'capsaicin', and it is this which makes them taste hot. The Scoville heat unit, named after Dr Wilbur Lincoln Scoville of the Parke Davis company, measures how hot these foods are in terms of the parts per million which are capsaicin. For example, the scores of some such foods are given in Table 5.1.

5.7 Measuring strength of attitude and opinion

There are many definitions of 'attitude'. A useful one, and one which has been adopted by many researchers, is that of Katz (1960, p. 168): '*Attitude* is the predisposition of the individual to evaluate some symbol or object or aspect of his world in a favorable or unfavorable manner.' Many authors have regarded attitude as having three components: cognitive, affective and behavioural. Thus, a representational model is being constructed, one based on a theoretical idea of what is meant by 'attitude', and from which relations to manifest variables can be derived. Inevitably, however, many of the decisions which will have to be made will be pragmatic ones.

The simplest approach is to measure attitudes using direct pragmatic scales such as visual analogue or semantic differential methods. Here one positions individuals on a continuum such as extent of agreement with a stated position or relative position between two extreme and opposite adjectives. Typically, however, multiple such direct scales are used simultaneously, and indirect inference methods such as the unidimensional scaling methods described in Chapter 3 are used to condense the result down to a single score.

Dawes and Smith (1985, p. 539) caution against too literal an interpretation of the response to an attitude rating scale, drawing attention to the sorts of issues mentioned in Section 5.2: the scale may not include a category for the respondent's true feelings, the respondent may have a tendency to avoid being too pejorative, and so on.

Attitude and opinion measurement may be particularly susceptible to the potential criticism that it sensitizes the respondent to the fact that they are being studied – the reactive effect mentioned in Section 5.2. This may distort their response, and has led to the development of indirect methods of measurement. Of course, as was remarked above, measurement procedures in which the respondent is misled may involve ethical issues.

5.8 Actuarial versus clinical methods

Chapter 1 described how there had been a gradual evolution towards quantification throughout science and technology. This has not always been a smooth transition. Although, in the modern world, the measurements and the empirical attributes of properties such as length and weight are seen as so close as to be inseparable (without intellectual investigation of the kind the reader of this book is pursuing), as Chapter 7 shows, the meaning and processes of measurement of even physical quantities have not always been obvious or clear. In historical times, the notion that one might represent the magnitude of such things in terms of numbers was often regarded as absurd. Moving on, in the modern world we now also measure such things as depression, quality of life, intelligence and achievement, where the distinction between the attribute and measurement is more striking, and in many of these cases there remains considerable controversy about what measurement means. This is perhaps especially true in psychological and social domains.

Many psychological and social attributes are involved in decision-making processes. For example, health-related quality of life scores are used in investigations of the efficacy of medical treatments, while intelligence, aptitude, and ability tests are used in selection for employment, schools and universities, and assessments of creditworthiness are used to decide to whom a bank should lend money. At an extreme, single instruments are constructed for such decisions, measuring the key attribute in question. These measurement instruments are built using past data to relate characteristics which may be used for prediction to the outcomes from previous patients, employees, customers, or whatever. However, such formal instruments are not the only way to make such decisions. An older, less formal approach has been in use for far longer. This is the interview. Because the two approaches, the one objective and quantitative, and the other subjective but based on skills and experience, are so different, there has been a natural debate about which is the more effective. In fact, as we illustrate below, there have been several parallel such debates occurring in distinct domains.

In psychology, the debate is sometimes called the clinical versus statistical (or actuarial) approach. Here 'clinical' implies an interview or investigation by a trained expert, while 'statistical' or 'actuarial' implies the construction of a formal predictive statistical model based on analysis of records of previous patients.

The clinical versus statistical debate can be traced as far back as one likes. Early papers include those by Freyd (1925) and Viteles (1925). Lundberg (1926, 1929) argued that the clinical approach was a simplistic precursor to the more scientific statistical approach. On the other hand, Allport (1937, 1942) presented a rather contrary perspective, proposing that the statistical approach and an approach based on general principles of behaviour were only appropriate for large groups, and if one wanted to make statements about individuals one had to adopt a clinical, subject-specific, approach. Horst (1941) struck a more balanced position, arguing that both perspectives had roles to play. A review of the early work is given by Hough (1962).

Perhaps the best known discussant of this issue in modern psychology is Paul Meehl (e.g. 1954, 1957, 1965). In his 1954 paper, Meehl summarized 20 studies, 19 of which suggested equal or superior predictive performance for the statistical approach, compared to the clinical approach. Later work by others seems to support this position (e.g. Carroll *et al.*, 1982; Wedding, 1983; Wormith and Goldstone, 1984; Holt, 1986; Sarbin, 1986; Wedding and Faust, 1989). Wedding's (1983) study, for example, compared diagnostic performance of 14 clinical neuropsychologists with a recursive tree classifier (though not described in those terms) and linear discriminant analysis. The flavour of Wedding's conclusions is captured by the following:

> Two of the clinical judges outperform the discriminant functions, one clinician ties the functions and eleven perform more poorly. … The results offer strong support for the efficacy of actuarial and (particularly) statistical approaches to discrete decision-making in clinical neuropsychology.

However, it is also important to put these and similar results in a broader context. As Wedding also says:

> This finding is consistent with an extensive literature on judgement in areas other than neuropsychology and in no way denigrates the skill or expertise of the clinician. Clearly the qualities of a good neuropsychologist include the ability to relate to patients, establishing rapport, understanding the phenomenology of the brain damaged patient, devising rehabilitation plans and programs and helping the patient reshape a shattered world. In addition, not all clinical decisions will be as 'tidy' as those reported here.

This last point is particularly important when one wants to move statistical methods from the laboratory to the clinic or the real world more generally. A fairly recent and extensive review of clinical versus statistical methods is given by Dawes *et al.* (1989).

In the statistical approach, one builds a model targeted squarely at predicting the value of the outcome variable for people drawn from the same distribution as those earlier used to construct the model. This thus presupposes that one has available a database of previous subjects for whom the value/class of the outcome variable is known. It also means it is a highly specific model, and this in particular can be a weakness. If new cases arise which are different from those seen earlier, then the subjective clinical interview may be better. An interviewer will rapidly observe that someone is slow at writing because they have a broken arm, but this will not be obvious merely from a self-completed questionnaire. Less dramatically, an interviewer will also be able to note (perhaps subconsciously) someone's behaviour when answering the questions (whether they are hesitant, careless, slow, etc.). More generally, if the distributional characteristics of new cases differ from the distributional characteristics of earlier cases, then the model may also show degraded performance. This will happen if the nature of the population changes.

In such circumstances it is entirely possible that the subjective interview approach may be better, but this should not be taken as a criticism of the formal measurement strategy *per se*. Rather it is a comment about the use of empirical models. If mechanistic models are used, that is, if the measurement models are based on some kind of theory about how the underlying process works, rather than a simple empirical model of what has been correlated with the outcome in the past, then one would not expect so rapid a deterioration in performance. Unfortunately, as it happens, in the psychological, social and behavioural sciences mechanistic theories are weak.

On the other hand, working against the advantages of mechanistic models (based on a sound theory of the underlying mechanism), is the *flat maximum effect*. The models discussed above are predictive models based on a retrospective set of data for which both predictor variables and outcome variables are available. This means that purely empirical or descriptive models can be built. One can, for example, decide to use a weighted sum of the predictor variables and then estimate the best weights for predictive accuracy using regression analysis, without having any 'theory' about how the potential predictors are related to the outcome. A large number of empirical predictive models are available (see, for example, Hand *et al.*, 2001), including highly complex and sophisticated methods such as neural networks, support vector machines, multivariate adaptive regression splines, recursive partitioning models and so on. In some domains, however, and typically those involving some kind of prediction of human behaviour, it is often found that a simple weighted sum performs almost as well as much more complicated methods. More than this, however, it is often found that the predictive performance is relatively insensitive to the choice of weights in the model. This observation has been called the flat maximum effect (von Winterfeldt and Edwards, 1982) because changing the weights away from those which maximize predictive accuracy does not substantially degrade performance: the performance surface is relatively flat near to the maximum. Further discussion of this phenomenon in the context of predictive rules is given in Hand (1997, Section 9.3).

Similarly, in the context of forming weighted averages to act as overall measures, summarizing multiple separate measures, it is often found that using equal weights does not detract substantially from using the optimal weights. An informal argument showing why this is so runs as follows.

Suppose we have a set of variables $(x_1, \ldots, x_p)$, and, for simplicity, assume that $E(x_i) = 0$ and $V(x_i) = 1$ for $i = 1, \ldots, p$. Let $\mathbf{R} = \{r\}_{ij}$ be the correlation matrix between these variables. Now define two weighted sums:

$$w = \sum_{i=1}^{p} w_i x_i \quad \text{and} \quad v = \sum_{i=1}^{p} v_i x_i$$

using respective weight vectors $(w_1, \ldots, w_p)$ and $(v_1, \ldots, v_p)$. Now, the correlation between w and v, $r(w, v)$, can take extreme values of $+1$ and -1 (e.g. take $w_1 = 1, w_i = 0, i \neq 1$, and $v_1 = 1, v_i = 0, i \neq 1$, to yield $r(w, v) = +1$; and $w_1 = 1, w_i = 0, i \neq 1$, and $v_1 = -1, v_i = 0, i \neq 1$, to yield $r(w, v) = -1$). However, suppose we restrict the weights to be non-negative,

$w_i, v_i \geq 0$ for $i = 1, \ldots, p$, and also require $\Sigma w_i = 1$ and $\Sigma v_i = 1$. Now, using the facts that $E(x_i) = 0$ and $V(x_i) = 1$, we have that $r(x_i, x_j) = E(x_i x_j)$, so that

$$r(v, w) = r\left(\sum_i v_i x_i, \sum_i w_i x_i\right)$$

$$= \frac{\sum_i v_i w_i + \sum_i \sum_{j, j \neq i} v_i w_j r(x_i, x_j)}{\sqrt{\sum_i v_i^2 + \sum_i \sum_{j, j \neq i} v_i v_j r(x_i, x_j)} \sqrt{\sum_i w_i^2 + \sum_i \sum_{j, j \neq i} w_i w_j r(x_i, x_j)}}$$

Since

$$1 = \left(\sum v_i\right)^2 = \sum_i v_i^2 + \sum_i \sum_{j, j \neq i} v_i v_j$$

and since $r(x_i, x_j) \leq 1$, the denominator in $r(v, w)$ is less than or equal to 1, so that

$$r(v, w) \geq \sum_i v_i w_i + \sum_i \sum_{j, j \neq i} v_i w_j r(x_i, x_j)$$

Now, putting $v_i = 1/p$, $i = 1, \ldots, n$, we obtain

$$r(v, w) \geq \frac{1}{p} \sum_i w_i + \frac{1}{p} \sum_i \sum_{j, j \neq i} w_j r(x_i, x_j)$$

$$= \frac{1}{p} \sum_i \sum_j w_j r(x_i, x_j)$$

$$\geq \frac{1}{p} \sum_i \sum_j w_j r(x_i, x_k)$$

where

$$k = \arg\min_j \, r(x_i, x_j)$$

From this,

$$r(v, w) \geq \frac{1}{p} \sum_i \sum_j w_j r(x_i, x_k) \geq \frac{1}{p} \sum_i r(x_i, x_k)$$

In words, the correlation between an arbitrary weighted sum (with positive weights summing to 1) and the simple average of the variables is bounded below by the average of the smallest row entries in the correlation matrix of the x variables. This means that if the correlations are all high, then the simple average will be highly correlated with any other weighted sum: the choice of weights will make little difference to the scores.

The formal statistical method has several obvious advantages over the clinical method. Changes in the scores at different times are more likely to be solely due to changes in the patient when a formal objective method is used, than when a clinical interview is used;

interviewers can tire, become bored, be worried about extraneous issues, and so on, so that their measurement error is likely to be greater. A statistical model, given the same input data, will always yield the same output score. Likewise, formal statistical models do not imagine relationships, based solely on prejudice or irrational belief: clinicians are well known to be poor at estimating prior base rates, so that they may have grossly distorted ideas about posterior probabilities.

Towards the end of their review, Dawes et al. (1989) discuss why the evidence showing the undoubted value of the statistical modelling approach 'has had little impact on everyday decision making, particularly within its field of origin, clinical psychology'. They describe several factors, including lack of familiarity with the scientific evidence, misunderstanding of that evidence, the perception that a formal method dehumanizes the patient (ignoring the 'human cost of increased error that may result'), the argument that group statistics do not apply to individuals (ignoring the basic probabilistic arguments in all of this), and incorrect opinions of the effectiveness of clinical judgment. They conclude (p. 1673): 'Failure to accept a large and consistent body of scientific evidence over unvalidated personal observation may be described as a normal human failing or, in the case of professionals who identify themselves as scientific, plainly irrational.'

In retail banking, prior to the widespread adoption of formal credit scoring methods, the decision to grant credit was based on subjective human assessment (here, instead of 'clinical' the word 'judgemental' was used for such subjective interviews). A bank manager would interview a loan applicant, and make a decision based on the 'five Cs' – character, collateral, capacity, capital, and condition – which were thought to represent underlying aspects of the creditworthiness of a requested loan. As the retail financial industry developed, however, so this personal method became impractical. In particular, with the advent of credit cards, with many millions of customers seeking the product, personal interviews were completely out of the question. Some quicker, but hopefully not too much less accurate, assessment method was required. The answer was the development of credit scoring methods (reviews of these developments are given in Rosenberg and Gleit, 1994; Hand and Henley, 1997; Hand, 2001; Thomas, 2000; and Thomas et al., 2002). These are essentially generalized linear models to predict some outcome (e.g. default on a loan), based on a sample of retrospective cases, for whom the outcome is known (e.g. they were given a loan some years ago). These models require the applicant to complete an application form giving details about their financial circumstances, but also about other factors thought to be predictive of creditworthiness – for example, such 'scorecards', as they are known, because they yield a creditworthiness score, often include the time living at the current address, the time employed by current employer, occupation and marital status. Some variables – sex, for example – cannot be used by law since this is thought to discriminate against certain groups (although, as we have mentioned above, there is some controversy about this – see Hand and Henley, 1997).

Naturally enough, and in parallel with the introduction of formal measurement procedures in just about every other domain, the introduction of these statistical tools encountered scepticism, and stimulated some research aimed at investigating the relative effectiveness of the informal interview and the formal scoring approaches (e.g. Myers and Forgy, 1963; Chandler and Coffman, 1979; Reichert, et al., 1983). Looking back, the early studies comparing the two approaches had weaknesses, but were useful in identifying the key aspects of the comparison. They showed that even the initial formal methods seemed to have the edge in terms of narrow predictive accuracy, though this edge was not great. However, the practical need for such systems meant that they continued to be used and developed, and herein lies one of their great advantages: a formal quantitative method can be refined, modified, amended and tuned in transparent ways, so that it evolves to improve its performance. Doing the same with an informal method such as an interview is extremely difficult – there is just too much flexibility in an interview, too many opportunities to deviate from a specified track (indeed, this flexibility is what many would argue is the merit of the interview process). Nowadays, there is no doubt that formal credit scoring models are superior to interviews for the purpose of retail banking decisions. On the other hand, an interview can be effective at detecting those occasional anomalous cases, which are different from the statistical mass. Furthermore, as Chandler and Coffman (1979) note, simple accuracy of creditworthiness predictions is not the only thing which must be taken into account. Other factors to consider are managerial ability to exercise effective control over the credit-granting process, the ability to forecast results and prepare relevant management reports, compliance with legal constraints, and social and political acceptability. Chandler and Coffman (1979) conclude: 'it seems that, on the whole, the empirical evaluation process has no serious deficiencies not also shared by judgemental evaluation. It also appears that empirical evaluation of creditworthiness has certain advantages that do not exist with judgemental evaluation. On the other hand, judgemental evaluation may have an advantage in dealing with individual cases that truly are exceptions from past experience.' Similarly, while not convinced of the predictive ability of scoring approaches, Reichert *et al.* (1983, p. 102) observe that 'their real benefit may relate not to any superiority in predictive power but to the highly consistent, objective, and efficient manner in which such predictions are made'. They might also have added the fact that scoring is typically cheaper than the alternative. Hsia (1978) also discusses the disadvantages of judgemental systems.

Nowadays all retail banking organizations use credit scoring approaches, and these types of models are also used to yield scores for making a whole host of other decisions (Which applicants are likely to switch to another supplier? Who should be offered a top-up loan? Which transactions might be fraudulent? Who will pay off early? etc.). Fundamental to such models is the availability of a large pool of data describing previous customers, which can be used to build the model. For analogous decisions about corporations, such data are generally not available (there are not a million large corporations to provide delinquency and default records over the last three years). It follows that the development of analogous models for large corporations would have to be based more on a mechanistic model than an empirical one, and for this reason decisions for large corporations still tend to be subjective. The intermediate domain of small and medium enterprises is increasingly adopting formal measurement models.

As a concluding note, perhaps it is worth remarking that, even though the retail financial community has confidence in objective statistical measurement models for credit-granting and other decisions, there is still some suspicion in the customer base. This stems in part from concerns over the accuracy of the data relating to the individual applicant but also in part from anxiety about the impersonal nature of the process. As before, and as in so many other areas, acceptance of the fact that measurement is possible is not always easy.

In general, in whatever context, if the predictor variables do contain information which can yield accurate values for the outcome attribute, then formal methods have the advantage. Paul Meehl (1986) puts it very nicely: 'When you check out at a supermarket, you don't eyeball the heap of purchases and say to the clerk, "Well it looks to me as if it's about $17.00 worth; what do you think?" The clerk adds it up.'

Looking more closely at the decision-making procedures above, we see that they involve two steps. The first is the collection and collation of relevant evidence, and the second is the condensation and summary of that evidence. Improvement in one's predictions and hence one's decisions could be made by improving either of these steps. In the context of medical decision making, McGuire (1985) says: 'studies of clinical reasoning all too often have revealed disquieting defects in the process, namely, that physicians often fail to collect the data they need, to pay attention to the data they do collect, to use their knowledge effectively in making interpretations of and inferences from the data they do consider, to incorporate a systematic consideration of alternative risks and values in the actions they take on the basis of those inferences' (these issues are discussed further in Hand, 1985, Chapter 5). The point is that one of the key distinctions between the interview and the objective statistical method is that the latter is generally relatively unstructured. The direction in which the interview goes can be chosen by the interviewer. Clearly, however, this key distinction is not a binary divide: there can be degrees of structure in an information elicitation process. In many situations – some psychiatric interviews and some market research provides examples – structured interviews have been developed, in which the interviewer has to follow particular lines of questioning and is given training on how to elicit further information, but not on the precise question wording that must be used.

5.9 Educational testing

The fundamental aim of educational testing is to measure how well the student understands the material they have been taught – although there may be subsidiary aims, such as assessing the quality of the teaching. Clearly, in order to measure a student's understanding, great care has to be taken in constructing the tests. One needs to ensure that they cover the range of material taught, that they test understanding of that material and not other skills or understanding. For example, a modern statistics examination should not require too much manual calculation, unless the aim is to test skill in mere arithmetic manipulation. Examinations are clearly indirect measurement procedures, in which manifest variables (the individual questions) may be quite complex. There are also important issues of reliability and consistency, both between different examiners and even within a given examiner over time. Because of this, standardization strategies have been developed, though these are far from foolproof, especially in disciplines which hinge upon subjective judgement.

Educational testing is very often mass testing. Madaus and Raczek (1996) write:

> testing in the United States is a very big, very lucrative, unregulated industry. Haney *et al.* (1993) estimate that the revenues from the sales of elementary and high school (secondary) tests and related services, such as scoring and reporting (called the Elhi market), is estimated to be between one-half and three-quarters of a billion dollars annually. To appreciate educational testing in the US therefore, one must clearly keep in mind that the institution of testing is at its root commercial and competitive.

As mentioned above, one of the earliest mass tests was that developed for use by the US military during the First World War. One version of this test had eight subtests, covering such things as arithmetical reasoning, practical judgement, and analogies, and was taken by almost 2 million men. Over the following years these tests were elaborated further, with extra tests being introduced aimed at measuring aptitude for particular types of job. These developments have continued (see, for example, Segall and Moreno, 1999; Waters, 1997) and such ideas are now applied very widely indeed. For example, tests of ability and knowledge are used to select students for university and college, before they are subjected to the rigours of that education. Such tests explicitly distinguish between aptitude and achievement, to measure ability rather than experience. Thus the American Scholastic Aptitude Test (SAT) (there seems to be some ambiguity: sometimes 'Scholastic Assessment Test' or even 'Scholastic Achievement Test', though this last seems to miss the point), first used in 1926, is used to aid university admissions decisions. More generally, the use of such tests to select people for jobs is now widespread. While interviews continue to play a role in such selection, how else would one winnow down the 500 applicants for two posts? Interviewing them all would certainly not be cost-effective.

Multiple-choice questions were a natural (indeed, perhaps inevitable) way to handle such large-scale tests and are a precursor to the computerized test, although the latter now goes far beyond such simple ideas. The first machine for automatically marking such tests, by optically detecting which of a set of answers had been marked, was developed by Reynold B. Johnson, a schoolteacher (Downey, 1965).

Since computers are ideal for automatically processing large quantities of data which occur in the same format, it is hardly surprising that one of their earlier uses was in scoring tests for large numbers of candidates for examinations or selection procedures. But computers and computer software have developed in an extraordinary manner. They are now used in a huge number of ways which would have been inconceivable at the start of the computer revolution. To take but one example, computer power permits examinations to be matched to the individual candidate, and not only that, but matched in a dynamic way: the candidate's answers to early questions can lead to an adjustment in the later choice of questions, so that a more precise evaluation of the candidate's ability may be obtained. The days are numbered in which every candidate, from the least to the most able, is forced to answer the same set of questions. If the aim of testing is to rank individuals according to (say) their ability, then the fact that ability generally seems to follow a normal distribution means that tests must be most sensitive in the middle of the range. It is in this part of the range that the scores will be most dense. Indeed, this is often very evident from college examination scores, where the top candidate may be streets ahead of the second candidate, even though the latter is also very good. (Of course, an examination which is poorly designed may lead to the ceiling effect coming into play, so that many candidates achieve the maximum available number of marks.) Since sensitivity in the middle of the range is important, the majority of the test items will be aimed at discriminating in this range. The consequence of this for traditional paper-and-pencil tests is that the less able students find themselves faced with a majority of questions which are too tough for them, and the exceptionally able students have to spend substantial time answering questions which merely indicate that they are very able, without usefully characterizing this ability. The solution to this problem is the *adaptive* test (Wainer, 2000), in which successive questions are chosen on the basis of the student's previous responses, using easier questions whenever a student gets one wrong and harder questions when a student gets one right. It is clear that paper-and-pencil implementation of such a strategy would be very difficult, but that computer implementation is relatively straightforward. There are other peripheral benefits to such a strategy, not least the fact that a student can receive immediate feedback on how they are doing. Of course, implementation of this strategy requires knowledge of how hard the questions are relative to each other. Item response theory, discussed in Chapter 3, tells us how to obtain that knowledge. Furthermore, since item response theory models position each item on a continuum of difficulty, they permit one to assess the ability of different students on the same scale, even if those students have not attempted any items in common. Note that, with adaptive methods, the 'standard' scoring strategy of the number of questions a student answers correctly cannot be used: if the adaptation has worked well, then each student will be correct on about half of the questions they have attempted. The point is that with adaptive testing, the questions attempted by each student will differ between students, depending on their individual ability.

A key issue in adaptive testing is how a candidate's response to the questions so far asked influences the choice of the next question to be asked. The basic strategy is to choose the next question so that it gives as much information as possible near the ability range currently estimated to be the candidate's ability. Associated with this is the issue of when to stop – when we believe we know the candidate's ability (or whatever) sufficiently precisely that further measurement is unnecessary. In item response theory, each estimate of a candidate's ability is accompanied by an estimate of the accuracy of that estimate; thus we can stop the procedure when sufficient accuracy has been attained.

5.9.1 Scholastic Aptitude Tests

The SAT includes a test which measures verbal and mathematical reasoning and also tests which measure understanding of specific areas, such as biology and history. The tests are such that they can be applied to large numbers of candidates and are machine-marked – for example, multiple-choice or where the student has to write a number as an answer. The verbal and mathematical test has three verbal and three mathematical subtests, along with a subtest which does not contribute to the overall score but is used for test equating (see Section 5.10).

The SAT has its origins in the late 1920s and early 1930s, when the President of Harvard University, James Bryant Conant, wanted to increase the diversity of social backgrounds of his students. The University eventually adopted the SAT, redesigned as a multiple-choice test. Originally aimed at the selection of a few able students from less privileged backgrounds, the use of the test evolved to become a major tool in deciding which students in general should be admitted to colleges across the United States. Of course, as is perhaps inevitable, an industry of coaching and guidance has grown up to assist students taking the test.

The ACT Assessment Program is also used to assist in college admission decisions in the United States. This is rather different from the SAT, as it is directly aimed at assessing abilities which are particularly relevant for success at college.

Moving up the educational chain, the Graduate Record Examination (GRE) serves a role similar to the SAT, but for admission to graduate school in the United States. This test is increasingly computer-based.

5.9.2 Grade inflation

A topic which stimulates a tremendous amount of debate in the media at various times of the year is that of *grade inflation*. This is a proposed explanation for the observed phenomenon

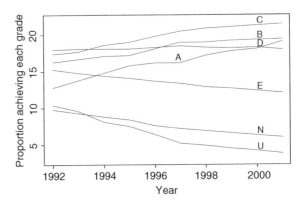

Figure 5.1 Proportions of UK students with each A-level grade, 1992–2001.

of a drift towards higher academic grades over time. Figure 5.1 illustrates the phenomenon for UK A-level examinations. These are examinations taken at age 18, and have traditionally served as the criterion by which UK universities decide which students to admit. In fact, things are rather more complicated than that. A-level examination results become known in August, after the universities have had to decide which students they would like. They therefore make offers based on predicted A-level grades, and conditional on the students achieving certain grades. Figure 5.1 shows that over the period 1992–2001 progressively higher proportions of students have achieved the higher grades (A, B, C) and progressively lower proportions of students have achieved the lower grades (E, N, U). The line for the D grade is almost constant.

The phenomenon is not limited to the UK. In the USA, grade point averages rose dramatically after the mid-1960s. For example, citing a report by Arvo E. Juola from the University of Michigan Office of Education Services (August, 1974), Birnbaum (1977) says: 'Juola, reporting on a survey form completed in whole or in part by 197 institutions, indicated an increase of .324 of a grade point over the eight year period from 1965 to 1973 in this nationwide sample'. Birnbaum went on to say: 'A study of 281 institutions conducted by Never and Kojahn (1973) indicated an increase in freshman grade average means in five categories of institutions from .28 to .46 of a grade point between 1966 and 1973. Suslow (1976) found an increase in mean grade point average between 1960 and 1974 from 2.47 to 2.94 at 15 research institutions.' More recently, also in the US, Levine and Cureton (1998) found that the proportion of A grades awarded in a survey of undergraduate college students increased from 0.07 to 0.26 between 1969 and 1993, while the proportion of C grades declined from 0.25 to 0.09. This is not simply something which affects lesser colleges. In an impressive review of grade inflation in the US which appeared in the *Boston Globe*, Patrick Healy (2001) said: 'Undergraduate honors [at Harvard] increased from 32 percent in 1946 to 91 percent in 2001, with the greatest growth in the 1960s and early '70s, and then again during the last 15 years.'

It is clear from these observations that there is a drift over time towards the higher grades. The question is, what causes this drift? In particular, is it caused by *grade inflation* – the gradual lowering of the grade thresholds so that successive cohorts of students have lower barriers to jump, or can it be explained in some other way?

Grade inflation is certainly one possible explanation, but others have been proposed. In the UK, these include:

- Students working harder and having a better understanding of the material.
- A reduction in the difficulty of the material taught. In fact, examination suggests that, although the material has changed over time, it is not necessarily any easier.
- A shift in assessment methods, away from straightforward terminal examinations to modular systems, with examinations after each module, to systems in which examinations are partially replaced by assessed coursework, and to systems in which overall grades can be improved by retaking examinations of modules.
- Population drift of students away from the tougher subjects towards those in which higher grades are easier to win. The UK recently changed its examination system, from one in which the norm was a two-year study period followed by examinations in three A levels, to a system in which a larger number of AS levels were taken for the first year, followed by pursuing just a few of these to the more advanced A-level examinations in the second year. This allowed the students to test the subjects in the first year, and drop those in which they were performing less well. This has been proposed as the explanation for the dramatic reduction in students taking A-level mathematics, from 66,247 in 2001 to 53,940 in 2002. This kind of inflationary effect will manifest itself on any kind of aggregate result, not merely overall proportions achieving certain grades, but also measures such as grade point average.
- A levels are a nationwide system, but there are several different examination boards. Inevitably, some of these are thought to provide easier examination papers than others, so there may well be a drift towards those which are thought to be easier.

The US college situation is rather different from the UK A-level situation, in that the grades in the colleges are not assessed on a national scale. But grade inflation still seems to be apparent. In the US proposed explanations include:

- The impact of the Vietnam war, when draft deferment was possible if one could get into graduate school.
- An increase in numbers of students from less traditional backgrounds, with these students being less familiar with and less prepared for the kind of assessment procedures they would be subjected to.
- The use of student evaluations of staff teaching performance. Perhaps not surprisingly, those teachers who award higher grades tend to receive better student evaluations (Wilson R., 1998).
- A high proportion of part-time and casual staff, who have a high workload and little vested interest in careful grading.
- There appears to be some evidence that graders adapt their thresholds to the ability of the cohort of students (e.g. Aiken, 1963; Juola, 1968; Baird and Flister, 1972).

If college grades are not based on national examinations, the SAT run by the Educational Testing Service at Princeton is. According to Rosovsky and Hartley (2002): 'During the time period in which grades increased dramatically, the average combined score on the Scholastic Achievement Test (SAT) actually declined by 5 percent (1969–1993).'

Underlying the entire issue is the fact that the grades are used for multiple purposes. They are used to evaluate students, but also to evaluate teaching methods, to evaluate teachers and, when a national system exists, to evaluate institutions. Even when used to evaluate students, they may be used for several purposes. Thus they may be a means by which the students can themselves assess their progress, or so that their teachers can assess their progress, or they may be a final evaluation, so that, for example, a potential employer or a college can see how well the student has grasped the material. In general, these different aims may require different things from a grading system. For example, from a student's perspective, a high grade is desirable if it is to be used when seeking a job, but an accurate reflection of ability is desirable if it is to be used to assess progress. Or, to take another example, a school might assess its own students on a criterion-referenced basis – that is, decide if the students meet some absolute level of achievement – whereas, in contrast, a prospective employer or college might prefer to use a norm-referenced basis – one in which a student's achievement is put on a quantile of scores in the distribution of all those being assessed in that year. In discussing the issue of grade inflation, one important aspect to consider is whether *criterion referencing* is appropriate or whether *norm referencing* should be adopted. In criterion referencing the scores are compared to some objective standard, whereas in norm referencing they are compared to the scores of a norm group. Which is appropriate will depend on the objectives.

Naturally the topic generates much heat. Employers and colleges are concerned that the criteria they use to make choices are becoming less discriminatory. Students, who study hard to pass the examinations, resent being told that their qualifications are not worth much.

Examination scores are not the only psychological measurements which show apparent improvement over time. IQ scores have also shown apparent dramatic improvements over the decades. It is not clear that our recent ancestors were that much more stupid than the present generations, so some other explanation is needed.

5.10 Test equating

As tests evolve and develop, one naturally wishes to establish comparability with previous tests. It is also often necessary to transform test results so that different tests are equivalent in some sense. Such procedures are known as *test equating*. Of course, equating of measurement procedures is not confined to educational testing. Economics, in particular, suffers much (unfair) criticism when methods for measuring underlying social and economic phenomena, such as unemployment, inflation, or retail spending, are altered to reflect the changed situation in which they are being applied. (Of course, I make no comment on unfortunate political manipulation which may use such adjusted measures to show, for example, that the crime rate has gone down when all that has happened is that the definition of what constitutes a 'crime' has been subtly adjusted.)

Lord (1980) identified four key characteristics that test equating procedures should possess:

1. Since a test positions someone on an underlying continuum, tests can only be equated if they refer to the same continuum; that is, if they measure the same construct. Since different tests will generally involve different items (else they would hardly be different) care is needed to ensure that they really are measuring the same thing.
2. Test equating must produce tests of equal difficulty, in the sense that subjects with the same ability will have the same conditional distributions of scores on the two tests.
3. The same equating transformation must apply across different populations. If different transformations are needed for different populations, then the tests are probably measuring rather different things.

4. The equating relationship must be symmetric, in the sense that if a score on test A is matched to a score on test B when A is transformed by the equating procedure, then the same scores must also be matched if B is transformed by the (inverse) equating procedure.

The symmetry requirement means that straightforward regression cannot be used to obtain equating transformations: the regression of A on B is not the inverse function of the regression of B on A.

The equating transformation will be based on samples of people who have been exposed to the tests to be equated. These samples will be used to construct distributions of test scores. It follows that these samples need to be as representative as possible of the population to which the tests are to be applied – so that, for example, the larger the sample used the better. Various designs are possible for constructing equating samples: the same sample could be used for the two tests, two different samples which are equivalent in that they are selected in the same way (e.g. simple random sampling) could be used, or two different samples which are not equivalent in this sense could be used. In the last of these cases, additional items are also answered, which can be used to equate the tests.

An example of a simple equating transformation is one which maps the percentiles of the distribution of sample scores on one test to the corresponding percentiles of the sample scores on the other test. If F is the sample distribution of scores on one test, and G is the corresponding distribution under the other test, then this transformation is simply $T(x) = F^{-1}(G(x))$. Often irregularities arising from the sampling nature are smoothed, and this has been done in a large variety of ways. If two non-equivalent samples have been drawn, then the equating procedures will involve two stages, each one equating a test to a common set of items which has been taken by both groups.

A simple approximation to this procedure is to match scores on the two tests in terms of how many standard deviations they are above or below the mean of that test. This yields a linear equating transformation. There are many versions of this approach based on the use of an equating test relating the two tests. They vary according to precisely what assumptions are made about the relationships between the three tests involved, and include the Tucker model, the Levine equally reliable model, and the linear frequency estimation model. Details of these are given in Kolen and Brennan (1995).

Yet a further alternative is to equate the tests via the item response theory model. In Chapter 3 we suggested that sometimes alternative transformed versions of the subject ability scores could be useful, and we illustrated with the example of the probability that a subject would get an item correct. In the simple two-parameter item response model this would be $\pi_i(y_r) = [1 + e^{-a_i(y_r - \beta_i)}]^{-1}$. Summing these scores over the items in a test A yields the item pool score for this subject:

$$T_A = \sum_i \pi_i(y_r) = \sum_i \left[1 + e^{-a_j(t - b_j)}\right]^{-1}$$

and this can be matched to the corresponding score, T_B, for test B.

6

Measurement in medicine

Such pretensions to nicety in experiments of this nature, are truly laughable!
They will be telling us some day of the WEIGHT of the MOON, even to
drams, scruples, and grains – nay, to the very fraction of a grain! – I wish
there were infallible experiments to ascertain the quantum of brains each
man possesses, and every man's integrity and candour: – This is a
desideratum in science which is most of all wanted.

Harrington (1804, p. 217)

6.1 Introduction

Medical measurement is a vast subject. One could easily write many books about it – and, indeed, many such have been written. One reason for its breadth is that it covers many different kinds of objects and conditions. Thus, for example, there are direct measurements of physiological systems, such as blood pressure, erythrocyte sedimentation rate, progesterone level, or size of a tumour or lesion. Then there are indirect derived measurements based on responses to questionnaires, such as depression and quality of life scales. Finally, there are measurements of aggregates of people: the overall health of a population, the mortality or morbidity rate, and so on. This chapter examines the nature of these scales rather than the precise details. The references will serve as guides to some of the thousands of measurements and scales which are used.

The representational–pragmatic distinction takes on a particular force in medical measurement. In Chapter 1 I quoted from Fayers and Hand (2002): 'The objective of the psychometric approaches outlined above might be characterised as *attempting to measure a single attribute using multiple items.* In contrast, clinimetric methods *attempt to summarise multiple attributes with a single index.*' The point is that in the first case, some underlying theory relates the observed or manifest variables to the underlying attribute to be measured, so that the value of the latter can be deduced from the observed values of the former. This is clearly a representational approach. In the second case, however, one explicitly defines the attribute to be measured in terms of the variables which can be directly observed. This is clearly a pragmatic approach. Alvan Feinstein explored these ideas in detail in the medical context (see, for example, Feinstein, 1987b). He was one of the first researchers, in a particular application domain, to articulate the distinction between the two aspects of measurement, though he tended to regard them as two distinct types of measurement procedure, for applications in distinct types of situation, rather than as aspects of a common problem.

6.2 The origins of medical measurement

6.2.1 The individual level

One can, of course, trace measurement in any domain back as far as one wishes (or, at least, back as far as recorded history will allow). In the case of medical measurement, the Ebers papyrus of around 1500 BC gives measures of quantities in prescriptions used in Egyptian medicine (Leake, 1952). Here we are talking about weights and volumes, so this is really physical measurement, but these sorts of use are the first in medicine. Something analogous was used much later when assessing the strengths of drugs by seeing how much of a drug was required to kill a cat. Porter (1995, p. 30) refers to Hatcher and Brody (1910):

> The 'progressive pharmacist' need only test each harvest of leaves by determining the minimal fatal dose per kilogram of cat. This should be called the 'cat unit.' Cats are easy to use, the authors explained, and their deaths 'do not affect the sentimental portion of the community to the same extent that the employment of dogs does.'

Problems arose, however, when, for example, it was found that digitalis was a mixture of active ingredients:

> It seemed that potential test animals were sensitive to different active ingredients. Already the 'frog unit' had fallen into disrepute because frogs tolerated the drug differently in summer and winter, and because they were often killed by its effect on their nerves rather than their heart. By 1931 there were more than seven hundred papers on the quantitative testing of digitalis, involving a variety of animals. Joshua H. Burn, one of the leaders of the field, remarked in 1930 that biological assay 'remains a subject for amusement or despair, rather than for satisfaction or self-respect. We have cat units, rabbit units, rat units, mouse units, dog units, and, latest addition of all, pigeon units. The field of tame laboratory animals having been nearly exhausted, it remains for the bolder spirits to discover methods in which a lion or elephant unit may be described' (Burn, 1930).

Methods of standardization paralleling work in physics and engineering were explored. The natural variability within an animal species was reduced by breeding standard strains, and preparations of drugs were created which could serve as the standard, with samples being sent to other laboratories. As with physical standards, such as the 'standard metre' described in Chapter 7, these standards were maintained in carefully controlled conditions. However, the length of even the standard metre changes over time, and the effect is even more pronounced with chemical and biological substances. Nevertheless, Joshua Burn was confident enough to write: 'The adoption of biological standards of this kind brings the estimation of biological properties into the same position as the measurement of length and weight' (Burn *et al.*, 1937).

Quantitative measurements of symptoms and physiological phenomena were slower to develop, although there are occasional early isolated examples, such as the number of days for which a fever lasts. Shryock (1961), in one of the few historical studies of measurement in medicine, refers to Erasistratus as measuring respiration and Heophilus as measuring pulse rate – using a water clock, apparently (Garrison, 1929).

As with many other domains, we find the early attitude that measurement cannot have much to say, in this case about health. Thus Cabanis (1798) quotes Hippocrates as saying: 'You can discover no measure, no weight, no form of calculation, to which you can refer your judgements in order to give them absolute certainty. In our art there exists no certainty except in our sensations.' On the other hand, this might be taken as implying that some, at least, had considered or were considering taking measurements – why else criticize the practice? The attitude that classification was the only level of 'measurement' which could be used in medicine seems to have persisted throughout the Middle Ages. However, an increased interest in quantification in areas such as physics and astronomy began to diffuse outwards – there was not the rigid division into different investigative areas in those days – and such ideas also began to be applied to the medical area. For example, Nicholas Cusanus in 1450 referred to pulse rate and respiratory frequency and the measurement of the weight of bodily fluids. It is interesting to note Shryock's (1961, p. 89) comment on such advances, written some 40 years ago, in the light of the discussion of indirect medical measurement, and in particular quality of life measurement, below. He says: 'These efforts were sometimes naïve, as when it was assumed that the "health" of an individual was a quality even as was weight, and that the former like the latter could be measured on a scale. We now speak of "positive health" as something more than "the mere absence of disease," but no one has yet come up with a means for measuring it.' Things have clearly progressed in the 40 years since that was written, and now people have come up with means for measuring health, though all such measures have a heavy pragmatic component, making explicit the definition of what, in any given context, is meant by 'health'.

Resistance to the application of measurement was not restricted to the clinical aspects. Keir (1789) remarked that 'if, instead of aiming at mathematical demonstration, and certainty, we be satisfied with examining the various modes, in which chemical phenomena may be viewed and arranged; and with comparing these … constantly distinguishing certainty from probability, and hypothesis from demonstrated truth … our minds will be ever open to receive the improvement that time and repeated experiments alone can produce' (quoted in Golinsky, 1995). Gradually, however, the power of the quantitative representation in the physical sciences and, of course in engineering as manifest in the industrial revolution, became apparent.

The success of quantitative methods in the physical sciences led at least some to predict similar success in medicine: 'As early as 1540, it was declared that medicine could be made as perfect as Copernicus had made astronomy' (Shryock, 1961, p. 89). Perhaps this is an early precursor of what has been termed the iatromathematical fallacy. For example, Feinstein (1977) wrote:

Iatromathematical enthusiasts could make substantial contributions to clinical medicine if the efforts now being expended on Bayesian and decision-analytic fantasies were directed to the major challenges of algorithmically dissecting clinical judgement, based on the way the judgements are actually performed. Instead, however, the enthusiasts usually become infatuated with the mathematical processes and with the associated potential for computer manipulations, so that the basic clinical challenges become neglected or evaded.

This is the same Feinstein who went on to establish a substantial reputation in medical measurement (e.g. Feinstein, 1987a, b).

The major obstacle in the path to more widespread use of quantification in medicine was the lack of understanding of physiological and disease mechanisms. This meant that only the crudest relationships between a measured attribute and an outcome could be contemplated. However, the impact of the success of measurement technologies in other disciplines had an increasing impact and this, combined with the increasing view of the body as a machine, naturally stimulated the quantitative perspective. Thus, for example, we find Galileo measuring pulse and body temperature, and later Boyle and Hooke carrying out experiments on blood circulation and respiration. We see that at least some were aware of the need for experimentation and the objective assessment of evidence: 'Giorgio Baglivi explained to physicians in 1696 the reasons for striking progress in astronomy. There had been in that field, he noted, various hypotheses on which astronomers differed. But they all agreed on the prediction of eclipses, since there was no refuting of verifiable, numerical data. Ergo, medical men should seek similar data, in order to minimise their divergencies in theory and to arrive at dependable knowledge' (Shryock, 1961, p. 92, referring to Baglivi, 1696).

Of course, there are limitations to what quantification can achieve or what can be quantified when the underlying theories are very weak – and this even more so at a time when it was uncertain precisely what should be involved in an underlying theory. For example, it was still not clear whether astrology should play a role. Inevitably, resistance to the possibilities of measurement continued: 'Measurement, declared so distinguished an authority as Goethe, could be employed in strictly physical science, but biologic, psychologic and social phenomena necessarily eluded the profane hands of those who would reduce them to quantitative abstractions' (Shryock, 1961, p. 93). Even today such attitudes are not unknown. I, of course, would argue that by measuring things we take the conceptual high ground – 'we hardly recognise a subject as scientific if measurement is not one of its tools' (Boring, 1929) – but those who have resisted quantification have asserted the opposite, claiming, as Shryock (1961, p. 93) again elegantly puts it: 'Here one detects the feeling that measurement somehow robs human phenomena of all mystery or beauty, and denies to investigators the satisfactions of age-old sense impressions and of intuitive understanding. Such feeling usually appears within any discipline when it is first threatened, as it were, by quantification. Dr Stevens terms it, in relation to current psychology, "the nostalgic pain of a romantic yearning to remain securely inscrutable".'

Because of this sort of attitude, even those medical measurements which could be taken, and which were usefully indicative, were often not made general use of, and it was not until the middle of the nineteenth century that more general objective measurements of bodily characteristics were employed. Thus graduated hypodermic needles were introduced in the 1840s, measurements of temperature and pulse rate were only adopted in general practice after 1850, and an instrument for measuring blood pressure was invented in 1887 (Shryock, 1961, p. 100).

The gradual encroachment of measurement into medicine at the individual patient level was assisted by the adoption of the idea that illness was caused by particular diseases, rather than being a general disorder. After about 1800, as more reliable and larger-scale data began to be compiled, it was also boosted by the clear successes of measurement at the aggregate level, discussed below, where epidemiological advances led to progress which was difficult to deny. The work of Florence Nightingale (1820–1910) in Scutari, for example, demonstrated, through statistical plots, how unhygienic conditions, rather than military action, were related to deaths amongst the soldiers.

Amongst the main players in promoting the value of statistical computations in a clinical context was Pierre Louis. Louis (1835) carried out what might best be termed a clinical trial of bloodletting in which he treated 78 cases of pneumonia, bleeding them at different stages of the disease. He found that bleeding in the first two days halved the duration of the disease although, unfortunately, it also doubled the risk of death. In fact, of course (I can see statistical readers raising their hands in horror), there are all sorts of statistical complications here, including non-compliance amongst those bled later (Freedman, 2002). Although compelling, inevitably not everyone agreed with the conclusions he drew: therapy was not a matter of aggregates, but of individuals, and, moreover, it was not clear how large a difference between the two treatment groups should be taken as indicative of a real difference, rather than a chance effect. We thus have two competing schools of thought. In one, promoted by Louis, doctors should based their decisions on careful analysis of many recorded medical facts, describing both the patient at hand and others. In the other, as promoted by Risueno d'Amador, for example, doctors should base their decision on expertise handed down from previous practitioners, perhaps modified by experience acquired over an extended period of time. This debate still rumbles on, in the form of the *statistical versus clinical* debate discussed in Chapter 5. We argued there that, despite the longevity of the debate, it is becoming increasingly clear that the formal approaches are superior.

Of course, it is a *post hoc* simplification to describe only two schools. Another (developed by Claude Bernard – see Murphy, 1981) argued that experiment should be used to remove all uncertainty about the cause of a disease – so that he was opposed to the notion of averages in medicine. Be that as it may, the proponents of rigorous data collection set the stage for formal applications of probability theory, though these applications were a long time coming, at least compared with similar issues in gambling. Perhaps this is not surprising: the throw of a six on a die is clear and unambiguous, whereas the occurrence of a case of a disease hinges on recognition of often ambiguous and unclear symptom patterns.

The gradually increasing enthusiasm for such methods is illustrated by another quotation from Shryock (1961, p. 103):

> Bouillaud of Paris declared in 1836 that statistical method was revolutionising medicine: a conjectural art was becoming an exact science. Henry Holland of London stated, four years later, that 'through medical statistics lies the most secure path into the philosophy of medicine'; and by 1842, J.F. Double of Paris claimed that all critics had come to admit the value of such data in therapeutic studies. Last, but not least, W.P. Alison of Edinburgh assured British scientists in 1855 that many of the most important questions in medicine could be investigated by statistics 'and in no other way'.

The English astronomer Herschell reported about 1850 that: 'Men began to hear with surprise, not unmixed with some vague hope of ultimate benefit that not only births, deaths and marriages, but ... the comparative value of medical remedies, and different modes of treatment of disease ... might come to be surveyed with the lynx-eyed scrutiny of a dispassionate analysis.'

Florence Nightingale was not alone.

6.2.2 The aggregate level

In certain medical situations, the objects of interest are complex organizational entities, perhaps properly described as aggregates of more elementary entities. Examples are medical practices, hospitals and populations. One might measure the properties of such aggregates by using summary measures of the elements. For example, one might measure the overall health of a population by means of a count or an average of the health of the individual members. The relationship between descriptions of the individual components and descriptions of the overall collective is discussed in Chapter 8. Desrosières (1998, p. 82) briefly describes this in a medical context:

> Illness and how a doctor treats it constitute a unique event, and this singularity was long monopolized by the medical community, reluctant to accept any form of categorization and summation that might violate the 'privileged communication' between doctor and patient. But this resistance soon gave way when disease became a collective problem calling for general solutions: epidemics, and how to prevent them. In this case, the sometimes urgent need for measures of public hygiene to prepare for or halt an epidemic involved viewing the epidemic as a whole, the constant causes of which – that is, the *factors* that encouraged it to spread – were being urgently sought. To this end, calculation of average death rates of segments of the populace classified according to different criteria (region of Paris, type of dwelling, level of income, age) provided data for possible preventive policies.

Governments nowadays collect a variety of health statistics, so that they can assess the well-being of the population and the quality of the services being provided. At the population level, the information collected includes details of births and deaths, including causes of deaths, notification of certain diseases, details of sickness absences and claims, details of industrial accidents, information about patient contact with the health service, and occasional surveys on a wide variety of health-related topics (e.g. dental health, infant feeding, use of contraception, smoking). At the hospital level, data are collected on services provided. Information on some topics is also collected in the General Household Survey. Further details are given in Kerrison and Macfarlane (1999). In recent years, as in so many other areas, there has been a tendency to set targets (see Section 8.2) which various measures should strive to reach at some specified point in the future. For example, in *Revitalising Health and Safety* (http://www.hse.gov.uk/revitalising/what_is/index.htm), the UK government and the Health and Safety Commission set three targets for the year 2010:

- [to] reduce the number of working days lost per 100,000 workers from work-related injury and ill health by 30% …;
- [to] reduce the incidence rate of cases of work-related ill health by 20% …;
- [to] reduce the incidence rate of fatalities and major injuries by 10%.

6.3 Direct measurement in medicine

At the low level of collecting data on physical symptoms, a wide variety of fairly direct measurements are made in medicine. These include things such as measurements of blood pressure, pulse rate, erythrocyte sedimentation rate, simple visual analogue scales for symptoms such as pain, severity scales with given categories (e.g. mild, moderate, severe), lesion or tumour size, temperature, and a host of other measures. We see, even from this brief list, that some of the direct measurements appear to be representational, in the sense that they are measures of clearly defined physical phenomena. However, this does not mean that they are representational in terms of the attribute being measured. The diameter of a tumour is a length measurement – a paradigmatic representational measurement – but it is really being used as a measure of how serious the growth is. Length here is being used as an indicator variable, and by (pragmatic) definition as a measure of severity.

This last example illustrates that, as in so many other measurement situations, many of the procedures for measuring medical phenomena have translated the magnitude of the attribute into a length (e.g. the length of a column of mercury for body temperature and blood pressure, or the position on a line segment for visual analogue scales). More recently, however, advances in electronic instrumentation have meant that direct digital readouts from electronic sensors are increasingly typical. This, coupled with direct recording from electronic measuring instruments, is leading to an improvement in data accuracy. For example, the phenomenon of digit preference in blood pressure measurements illustrated in Chapter 4 will be less likely with direct electronic readouts and recordings.

Many elementary direct measurements in medicine are subjective responses to simple questions: 'On a numerical scale on which 0 represents no pain at all and 100 represents intolerable pain, how would you rate the pain?' These are often supplemented by interpretive aids, such as visual analogue scales or semantic differential scales. For example, ask someone to rate the degree of comfort with some medical device by placing a mark on a line which runs from 'extremely uncomfortable' at one end to 'completely comfortable' at the other end. Another alternative is to give a categorical scale with several points: not at all, mild, moderate, severe, intolerable, for example. On the subject of such strategies, McDowell and Newell (1996, p. 15) say:

> Subjective measurements are, of course, little different from the data collected for centuries by physicians when taking a medical history. The important difference lies in the standardisation of these approaches and the addition of numerical scoring systems ... Despite these potential advantages of subjective indicators, several problems delayed their acceptance. Compared to the inherent accuracy of mortality rates as a source of data, asking questions of a patient seemed to be abundantly susceptible to bias.

As we have seen in Chapter 3, there are strategies for making such simple subjective scales more reliable, but scales based on multiple questions are generally preferable, for the reasons described in Chapter 3.

A fundamental issue with subjective scales is the distinction between (a) the idea that a single scale is appropriate for everyone and (b) the idea that different scales are appropriate for different people. It is a common experience that different people will interpret even the same question differently. This can be overcome to a large extent when there is the opportunity for training and calibrating the respondents. For example, in many situations involving interviews (e.g. the Present State Examination in psychiatry) it is possible to explain in detail how different types of responses should be coded. However, in many other situations, and especially if the patient is responding directly, this will not be possible.

6.4 Indirect measurement in medicine

Indirect measurement is representational to the extent that it is based on an underlying theory relating the observed variables to the underlying attribute which is the target of the measurement exercise, or to the extent that it is based on observed empirical relationships in data. Indirect measurement is pragmatic to the extent that external criteria are used to constrain the way the observed variables are combined to yield the measurement of the attribute of interest. Thus both representational and pragmatic aspects feature in indirect measurement. It is probably fair to say that in medical measurement the emphasis is on the pragmatic aspects or on empirical relational aspects, rather than on theoretically based representational aspects. This is in contrast to, for example, psychology and education, with their representational tools such as factor analysis and item response theory designed to extract the value of the underlying attribute from the observed values of individual item scores.

A particular example of a scale developed on an empirical representational basis is Leavitt's Back Pain Classification Scale. Of this scale Leavitt (1983) says: 'Why this particular set of verbal pain descriptors works as discriminators and others do not is unclear from research to date. The shared variance of pain words with MMPI items is only 21%, and does not seem to fit any particular pattern in terms of the sensory and affective divisions of pain experience.' More generally, those branches of medicine which have developed in terms of syndrome constructs (constellations of symptoms which tend to occur together) are based primarily on empirical representational measurement: one will try to condense the symptoms comprising the syndrome into a single overall measure which represents that syndrome. Traditional psychiatry is the classic example.

The Apgar score (Apgar, 1953) of the clinical condition of a newborn baby provides a nice illustration of an explicitly pragmatic indirect measurement in medicine. This scale takes values 0, 1, or 2 for each of colour of complexion, heart rate, respiration rate, reflex response to nose catheter, and muscle tone, and sums these to yield a total score. The scoring procedure is repeated a specified number of minutes after birth, with scores in the range 7 to 10 being considered normal.

Many concepts in medicine and health must, of course, of necessity be measured indirectly. This is because, at the higher level, complex phenomena with many interrelated factors and influences are involved. With sociomedical measurement, such as social functioning, physical disability, and quality of life, there is no alternative to an indirect and largely pragmatic approach. An extreme example of this is the measurement of 'health' itself, and in this context we should perhaps make the cautionary remark that the freedom inherent in the pragmatic approach should not encourage one to expect miracles: 'The WHO's definition of health in terms of "physical, mental, and social well-being, and not merely the absence of disease and infirmity" has been criticised as unmeasurable and unrealistic' (McDowell and Newell, 1996, p. 12).

In common with other indirect measurement situations concerned with humans (e.g. psychological instruments), often profiles of scores are used rather than single indices. Thus, for example, many regard quality of life as fundamentally multidimensional (see Section 6.5.4), and question the wisdom of combining the dimensions into a single scale. In fact, the two issues – the multivariate nature of a phenomenon, and the wisdom of measuring it by a single index – are not really contradictory. As we have seen, in pragmatic measurement the aim is explicitly to combine different qualities into a single scale. The structure of this single scale defines the aggregate of the different qualities, as well as measuring it. And this single scale can then be used for administrative, monitoring and other purposes. Again, the Apgar score is a perfect illustration, being used to indicate whether resuscitation attempts are necessary. There are clear advantages in adopting particular scales by convention; then, at least, everyone is talking about the same thing. On the other hand, one is at liberty to disagree about the ways in which the individual components are combined, and to modify scales, produce alternatives and investigate their properties. There is certainly a tendency for researchers to invent their own measures, partly because there is often a lack of awareness of earlier work, partly because it appears so easy (though it is not – as Chapter 4 shows), and partly because some see merit in retaining some ambiguity about the concept being measured in a world where those concepts are in fact changing. Indeed, McDowell and Newell (1996, p. 10) remark that 'there is a historical tension in approaches to health between those who prefer to keep the concept somewhat imprecise, so that it can be reformulated to reflect changing social circumstances, and others who define it in operational terms, which often means losing subtle shades of meaning.' In general, different objectives are likely to require different ways of combining the simple measurements and raw observations. It would not be surprising if rather different scales were appropriate for measuring the severity of a disease, for measuring the disease with a view to prognosis, or for measuring the disease with a view to recommending treatment. On this sort of basis, indirect measurement constructs in medicine may be categorized in various ways. One is on the basis of the function to which the instrument is to be put. For example, Bombadier and Tugwell (1982) distinguish between the functions of diagnosis, prognosis and evaluation. A second kind of categorization is on the basis of the domains with which they are concerned: for example, ability to accomplish daily tasks, or extent of depression, or quality of visual function, or quality of life. Yet a third kind classifies measures on the basis of the ways the instruments have been constructed – is the instrument a rating scale, is it based on subjective or objective measures, is it fundamentally unidimensional or does it determine position on multiple dimensions? These various kinds of classification will be useful to different people using them.

There is another important approach to indirect measurement in medicine. This is based, perhaps not explicitly, on the notion of a gold standard. One could, for example, observe the effect of an illness on multiple functions, and integrate these to yield an overall measure of severity.

One should be aware that even relatively simple indirect measurements can produce unexpected complications. Thompson *et al.* (1991) describe a study of the perception of pain following wisdom tooth extraction. Initially, two direct measures of pain were taken:

(i) A visual analogue scale consisting of a 10 cm horizontal line with the words 'complete pain relief' on the left, and 'no pain relief' on the right. The pain relief score was the length in millimetres, from the left hand end of the line.

(ii) A categorical rating scale, with five categories, scored 0 to 4, and corresponding to no pain relief, a little relief, some relief, a lot of relief, and complete relief respectively.

Both measurements were taken for each of 71 patients, and the correlation was calculated. This was -0.815, described as 'high, as the researcher who had conducted the study expected, and negative because the categorical scale was scored 'backwards' relative to the visual analogue scale (with a low score corresponding to no pain relief)'. The researcher was also interested in the effect of the treatment over time, and defined this in a pragmatic and clinimetric way as the area under the curves of the pain scores over time. The time for which each subject gave pain measurements was a random variable (details are given in the paper). Now the area corresponding to the visual analogue scale is a measure of total pain and that corresponding to the categorical rating scale is a measure of total pain relief, so the researcher again expected to find a negative correlation. However, it turned out that although this correlations was indeed negative, it was puzzlingly small, being -0.039.

The explanation for this puzzle lies in the inverse relationship between the two pain scores used to produce the indirect 'total' values. It is most easily grasped by way of an extreme example. Suppose, firstly, that the two pain measures are perfectly negatively correlated such that the scores, if plotted, lie on a line with slope $-45°$. When we calculate total values over time, the two coordinates of each point will be multiplied by a common value, different for each point (because of the different times for which each patient is measured). Apart from certain very special cases, this will have the effect of destroying the perfect negative correlation: the transformed points will no longer lie on a straight line. Now suppose that the two pain measures are perfectly positively correlated, such that the scores, if plotted, lie on a line through the origin with slope $+45°$. Now, when we transform the points by multiplying them as above, they still lie on this line. That is, the perfect correlation is preserved. The configurations here were deliberately chosen to illustrate the ideas, but the same effect is to be expected for less extreme configurations. Essentially, what happens is that a negative correlation is likely to be reduced in absolute value, while a positive one may not be.

6.5 Some examples of indirect medical measurement

6.5.1 Physical disability

Measures of physical disability play an important role in legal claims for compensation following injury. Some such measures are highly specific – for example, the angle through which an injured knee can be bent. Others are more general, and are necessarily indirect measurements. Thus, measures of *Activities of Daily Living* (ADL) are illustrative of functional disability scales, and are generally aimed at fairly severe levels of disability, for example for long-term inpatients. During the 1970s the ideas were extended to tools for measuring less severe disability, for example as experienced by those who can manage in the outside community. Scales of this latter kind are sometimes called *Instrumental Activities of Daily Living* (IADL) scales, and are used, for example, in surveys of minor disability in the community. IADL scales are thus intermediate between ADL scales which measure physical disability and scales which measure social functioning. Because IADL scales measure how well subjects can perform everyday activities they often involve culture-specific and social role-specific questions. One also has to be careful about how questions are phrased: 'do you wash the dishes?' may well elicit a different response from 'can you wash the dishes?'. Both forms of question have strengths and weaknesses, and either may be an appropriate form.

Most scales whose use persists have several or even many variants, as researchers refine them, adapt them to slightly different purposes or target groups, or respond to criticisms of the earlier versions. This can make comparison difficult. Examples of ADL scales are the *Barthel Index* (Mahoney *et al.*, 1958) and the *Index of ADL* (Katz *et al.*, 1963, 1970). An example of an IADL scale is the *Functional Activities Questionnaire* (Pfeffer *et al.*, 1982).

The Barthel Index measures the level of nursing required by long-term hospital patients, and is based on medical records, direct observation or self-completion. It is thus at the more severe end of the scale and cannot be used to detect minor disability. The original version had ten items covering personal care and mobility (e.g. feeding, washing, dressing), each indicating the extent of assistance required to perform some task, with the final score being a sum of the separate scores of these items. The scores are accompanied by an extensive set of instructions for how the separate items should be interpreted. For example, under 'dressing', a score of 5 is described as: 'Patient needs help in putting on and removing or fastening any clothing. He must do at least half the work himself. He must accomplish this in a reasonable time. Women need not be scored on the use of a brassiere or girdle unless these are prescribed garments.' There are many variants of the Barthel Index, with popular versions of the basic form including those of Collin *et al.* (1988) and Granger *et al.* (1979).

Katz based his Index of ADL on the observation that, as activities are lost, the most complex are lost first, and when they are regained, the simplest are regained first. Thus, to a limited extent, this scale is based on a theoretical model (or, at least, an empirical relational system). Six activities are combined in the index (bathing, dressing, toileting, transferring from bed to chair, continence, and feeding), each being rated on a three-point scale. The level of accomplishment to achieve each possible score on each activity is described in detail (e.g. Dressing: 'gets clothes from closets and drawers – including underclothes, outer garments and using fasteners (including braces if worn)' with the middle of the three possible scores being described as 'Gets clothes and gets dressed without assistance except for assistance in tying shoes'). Performance on each of the six scores is classified as 'dependent' or 'independent' (according to which of the three scale points the patient falls into) and then either the number of activities scoring 'dependent' is reported (hence a scale 0 to 6) or an ordinal categorization is used.

Like the Index of ADL, the Functional Activities Questionnaire (Pfeffer *et al.*, 1982) is also based on the notion that different activities have some kind of intrinsic order of complexity. However, the Functional Activities Questionnaire is an IADL scale, and is therefore based on ten higher-level activities, including such things as 'Writing cheques, paying bills, balancing chequebook, keeping financial records' and 'Remembering appointments, plans, household tasks, car repairs, family occasions (such as birthdays or anniversaries), holidays, medications'. Each item has an associated list of possible levels of response, with precise descriptions of these being given. A score ranging from 0 (independence) to 3 (dependence) is given for each item and the total scored is given by their sum. Many of these items in the Functional Activities Questionnaire illustrate the culture-specific nature of the questionnaire. The *OECD Long Term Disability Questionnaire* (McWinnie, 1981) represented an international effort which sought to overcome this, at least partially, by using questions of a more general nature, such as 'Is your eyesight good enough to see the face of someone from 4 metres? (with glasses if usually worn)' and 'Can you speak without difficulty?'.

6.5.2 Psychological health

There are many measures of psychological health, and measures of many different components of it, including such things as depression, anxiety, minor psychiatric morbidity, and so on. One broad distinction is between measures which cover symptoms, measures which ask how the subject feels, and measures which seek to combine these. Inevitably, all such measures are necessarily indirect. Sometimes a gold standard is available, in which one uses external criteria (e.g. an intensive psychiatric interview) to classify subjects into classes or assign subjects a severity score and then uses tools of supervised classification or regression to construct a rule which gives the probability that a new subject will belong to each class or predict their severity score, based on easily measured variables. For simplicity, we shall talk in terms of classification, but the regression context, with a continuum of severity scores, is very similar. In the second type, there is no gold standard classification, so that the measurement must be constructed from the available variables. For this second type one has to have some notion of an underlying theory, even if it is a very weak one, in order to decide which variables to include and an appropriate model form for combining them, or else one has to have pragmatic motivations for choosing what items to use and combine. For methods based on a gold standard it is not necessary to have such an underlying theory, and such methods can be entirely empirical. In principle, measures of this type can include predictor variables which seem to have a peripheral relationship to the target classification or score if these variables turn out to have predictive power, and this can make such models unappealing. A further potential problem with the first type of strategy is that one does need clear notions of the classes. If, for example, the two classes are, on the one hand, a healthy class, and, on the other, patients who are depressed, or anxious, or have other emotional problems, then it may not be clear precisely what the scale is measuring. (A similar issue arises in the rather different context of credit scoring (Section 5.8), where the definition of what constitutes default on a financial product is to some extent arbitrary.) In this regard, Dohrenwend *et al.* (1980, p. 1229) comment: 'As might be expected given the actuarial procedures and undifferentiated patient criterion groups used to construct them, none of the screening scales reflects a clearly specified conceptual domain. Thus, there is no ready correspondence between the content of the scales and conceptions of major dimensions or types of psychopathology such as mania, depression, hallucinations, or antisocial behaviour.' In view of this, one interpretation of such scales is that they are analogous to body temperature or erythrocyte sedimentation rate – simply a general measure of health, and not a diagnostic tool.

One popular example of such a tool is the *General Health Questionnaire* (Goldberg, 1972). This is a self-completed instrument aimed at detecting acute minor psychiatric illness, primarily in surveys of the general population or outpatients. There are many versions (both in English and in translation into many languages), and many evaluation studies have been carried out. The main version has 60 items, including questions such as 'Have you recently been feeling perfectly well and in good health?', 'Have you recently been getting up feeling your sleep hasn't refreshed you?', 'Have you recently spent much time chatting with people?', and 'Have you recently felt that life is entirely hopeless?'. Each item is scored on a four-point scale (better than usual, same as usual, worse than usual, much worse than usual), scored either 0, 1, 2 and 3, respectively, or 0, 0, 1 and 1, respectively, and then summed to give a final overall score. The items were chosen by a complex item selection process from a large database of questions. Having so many as 60 items means that the instrument has good reliability, but it also means that there are circumstances where a reduced version would be preferable, and versions with 30, 20 and 12 items are also in common use.

Turning to depression, the ambiguity of the term is indicated by Mayer (1978), who remarks that it can be thought of as 'an affect, emotion, mood, symptom, disease, or diagnostic label, and may refer to both normal and pathological events that can be transient or enduring'. The term also spans the range from minor feelings of sadness to major depressive periods (even life-threatening – most suicides are attributable to depression). There are also various types and categorizations of depression (Dunn *et al.*, 1993). In general, depression may have multiple causes, so that it is defined as a symptom constellation, a syndrome, rather than in functional terms. Moreover, a wide range of different symptom patterns are taken as indicative of depression, so that an effective scale must cover multiple dimensions. This gives scope for many different scales of depression measurement. Depression scales can have particular practical measurement difficulties, with a subject's ability or motivation to respond being affected in severe cases, and also with distortion arising from other cognitive deficits, with cultural gender bias, and so on.

Both self-assessment instruments and clinician-assessment instruments have been developed. The self-assessment scales are generally quick and rough, useful for detecting that someone is exhibiting the symptoms of depression but not distinguishing depression from other causes with the same manifestation. They are used in epidemiological contexts.

The *Beck Depression Inventory* (Beck *et al.*, 1961, 1979) is an example of a self-administered instrument (though it was originally designed to be completed by a trained interviewer). It takes an inclusive perspective on depression, covering 'affective, cognitive, motivational, and behavioral components' (Beck, 1967). Each of 21 symptoms is rated on a scale from 0 to 3 (with the last being the most severe). For example, the symptom 'social withdrawal' has response categories 0 = 'I have not lost interest in other people', 1 = 'I am less interested in other people than I used to be', 2 = 'I have lost most of my interest in other people', and 3 = 'I have lost all of my interest in other people'. These individual scores are then added to yield a final score.

The *Hamilton Rating Scale* (Hamilton, 1960) is perhaps the most popular clinician-completed depression instrument. The depression score itself is comprised of 17 symptoms, with scores of 0 to 2 used for aspects that cannot be finely quantified, and scores of 0 to 4 being used for the others. Again higher scores are more severe (e.g. the 0–2 scale has 0 = symptom absent, 1 = slight or doubtful, and 2 = clearly present) and the final score is the sum, which ranges from 0 to 52. Administering the instrument can be an elaborate procedure, sometimes using two raters, who later discuss and resolve differences, and sometimes also including information from others close to the subject. As with many of the scales which were developed early and which have persisted over time, there are many variants.

6.5.3 Pain

The measurement of pain is particularly challenging. Like depression, pain is subjective and also has multiple dimensions (intensity, duration, etc.). Depression, however, is generally seen as a response, while pain is seen as a cause. That is, we try to infer the strength of the pain a person is experiencing by studying their response to that pain, whereas depression is itself seen as the response, the depth of which is to be measured. Early instruments for measuring pain were simply based on the subjective response of the patient but more recent ones are based on the theory that pain has both physiological (sensory) and psychological (reactive) components. This illustrates the fact that attempts to develop instruments for measuring pain have led to careful thought about the nature of pain, providing an example of the synergistic effect in which refining measurement tools leads to deeper theoretical development and vice versa.

There are two main classes of pain-measuring instruments: those that match the pain to a description of it and those that determine its effect on the subject. Tools in the former class may be in terms of simple three-point intensity scales (mild, moderate, severe), or be based on visual analogue scales, or ask the subject to rate the pain on a numerical scale, or even ask the subject to match their pain against a brief description from an ordered list of descriptions. Tools in the latter class include measures of the extent to which pain prevents some action being carried out, the quantity of medication taken, and observation of behaviour patterns indicative of pain, such as grimacing, gasping, or massaging an injured spot.

The *McGill Pain Questionnaire* (Melzack, 1975), the most popular pain-measuring instrument, is an example of the first kind of scale. It is based on Melzack's characterization of pain as having three components: sensory-discriminative, motivational-affective and cognitive-evaluative (Melzack and Wall, 1988). Thus it has a sound theoretical foundation. The scale consists of 78 words (selected from a larger group when the questionnaire was designed) describing pain, sorted into 20 groups, with the words in each group ordered according to increasing intensity of the aspect of pain they are describing. For example, three of the groups are {flickering, quivering, pulsing, throbbing, beating, pounding}, {sharp, cutting, lacerating}, and {nagging, nauseating, agonizing, dreadful, torturing}. Each word is assigned a weight (based on an analysis of previous patient ratings). A new subject then selects the single word from each group which best describes their pain, choosing none if no word in the group fits. The overall score is the sum of the weights (or, sometimes, the sum of the rank of the word in its group, or other variants).

The *Oswestry Low Back Pain Disability Questionnaire* (Fairbank *et al.*, 1980; Baker *et al.*, 1989) is an example of the second type of pain-measuring instrument, one which determines the level of pain by the effect it has on the subject's functioning. The scale comprises 10 six-point subscales, each scored 0 to 5, with the first measuring pain intensity directly (on a scale ranging from 'I have no pain at the moment' to 'The pain is very severe at the moment'), and the other nine measuring the extent to which activities are disabled. For example, the 'Lifting' subscale has response categories from 'I can lift heavy weights without extra pain' to 'I cannot lift or carry anything at all', and the 'Sleeping' subscale ranges from 'My sleep is never disturbed by pain' through to 'Pain prevents me from sleeping at all'. The single item which most matches the pain level is chosen from each subscale, and the overall score is the sum of the scores of the ten subscales.

Coincidentally, while I was writing this section of the book, the death was announced of Patrick Wall: 'widely acknowledged to be the world's leading expert on pain; his research over more than 50 years transformed thinking in the field and led to the development of new forms of pain relief' (*The Daily Telegraph*, 23 August 2001, p. 27). Wall founded the journal *Pain* in 1975, of which McDowell and Newell (1996, p. 377) say: 'Finally, the reader gains the impression that the many researchers working on the problem of pain measurement, who come from widely differing disciplines, benefit from the existence of a dedicated journal (*Pain*) that has published many of the leading articles on pain measurement. By contrast, other fields of measurement lack a clearly identifiable journal that deals comprehensively with measurement and conceptual issues.'

6.5.4 Health-related quality of life

In recent years there has been a growing interest in quality of life measures. The reasons for this growth in interest are various. Firstly, the success in other areas of medicine has led to the recognition that merely prolonging life, no matter how miserable, is not enough.

Secondly, with chronic conditions in which the symptoms cannot be cured, a more global assessment of the merits of treatment is needed (Ebbs *et al.*, 1999). Thirdly, at the very least, quality of life measures are useful as secondary instruments for comparing treatments which are otherwise equivalent, in the same way that one may choose between equally efficacious treatments which have different numbers of side effects. They are particularly relevant for those patients with a terminal illness and for whom extending life is not possible, so that improving the quality of the remaining life becomes paramount.

We should distinguish between *health-related quality of life*, which describes the social and emotional aspects of health, and simple *quality of life*, which simply describes how satisfied people are about their circumstances. Of course, those working in health-related quality of life will often abbreviate the term simply to 'quality of life' (e.g. Fayers and Machin, 2000), and we will adopt this abbreviation here.

Quality of life, perhaps even more than many of the other attributes described in this section, lacks a convincing and generally agreed definition. Rogerson *et al.* (1989) have noted that no universally agreed list of components of quality of life has been (or probably could be) agreed, let alone a way of combining these components. Various broad definitions have been given. In 1834 the utilitarian philosopher Jeremy Bentham defined well-being as 'the difference in value between the sum of pleasures of all sorts and the sum of pains of all sorts which a man experienced in a given time' (Bentham, 1983). This definition clearly leaves a huge number of pragmatic choices to be made before it will yield a useful measurement! The World Health Organization Quality of Life Group definition of quality of life is: 'Quality of life is defined as an individual's perception of their position in life in the context of the culture and value systems in which they live and in relation to their goals, expectations, standards, and concerns. It is a broad ranging concept affected in a complex way by the person's physical health, psychological state, level of independence, social relationships, and their relationships to salient features of their environment' (WHOQOL Group, 1993). Bowling (1995) gives: 'health-related quality of life is defined here as optimum levels of mental, physical, role (e.g. work, parent, carer, etc.) and social functioning, including relationships, and perceptions of health, fitness, life satisfaction, and well-being'. There is also the issue of the subjective/objective aspects of any definition. Ziller (1974) and Rosenberg (1992), for example, are amongst authors who have argued that subjective aspects of quality of life should be included in any instrument (and see, for example, Guyatt *et al.*, 1989a, b; Ruta *et al.*, 1994). Formal methods of quantifying subjective feelings have been developed. For example, patients may list the areas of quality of life they regard as most important (prompted by open-ended questions), rate each of these on a visual analogue scale, and also indicate the relative importance, to overall quality of life, of each of the components. The extent to which the scale should tap subjective aspects depends on one's aims.

It is clear from this that many different pragmatic definitions can be (and have been) created. However, the one thing that most researchers would agree on is that quality of life is a multidimensional concept. Because of this, many prefer to use a profile summarizing its different aspects, especially in a clinical context. This is all very well, but if we are to be able to use the concept to make overall social or economic decisions, it is necessary to reduce it to one dimension – essentially using a pragmatic (or, one might say in this context, clinimetric) approach. In the context of such economic decisions, a related measure which is sometimes used is the 'quality of life year' or 'quality-adjusted life year' (QALY), which is a year of maximum quality of life, scaled to equal 1. Then an overall measure of quality of life benefit from a treatment is given by the number of years of life gained through using the treatment multiplied by the average quality of life over those years – very much a pragmatic definition. As might be expected, such a measure is not without its critics. Bowling 1995, p. 13) comments: 'There is no evidence that the judgements determining the QALY for a particular condition bear any relationship to real judgements faced by patients suffering from that condition.' More direct methods have also been developed for QALY measurement, in which information on the quality–length trade-off is elicited from the patients (using standard betting techniques used for eliciting utility measures).

An illustration of such scales is given by the *Quality of Life Index* (Spitzer *et al.*, 1981), which provides a single overall summary measure of the quality of life of patients with cancer or other long-term diseases. The self-administered version consists of three statements in each of five areas: activity (whether the patient works, studies, etc.), daily living activities, health, social support, and outlook on life. For example, the statements on 'support' are: 'I have good relationships with others and receive strong support from *at least one* family member and/or friend', 'The support I receive from family and friends is limited', 'The support I receive from family and friends occurs infrequently or only when absolutely necessary'. The three statements in each area are numbered 0, 1, or 2 consecutively, and the responses summed to yield the overall score.

6.5.5 Measuring spinal deformity in osteoporosis

We would not want to give the impression that indirect models only arise in measurement problems which have a psychological component, so here is a very physical example. It is indirect because the overall measure is an aggregation of many constituent measurements.

Osteoporosis is a condition in which the bones weaken and crumble. It can be diagnosed by bone density measurements or by noting vertebral crush fractures. Originally such fractures were detected on a judgemental basis, by an expert radiologist studying an X-ray. However, as seems to be a theme running through this book, the judgemental process was subject to considerable unreliability, both within individual radiologists and between different radiologists, and was usurped by a more formal process in which measurements taken on the X-rays were used in a formal rule.

The vertebrae of most interest in this context are the lower thoracic vertebrae and the lumbar vertebrae, since these carry most weight and are more susceptible to crushing. Various measures may be taken on each vertebra, including the anterior height, the posterior height, and the mid height, and elaborate procedures are defined for measuring these as accurately as possible – any layperson who has looked at an X-ray will know that interpretation is not easy, so that precisely defined measurement procedures are important. For example, because of overlapping images from the edges of the endplates of the vertebrae, the mid height might have to be calculated from one, two or three distinct lines in the X-ray at each of the top and bottom of the vertebrae. There are other problems, such as subtle distortions arising from the position of the spine when the X-ray is taken.

Once the various characteristics of each vertebrae have been determined, various statistical models can be built to decide if any given vertebrae is abnormal. After all, 'small' alone is not enough: it has to be in the context of the other vertebrae in that spine. These models also have to take account of the natural shape of the spine: for example, the fact that the lower vertebrae are larger than the upper ones, that the mid height is typically smaller than the anterior and posterior heights, and the anterior height is smaller than the posterior height (except for lumbar vertebrae L4).

With all these different measurements to integrate into a single index of deformity, an indirect model is necessary. Various such models have been developed. An example is that of Minne *et al.* (1988). They standardized all the vertebrae by the corresponding (anterior, posterior, mid) height of the fourth thoracic vertebrae and fitted cubic curves to the standardized height versus vertebrae number. A deformation was defined as occurring if any of the standardized heights was less than the minimum value in a sample of spines from 110 healthy subjects. Summing the difference between this minimum and the observed height, over all such deformed heights, yielded the *spine deformity index.*

A wide variety of other models have been built, many of them with a strong *ad hoc* component, as modifications are introduced to rectify weaknesses of earlier models.

Clearly, in these models, an underlying empirical reality is being measured, so that the approach has a major representational component (we would hope that someone whose spine is manifestly damaged by fractures would have a larger spine deformity index than someone without such damage). On the other hand, many fairly arbitrary decisions have been made in reaching the final model. Thus the indices also have a major pragmatic component.

6.6 Measures of diagnostic accuracy

Diagnostic accuracy is a fundamental issue in medicine: without being able to measure diagnostic accuracy, one cannot expect to be able to choose appropriate treatments. The most important special case, and the only one we will discuss here, is when we have two classes, and the aim of the diagnostic instrument is to assign people to one or the other class. Often the classes will be diseased and non-diseased (e.g. peptic ulcer versus no ulcer), where a particular disease is of interest, but sometimes the aim will be to distinguish between two distinct diseases (e.g. benign or malignant tumour). For convenience and generality, we will refer to the two classes by the letters A and B. Often in medicine it is useful to regard the two classes in an asymmetric way, and when this is appropriate we will designate class A the 'case' class and class B the 'non-case' or 'normal' class. The word 'positive' will then signify membership of class A, and the word 'negative' membership of class B. Our interest, then, lies in measuring the effectiveness of a diagnostic instrument in assigning people to the correct class.

The two-class situation is summarized in the *confusion matrix* shown in Table 6.1. This shows the probabilities of falling into each cell of a cross-classification of true class (column) by predicted class (row). 'Predicted class' is the class assigned by a diagnostic instrument, and 'true class' is the class assigned by a 'gold standard'. Often the gold standard is

Table 6.1 Confusion matrix for two-class diagnostic problems.

Predicted class	True class	
	A	B
A	P_{AA}	P_{AB}
B	P_{BA}	P_{BB}

a strategy which cannot be applied for practical reasons; for example, it could be something which can be determined only by a post-mortem examination.

Our aim is to reduce the four numbers shown in Table 6.1 to a single number which can serve as a measure of the effectiveness of the diagnostic prediction. Given that there are four numbers in the table, it will come as no surprise to hear that this summary may be accomplished in various ways, and that pragmatic choices have to be invoked to lead to a unique value.

There are two immediately obvious ways of reducing the four numbers to two. These are to look at the conditional probabilities. Thus, on the one hand, we have:

- *sensitivity*, defined as the probability of a positive test result when the condition is present, $p_{AA}/(p_{AA} + p_{BA})$;
- *specificity*, defined as the probability of a negative test result when the condition is absent, $p_{BB}/(p_{AB} + p_{BB})$.

On the other hand, we have the row conditional probabilities:

- the probability of being a case, given a positive test result, $p_{AA}/(p_{AA} + p_{AB})$.
- the probability of not being a case, given a negative test result, $p_{BB}/(p_{BA} + p_{BB})$.

One must think carefully about which measures are appropriate for the problem. Williams *et al.* (1982) give the example of an instrument for screening for eating disorders. One of the confusion matrices they present is reproduced in Table 6.2. From this, the sensitivity is 91.7% (this percentage of the anorexics are correctly diagnosed as such) and the specificity is 66.1% (this percentage of normals are correctly diagnosed as normal). However, amongst those that the instrument diagnoses as positive, only 15.9% (11 out of 69) really are anorexics. The sensitivity and specificity look good, but, as measures of diagnostic accuracy, they are clearly missing something. (Closer examination shows that the apparent discrepancy between the sensitivity and specificity on the one hand, and what is sometimes called the 'positive predictive value' on the other, arises because of the low prevalence: only 12 of the 183 subjects are anorexic.)

There are various ways of combining sensitivity and specificity to yield single measures of diagnostic performance. An obvious way is to use a weighted average, where the weights are the proportions of the population which belong to the two classes. This tells us the overall proportion of the population which will be classified correctly by the instrument. Its complement tells us the misclassification rate or error rate.

This simple weighted average is all very well, but it is not always appropriate. Indeed, some would say that it is hardly ever appropriate, because implicit in this choice of weights is the assumption that the two types of misclassification are equally serious, whereas usually one type of misclassication is more serious than the reverse. To take an extreme example, misdiagnosing someone suffering from an easily treatable but otherwise fatal disease as

Table 6.2 Confusion matrix for the two-class diagnostic problem for eating disorders (Williams *et al.*, 1982).

Predicted class	True class	
	Anorexics	Controls
Anorexics	11	58
Controls	1	113

healthy will generally be much more serious than the reverse misclassification. To overcome this, the two types of misclassification must be weighted by the relative misclassification costs. If the cost of misclassifying someone from class A is c_A and the cost of misclassifying someone from class B is c_B, then the overall misclassification cost of a diagnostic rule is

$$c = c_A p_{BA} + c_B p_{AB} \tag{6.1}$$

In terms of sensitivity (Se) and specificity (Sp) this is

$$c = c_A \times (1 - Se) \times \pi_A + c_B \times (1 - Sp) \times \pi_B$$

where π_A and π_B are the respective proportions in the two classes. To be useful, however, this measure requires agreement about the relative sizes of the costs, and this may be difficult to obtain.

Often diagnostic instruments take the form of a score, S (a 'measurement', one might say), on some continuum, which is compared with a threshold in order to reach a diagnosis. In fact, the predicted class in Table 6.2 was given by comparing the score on the *Eating Attitudes Test* (Garner and Garfinkel, 1979) with a threshold of 30: less than 30 and the subject was predicted normal, and otherwise as anorexic. When this is the case, one has the option of adjusting the threshold. If one does know the relative misclassification costs, then it is easy to find the optimal threshold which minimizes the cost-weighted misclassification rate in (6.1). However, if the relative costs are unknown, so that the optimal threshold cannot be chosen, some other strategy is needed.

A common strategy is to integrate the cost-weighted misclassification rate over all possible values of the ratio of the two misclassification costs, using, at each of the possible values of the cost ratio, the threshold which is optimal for that ratio. Of course, one must decide which values of the cost ratio to weight most heavily in this integration; that is, one must decide on a distribution over cost ratios in the integration. It turns out that, if one adopts, as this weighting distribution, the overall mixture distribution

$$f(s) = \pi_A f_A(s) + \pi_B f_B(s) \tag{6.2}$$

then the result is a simple linear transformation of the test statistic used in the Mann–Whitney–Wilcoxon nonparametric test of two samples. This is the probability that a randomly chosen member of class A will have a smaller score than a randomly chosen member of class B. This statistic is very important, and can also be derived in other ways. One way is from the *receiver operating characteristic* (ROC) curve. Suppose that low scores mean a person is more likely to belong to class A and high scores that they are more likely to belong to class B. We can adopt some score for use as a decision threshold and predict that patients with lower scores will belong to class A and those with higher scores to class B. Such a threshold will have associated values of sensitivity and specificity, and these values will change as we change the threshold. In fact, the (sensitivity, specificity) pair will describe a curve as the threshold is changed: this is the ROC curve. An example is given in Figure 6.1, showing the typical shape of such a curve. The better the separation between the two score distributions, the nearer the curve will lie to the top left-hand corner of the square. If the two score distributions were identical, the curve would lie along the diagonal from $(0,0)$ to $(1,1)$. The area under the curve can thus serve as a measure of separability between the two score distributions – as a measure of effectiveness of the score in distinguishing between the two classes, but without adopting any particular decision threshold. It turns out that the area under

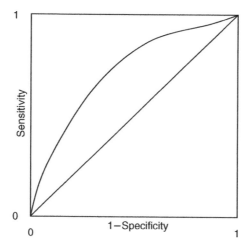

Figure 6.1 An example of an ROC curve.

the ROC curve (the 'AUC') is also a linear transformation of the test statistic used in the Mann–Whitney–Wilcoxon nonparametric test of two samples.

Variants of the AUC have also been proposed as improved measures of diagnostic performance. For example, the implicit integration over all possible cost ratios is unrealistic. Even if one cannot say exactly what the ratio is, one generally has some idea of a ball-park figure. This suggests using a weight distribution other than the simple mixture in (6.2). Such ideas have been explored in McClish (1989), Lloyd (1997), and Adams and Hand (1999).

The measures of diagnostic accuracy described above by no means exhaust those which have been proposed. Other popular ones include the odds ratio

$$\frac{p_{AA}/p_{BA}}{p_{BA}/p_{BB}} = \frac{p_{AA}p_{BB}}{p_{BA}p_{BA}} = \frac{Se/(1-Se)}{(1-Sp)/Sp},$$

Youden's index,

$$Se + Sp - 1$$

and the *positive* and *negative* likelihood ratios:

$$\frac{p_{BB}/(p_{BA}+p_{BB})}{p_{AB}/(p_{AA}+p_{AB})}, \frac{p_{BA}/(p_{BA}+p_{BB})}{p_{AA}/(p_{AA}+p_{AB})}$$

Measures of diagnostic accuracy are explored in detail in Hand (1997), Zhou *et al.* (2002) and Pepe (2003).

6.7 Measures to compare treatments

A common aim in medical research is to compare the effectiveness of two different treatments. Suppose that treatment 1 is such that a proportion π_1 of the patients so treated recover, and treatment 2 is such that a proportion π_2 of the patients so treated recover. There are many measures of relative performance, including:

- absolute risk reduction, $\pi_1 - \pi_2$;
- relative risk, π_2/π_1;
- odds ratio, $\pi_2(1 - \pi_1)/\pi_1(1 - \pi_2)$;
- log-odds ratio, $\log[\pi_2(1 - \pi_1)/\pi_1(1 - \pi_2)]$;
- relative risk reduction, $(\pi_1 - \pi_2)/\pi_1$.

Of course, all of these measures seek to condense two numbers (the success rates under each of the treatments) into a single measure. Such reduction necessarily implies some sacrifice of information. This point is discussed, in the more general context of classification rules, in Hand (1997, especially Chapter 8). Since, generally, the representational constraints will not completely determine which of these measures should be used, it is common to adopt a scale which is convenient for practical reasons, and define the pragmatic constraints as those which lead to this scale. For example, the log-odds ratio has the attractive property of covering the entire real line, and treating π_1 and π_2 symmetrically, as does the odds ratio. On the other hand, absolute risk reduction is expressed in the original success rate scale. Whether absolute risk reduction is preferred to relative risk reduction will depend on the type of statement one wants to make (see also Section 2.4.3 for a discussion of the relationship between additive and multiplicative models).

Another measure, originally strongly motivated by pragmatic considerations, is the *number needed to treat* (*NNT*) measure (Laupacis *et al.*, 1988). If a number *NNT* of patients are given treatment 1 then $\pi_1 NNT$ of them will recover, whereas if *NNT* patients are given treatment 2 then $\pi_2 NNT$ will recover. The difference between the numbers recovering under the two treatments, if I apply each treatment to *NNT* patients, is thus $(\pi_1 - \pi_2)\,NNT$. This will be equal to 1 patient if

$$NNT = \frac{1}{\pi_1 - \pi_2}$$

At first glance, this looks like an attractive and easily interpretable measure: it tells us the number of patients *NNT* who would need to be treated by each method in order that one method would yield one extra success (on average, of course, all subject to sampling variability). However, not everyone agrees: 'Physicians, patients and students consistently found results expressed in terms of the *NNT* difficult to interpret' (Grieve, 2003, p. 93), which rather undercuts the pragmatic argument. Hutton (2000) argues that absolute risk reduction 'is preferable to the number needed to treat, for both theoretical and practical reasons.'

6.8 Conclusion

Medical measurement is a vast subject. It is also an area which is changing dramatically with technological progress in other areas, perhaps especially electronics and computers. Computerized versions of questionnaires have many merits, not least their inability to be influenced by irrelevant aspects of the patient, and the fact that patients often prefer the anonymity such instruments provide when responding to questions on sensitive topics. It is entirely possible that more truthful answers are given to questions about smoking and sexual practices when it is not a person asking those questions. Of course, use of such systems presupposes a certain level of literacy in the patient – though even that may change in the near future, with progress in speech recognition software.

At the opposite end of the continuum, scanners of various kinds have revolutionized medicine. At a low level, they require highly sophisticated indirect formulae to extract the measures of interest (e.g. inverting large and ill-defined systems of equations to deduce what the inside of a head or body looks like from marginal projections). But at a high level, they give relatively simple measures which a clinician may interpret.

We have already mentioned the change which electronics has brought to even basic medical measurement procedures such as temperature and blood pressure measurement: no longer do these have to be translated into lengths of a column of mercury, relying on the nurse to measure this length accurately. Instead, the machines give numbers directly.

7

Measurement in the physical sciences

The history of science is largely coextensive with the history of measurement.
Tryon (1991)

7.1 Dimensions in physics

7.1.1 The history

This chapter discusses the physical sciences, an area in which most of the variables take ratio scales. Examination of these variables and the relationships between them immediately shows that they have a very striking form of relationship: we can express each (of the over 100 useful variables) as a monomial combination of certain basic variables or 'dimensions'. Thus, we might choose charge (C), temperature (R), mass (M), length (L), time (T), and angle (A) as the basic dimensions, so that, for example, speed has dimension LT^{-1}, density has dimension ML^{-3}, and frequency has the dimension AT^{-1}. It seems that, once we have chosen units for these six basic dimensions, the units of all other dimensions follow automatically. Relationships of this kind are very familiar: they are part of the backdrop against which our everyday life takes place. But if one confronts them squarely one sees that they pose something of a mystery. After all, why should all useful dimensions be simple monomial combinations of a small set of basic dimensions? Why should only six basic dimensions suffice? Are all six necessary? Why, in the monomial expressions, do only small integer exponents (only 1, -1, and -3 in the examples above) occur? And so on.

This section provides some historical background to such questions, and Section 7.1.2 examines them in more detail. Section 7.1.3 provides practical illustrations of how the theoretical ideas can be useful in genuine scientific investigations. Section 7.2 describes practical physical units of measurement, as well as something of their history, in a wide range of areas.

The concept of a measurement dimension grew up gradually. Although, as we mentioned in Chapter 1, it had its origins in geometric notions of length, even Descartes (1596–1650) in the seventeenth century had a broader interpretation: 'By dimension we mean simply a mode or aspect in respect of which some subject is considered to be measurable. Thus length, breadth and depth are not the only dimensions of a body; weight too is a dimension – the dimension in terms of which objects are weighed – speed is a dimension – the dimension of motion; and there are countless other instances of this sort' (Descartes, 1985).

Some (e.g. François Viète, 1540–1603) had even recognized by this time that the different elements of an equation should have the same dimensionality, if only of spatial dimension. Fourier later introduced this dimensional homogeneity requirement into physics more generally. In doing this, he presented a method for deciding the dimensionality of physical quantities. He took mass, length, time, temperature, and quantity of heat as defining base dimensions, and determined the dimensions of any indirectly measured quantity from its definition using these base quantities. His relationships between dimensions led to relationships between the units of the base dimensions and the units of the indirectly measured quantities which forced the preservation of structural form under changes of units of the base dimensions. Of course, if a law involves quantities all of which have already been defined, then changing the units can lead to numerical factors appearing. If Mach 1 (see below for a discussion of Mach numbers) denotes the speed of 340 metres per second, then if we convert from metric to Imperial units, we find we have to multiply by 3.28 to yield a speed of 1115 feet per second, that is,

$$\text{Mach } 1 = 340\,\text{m s}^{-1} = 340 \times 3.28\,\text{ft s}^{-1}$$

Since the value of these numerical factors will depend on the choice of base units, they too have dimensionality in terms of the base quantities – which can be determined from the equation. For example, in the law $F = Gm_1m_2/r^2$ describing the mutual gravitational attraction between two bodies, G takes the value 1 for certain units only. More generally, since all of F, m, and r have defined dimensionality, G changes with the choice of units and, by inverting the equation, we can determine the dimensionality of the constant G: it is ML^3T^{-2}. Roche (1998, p. 195) describes these Fourier dimensions as 'essentially a signature of quantity – determined by its measuring algorithm'. A system of units in which the units of derived quantities are defined in terms of a set of base quantities, so that the forms of relationships are preserved under changes to the base units, is called a *coherent* system.

It is clear from this sort of thing that measurement dimensions convey something important about attributes, their magnitudes, and their relationships. Recognizing this, some researchers attempted to take things further. Weber, for example, described electromagnetic quantities in terms of the mass, length and time units needed to measure them, and this encouraged the view (held, for example, by Maxwell) that electromagnetic phenomena were fundamentally mechanical (as heat had turned out to be) and led to the hope that the dimensions could explain some fundamental truths about reality. For example, 'It is by many considered that a dimensional formula has some esoteric significance connected with the "ultimate nature" of an object, and that we are in some way getting at the ultimate nature of things in writing their dimensional formula' (this is from Bridgman, 1922, although he was, in fact, criticising this position). Isaacson and Isaacson (1975, p. 195) comment: 'It has been suggested, in effect, that the dimensional formula represents the physical identity of a quantity in a manner analogous to the way in which chemical formulae indicate the chemical identity of the corresponding compounds'.

These hopes gradually dissipated. Isaacson and Isaacson later go (too far?) to the other extreme: 'the dimensional representation of a quantity merely tells us the manner in which that quantity is measured and nothing as to its ultimate reality'. Be that as it may, at least one important advance was stimulated by the perspective that the dimensional representation had something fundamental to say. Campbell (1920, p. 434) reports:

> Bohr has confessed that he was led to the fundamental ideas of his theory of atomic structure by noticing that the constant h, in the numerical law $E = hv$, has the dimensions of a moment of momentum. Up to that time it had always been called 'an element of action' and such theories as had been put forward to 'explain' it had attempted to represent it as an 'action'. These theories are not disproved by the new view of matter; but there is no doubt that the newer view that h 'is' a moment of momentum has led to theories and hypothetical ideas of great fruitfulness.

Confusion is often rife. Roche (1998, p. 217) quotes N.R. Campbell: 'Writers on "dimension" usually ignore almost the whole of the vast literature on the subject.' Roche goes on to say: 'Perhaps, also, some of those accustomed to a mathematical manner of thinking assume that problems of interpretation can be resolved by pure reasoning without the need to study origins or the literature of discussion.' Roche outlines some of the other difficulties with which dimensional analysis has had to contend. They include the fact that 'some of the arguments used are quite subtle and involve a kind of interpretative analysis not widely practised in physics or in engineering', as well as the long history of ambiguity in the terminology.

We should note that, even in the context of the present discussion, the term 'dimension' itself is used in different ways. Sometimes it means the 'mode' of measurement (length, mass, etc.), at other times it means the exponent to which each mode is raised in an expression (two-dimensional, in L^2), and sometimes it means the aggregate 'dimensional expression' (MLT^{-2}). My *Concise Oxford English Dictionary* defines a dimension both as (amongst other things) 'a measurable extent of any kind, as length, breadth, depth, area, and volume' and as 'the product of mass, length, time, etc., raised to the appropriate power, in a derived physical quantity'. On the other hand, as Roche (1998, p. 207) notes, 'despite the severe problems of interpretation, practitioners have had few difficulties in using the formalism of dimensional theory effectively. After Palacios [1964] most writers on the subject seem to have contented themselves with technicalities and with the admission that many problems of interpretation have not yet been resolved'. In this, at least, there appears to be a similarity to quantum mechanics.

7.1.2 The theory

Demonstrations that derived variables should take the form of monomial combinations of base variables often take the lines described in Chapter 2, when we were discussing Luce's principle. We shall repeat the derivation in a slightly different form here, for convenience. Such derivations begin with a demonstration that a function of a ratio scale is itself a ratio scale if and only if that function is proportional to a power of its argument. More formally, let x be a ratio scale – so that the ratio x_1/x_2 is invariant to all legitimate transformations (i.e. similarity transformations) of x. Then we wish to show that the derived variable $f(x)$ is also a ratio scale if and only if f has the form $f = ax^b$. That is, $f(x_1)/f(x_2)$ is invariant to all legitimate transformations of f (which are those obtained by rescaling x: $f(\alpha x)$) if and only if f has the form $f = ax^b$.

First, suppose that the function f takes the form $f(x) = ax^b$ for some constants a and b. Then for two quantities, giving rise to the raw measurements x_1 and x_2, the ratio of the values of the derived variable is ax_1^b/ax_2^b. If the units of the raw measurement are now changed, by rescaling them by a factor α, then the ratio of the values of the derived variable becomes $a(\alpha x_1)^b/a(\alpha x_2)^b = ax_1^b/ax_2^b$. That is, the ratio remains unchanged, so that the derived variable also has a ratio scale. (Formally, perhaps we should have written $x(X)$ to indicate that x is a numerical value obtained from the physical quantity X, and $f(x(X))$ to indicate that f is a derived numerical value obtained from the quantity X, but the extra rigour seems unnecessary here.)

Now suppose that f is a ratio scale. This means that the only legitimate transformations are rescaling transformations. Since the legitimate transformations arise from rescaling transformations of the raw variable (these being the admissible transformations of this variable), this means that if the raw variable is rescaled the derived variable f will have constant ratio:

$$\frac{f(x_1)}{f(x_2)} = \frac{f(\alpha x_1)}{f(\alpha x_2)} = k(x_1, x_2)$$

with $k(x_1, x_2)$ independent of α. From this we see that $\log k = \log f(\alpha x_1) - \log f(\alpha x_2)$ so that

$$0 = \frac{\partial \log k}{\partial \alpha} = x_1 \frac{f'(\alpha x_1)}{f(\alpha x_1)} - x_2 \frac{f'(\alpha x_2)}{f(\alpha x_2)}$$

Putting $\alpha = 1$ gives

$$x_1 \frac{f'(x_1)}{f(x_1)} = x_2 \frac{f'(x_2)}{f(x_2)}$$

That is, $f(x) = ax^c$ for some constants a and c, so that f has the required form.

Having shown that a function, f, of a ratio scale is itself a ratio scale if and only if that function is proportional to a power of its argument, it follows that if f is a function of several ratio variables, $x_1, \ldots, x_n$, then, f will be a ratio scale if and only if it is proportional to a power of each of them, so that it has the overall form $f(x_1, \ldots, x_n) = ax_1^{c_1} \ldots x_n^{c_n}$.

Of course, this takes as a starting point the fact that the derived variable has to be ratio. Krantz $et\ al.$ (1971, p. 516) suggest that the statement that the derived variable should also be ratio is merely a statement of dimensional invariance for the case of $d + 1$ variables. They suggest that, in fact, there are two fundamental questions:

(i)   When is a numerical representation of the qualitative physical attributes possible?
(ii) What properties must a relationship between qualitative attributes possess in order that it might be a physical law?

Observation shows that physical quantities are described in terms of products of powers of a few base fundamental quantities each of which is measured on a ratio scale. Furthermore, observation also shows that one can choose a basis of fundamental quantities such that no one is proportional to a product of powers of the others. The overall structure seems similar to a vector space structure, with multiplication replacing addition. If we take this as a foundation, then to answer the first question it is necessary to obtain empirical conditions on the qualitative structure which map into this vector space structure. Krantz $et\ al.$ (1971, Chapter 10) (see also Luce $et\ al.$, 1990, p. 309) describe such an axiomatization of a qualitative structure. They call systems satisfying their axioms *structures of physical quantities*. While such axioms map the system into the requisite vector space, they say nothing about the numerical measures produced to measure each dimension separately. These are produced by extensive and conjoint measurement procedures. Somehow, then, it is necessary to link the individual measurements to the structure of physical quantities. This link must model not only extensive measurements such as length and weight, but also intensive measurements such as density. Krantz *et al.* (1971) describe such a link via *laws of similitude* and *laws of exchange*, relating triples of qualitative attributes. These are qualitative properties which must be satisfied for the conjoint measures (arising from independence properties assumed in the structure of physical attributes, and which lead to multiplicative combination of measures) to be related to the extensive measures (which arose from the basic additive nature of some qualitative attributes). Narens (1976) and Narens and Luce (1976) describe an alternative approach based on *distributive triples*.

Turning to the second question, a basic starting point of dimensional analysis is that laws of physics are dimensionally invariant: that the *form* of a law does not depend on the choice of units in which the qualitative variables are measured. Buckingham's pi theorem (see below) shows that dimensional invariance means that the dimensions in a law can be combined into dimensionless products of powers, and that the law is a function of only these dimensionless products. It is clear that the value of this theorem hinges on the assumption that the laws of nature are dimensionally invariant, but why should natural laws be so? It is not enough to observe that, in physics, the laws generally are so. Something rather more fundamental is required. What is needed is a definition of what a physical law is, from which dimensional invariance can be shown to follow.

Luce (1978, p. 3) provides such a definition: a relationship, described in terms of the structure of qualitative attributes, can be a scientific law if and only if it is invariant under the automorphisms of the qualitative structure. This is very similar to the notion that statements about a single measured attribute are meaningful only if they are invariant to permissible changes in the representation, and, indeed, Luce (1978) shows they are both special cases of a general concept of meaningfulness in relational structures. Luce (1978, p. 15) also cautions that the system of interrelated extensive and conjoint structures and the distribution laws connecting them, on which the above hinges, 'is adequate for classical physics, but it is too restrictive for modern physics (e.g. relativistic velocity does not relate to distance via a distribution law) and probably for other sciences'. In general, however, the notion of meaningfulness as being invariance to automorphisms of the qualitative structure seems a good working definition. It means that the fact that a physical law is dimensionally invariant does not impose any additional restrictions on the qualitative relationship it is modelling. Conversely, a law which is not dimensionally invariant – one for which the structure does change with the units – imposes extra constraints on the qualitative attributes. These extra constraints come from pragmatic considerations.

7.1.3 The practice

Perhaps the most familiar use of dimensional theory in physics has been to check equations for internal consistency. However, it can also be used a priori to restrict the form that relationships can take. We consider the latter first, illustrating with a popular example, which leads us to the fundamental Buckingham's pi theorem, mentioned above.

7.1.3.1. *Finding the forms of equations*

Suppose we want to find the relationship between the period, t, of swing of a simple pendulum, its length, l, the mass, m, of the weight at its end, and the acceleration, g, due to gravity, when it swings in short arcs. Since we have observed that the monomial relationship between variables described above generally holds, we will assume initially that the relationship is of the form $t = k \times l^a m^b g^c$, where k is an unknown constant. The dimensions of t, l, m, and g are, respectively, T, L, M, and LT^{-2} so we have the dimensional equation $T = L^a M^b L^c T^{-2c}$. The notation X^y signifies that the dimension of X appears with exponent y in the equation.

From this, in order to make the two sides of the equation balance in terms of their dimensions, we derive the system of equations

$$1 = -2c$$
$$0 = a + c$$
$$0 = b$$

from which $b = 0$, $a = -c = 1/2$. This leads to $t = k\sqrt{l/g}$.

The above is already an impressive result. Without knowing anything about the physics involved, we have apparently derived the functional relationship between the variables in the equation.

Now let us take this example further. In the above we assumed that the arc lengths of the swings were only small. If they are not small, then experiment shows that the above relationship breaks down. We need to include s, the arc length, in our calculations.

So, let us begin with the postulated relationship $t = k \times l^a m^b g^c s^d$. The dimensional equation now leads to the system

$$1 = -2c$$
$$0 = a + c + d$$
$$0 = b$$

Since there are four unknowns here, but only three equations relating them, we cannot uniquely solve the system. We have $c = -1/2$, $b = 0$, and then $a + d = 1/2$. Looking back at the previous equation derived for the simpler case, we see that we can write the solution for this new case as

$$t = k\sqrt{\frac{l}{g}}\left(\frac{s}{l}\right)^d$$

Now, the ratio s/l is dimensionless, so any power of it or any series of powers of it will satisfy the dimensional requirements. That is, we have solution

$$t = k\sqrt{\frac{l}{g}}f\left(\frac{s}{l}\right)$$

where f is an arbitrary function of the ratio s/l. Determination of the function f has to be made by some means other than dimensional analysis.

As a second simple example, consider another vibrating system, the tuning fork. Here the period is a function of the linear dimensions, l, the density of the material of which the fork is made, τ, and Young's modulus, Y, a measure of its elasticity. Linear dimension has unit L, density has units ML^{-3}, and Young's modulus has units $MT^{-2}L^{-1}$. Following the same process as before, and assuming a relationship of the form $t = k \times l^a Y^b \tau^c$, we obtain $t = k \times lY^{-1/2}\tau^{1/2}$.

The argument of the function f in the first example was dimensionless. It was this that made it impossible for us to determine the form of f by dimensional arguments. The converse of this shows that arguments of transcendental functions, such as logarithms, exponentials or trigonometric functions must be dimensionless. This is easily seen from the Taylor expansion of such a function (see below), since this will involve increasing powers of its argument, so that the equation will not be dimensionally homogeneous unless the argument is dimensionless. Note, however, that the dimensionless nature of the argument may sometimes be disguised. For example, an equation might involve $(\log m + \log a - \log F)$, where F is force, m is mass, and a is acceleration. Superficially this equation appears to be dimensionally meaningless, but it is of course equivalent to $\log (ma/F)$, in which the argument of the log function is dimensionless.

Further examples from physics may be found in any book on dimensional analysis, such as Bridgman (1922), Birkhoff (1950), Palacios (1964), Porter (1958), Isaacson and Isaacson (1975), Taylor (1988), and Szirtes and Pal (1997).

The ideas illustrated informally in the above examples may be expressed in a more formal and algorithmic framework. They were explored by Rayleigh (1842–1919), Bertrand (1822–1900), Vaschy (1857–1899) and others, culminating in the formal expression by Buckingham (1867–1940) of what is now known as *Buckingham's pi Theorem* (Buckingham, 1914):

> If there is a dimensionally homogeneous equation relating n quantities defined in terms of r reference dimensions, then the equation may be reduced to a relationship between $(n - r)$ independent *dimensionless products*, provided that the members of the reference set be themselves so chosen as to be independent of one another.

Or:

> If an equation is dimensionally homogeneous, it may be reduced to a relationship between a complete set of dimensionless products.

This theorem arises from the following considerations. We can express the relationship between the n quantities in the form $f(x_1, ..., x_n) = 0$, so that each of the reference dimensions must separately combine to yield a dimensionless result. For example, if some of the variables involve time, with dimension T, then the overall exponent of T must be 0. Since there are r reference dimensions, this means that there are r relationships to be satisfied – leaving $n - r$ degrees of freedom.

To illustrate, $s = vt + at^2/2$, relative distance s, velocity v, time t, and acceleration a, may be rewritten as $1 = vt/s + at^2/2s$. There are four original variables but only two reference dimensions here, L and T, and the relationship has been reduced to a relationship between two dimensionless products, vt/s and at^2/s. These are independent, as may be seen from the fact that v appears in one and a in the other.

Buckingham's pi theorem was a terribly exciting development: '[the student of physics] is astonished that so large a proportion of his equations, which he imagined were based solely on experiment, are in fact unavoidable consequences of the nature of the physical quantities involved, as observed in the dimensional formulae' (Huntley, 1952). On the other hand, exciting though it may be, it is also deceptive. To apply the methods effectively, substantial knowledge about the domain is required: dimensional analysis, for example, cannot tell you that you have omitted necessary terms from an equation. Dimensional analysis can tell you nothing about dimensionless ratios (so, for example, it cannot say whether such expressions enter via sin, exp, log or other functions). Dimensional analysis cannot perform miracles.

As a parenthetical note, if one believes that the relationship between variables $v_1, \ldots, v_4$ takes the form of a monomial combination, perhaps $v_1 = k v_2^\alpha v_3^\beta v_4^\gamma$, then standard statistical model-fitting methods can be used to estimate the exponents, again without any understanding of the physics involved. In fact, if we take logs, we can use regression to fit the model

$$\ln v_1 = \ln k + \alpha \ln v_2 + \beta \ln v_3 + \gamma \ln v_4$$

to the data. The formula then is a direct consequence of experiment.

7.1.3.2 Detecting errors

If dimensional analysis cannot perform miracles, it is nevertheless extremely useful in detecting malformations in equations. This does not apply solely in physics, and this section gives some examples.

7.1.3.2.1 Examples in mathematics

Many mathematical equations admit a geometric interpretation, and in such cases length, with dimension L, plays a key role. Consider the equation of a parabola, written in the form $y = p + qx + rx^2$. Here y is the ordinate (dimension L), x is the abscissa (dimension L), p is the intercept, given by the value of y when $x = 0$, and hence also has dimension L, q is the gradient at $x = 0$ (dimension L/L) so that qx has dimension L, and r is the second derivative of y (dimensions $L/L^2 = L^{-1}$), so that rx^2 has dimension L. This geometric interpretation leads to the recognition that mathematical manipulation must preserve dimensional homogeneity – a fact which permits an easy test of such manipulations. Sometimes care needs to be exercised because coefficients may not be explicitly stated. Thus, for example, the apparent dimensional inhomogeneity of $(1 - x)^{-1} = 1 + x + x^2 + \cdots$ is resolved either by taking x to be dimensionless or by regarding the 1 in $(1 - x)^{-1}$ as an expression of unit magnitude (e.g. length). Then $(1 - x)^{-1}$ is a special case of $(a - x)^{-1}$, with a having the same dimension as x, which maintains dimensional homogeneity in its expansion:

$$(a-x)^{-1} = \frac{1}{a}\left(1 - \frac{x}{a}\right)^{-1} = \frac{1}{a}\left(1 + \frac{x}{a} + \frac{x^2}{a^2} + \cdots\right)$$

We have already remarked that a similar issue arises with transcendental functions such as exp, log, and sin. For example,

$$\exp(x) = 1 + x + \frac{x^2}{2!} + \frac{x^3}{3!} + \cdots$$

so that x has to be dimensionless. In general, an expression in any transcendental function, such as $\exp(6x)$, with x a dimensioned variable, conceals the fact that the '6' has dimensions the reciprocal of those of x, so that it will change value if the units of x are changed.

7.1.3.2.2 *Examples in statistics*

In statistics, any model formula must obviously conform to the constraints of dimensional homogeneity. Likewise, any test statistic must be invariant to changes of the units in which the measurements are expressed. In general, in statistics, however, we will not simply be concerned with physical dimensions, and this demonstrates the point that there is some freedom in what one considers to be a 'dimension'. For example, although, usually, length is regarded as a single dimension, regardless of direction, it can sometimes be advantageous to regard it as comprising three dimensions: L_x, L_y and L_z. This is illustrated in the context of the measurement of angle. Angle is a ratio of lengths (length of subtended arc divided by radius), and hence is dimensionless. And yet we can concatenate angle, so that it does have a unit. One way to resolve this is to treat the three spatial components of length as separate dimensions.

If we define the probability of an event as the limiting relative frequency of occurrences of the event as the total number of trials increases, we see that, as a ratio of like things, probability is dimensionless. This means, in turn, that the dimensionality of a probability density function for a variable x with dimension D is D^{-1}. Finney (1977) refers to a textbook which contained the formula

$$\text{const} \times \left(\frac{n-1}{2\sigma} \right)^{(n-1)/2} (s^2)^{(n-3)/2} \exp\left[-(n-1)s^2/\sigma^2 \right]$$

for the density function of $s^2 = \Sigma(x - \bar{x})^2/(n-1)$, the estimate of variance for a sample of size n from a normal distribution. This should, of course, have dimensionality D^{-2}. However, Finney points out that although σ and s have dimensions D, so that the exponent correctly has dimension D^0, the other factors lead to dimensions $D^{(n-5)/2}$. The simplest adjustment, to give an overall dimensionality of D^{-2}, is that the power of σ should be $-(n-1)$ instead of $-(n-1)/2$. ('Of course, no check on the dimensionless multiplying factor is possible without verifying the integration.')

In general, of course, regression coefficients have the dimensions of the ratio of response dimensions to predictor dimensions and, indeed, even if these are measured in the same units it may be useful to regard them as having distinct dimensions (so that, for example, a regression coefficient in an equation predicting a person's height from their forearm length has dimensions $L_{\text{response}}/L_{\text{predictor}}$). Note also that, in a regression with x as a potential predictor, fitting $\beta_1 x$ without a constant term β_0 means one is fitting a restricted form of model. In particular, it assumes that the response is 0 when x is 0. For some types of measurement scale this will be reasonable. For example, with a ratio scale, any rescaling will preserve the property that the rescaled value αx is 0 when x is zero, so if the relationship between the response and original x (that the response is 0 when x is zero) is true then it is also true for the rescaled x (that response is 0 when rescaled x is zero). This means that the (restricted) model form remains the same. Of course, one may not wish to contemplate such a special class of models, even for variables for which they retain their special form under legitimate transformations. If the special class is appropriate, then β_0 will be found to be zero.

For other types of scale it will not be reasonable to contemplate such a restricted form of model. For example, with a scale which permits translations (e.g. temperature) the x is transformed to $\alpha + x$. If the response is 0 when the untransformed x is zero, it will not be 0 when the transformed x is 0. This means that the transformation has induced a change in the shape of the model. This is overcome by relaxing the restriction and by introducing a constant into the model. In the context of this example, with scales which permit translations, suppose we have a particular set of measurements (measured according to the 'untransformed' x variable). It could well be that, when we fit the model, we find that the constant term is zero: $y = \beta_1 x$. This is not a legitimate model since legitimate transformations of the x variable will lead to models of the form $y = \beta_0 + \beta_1 x$. Put another way, one cannot report the model shape as $y = \beta_1 x$ and say 'all one has to do is choose the value of β_1 according to the particular scale used'. Rather, one has to report the model shape as $y = \beta_0 + \beta_1 x$, and choose both parameters according to the particular scale used. The point here is that the special case of the model holds if measurements are given in a particular scale, but it cannot be a scientific law because the shape of the model depends on the particular scale chosen. In general, given a proposed model form, we should include all terms which arise when the raw variables are transformed by legitimate transformations. We give below a real illustration of what happens when this condition is breached, but a simple example is the following. Suppose that we postulate a law which says that the masses of organisms in a particular class are proportional to the cube of a linear dimension: $M = kH^3$. For an appropriate choice of units, this will reduce to $M = H^3$. Now define $m = \log M$ and $h = \log H$, so that $m = 3h$. If, by chance, we were using this particular choice of units, then this, $m = 3h$, would be the law we propose. *But this is not a legitimate model.* Legitimate transformations of the underlying variables (which are, in this case, rescaling transformations – by α, say) lead to the introduction of a constant:

$$m = 3\log(\alpha H) = 3\log\alpha + 3\log H = \beta + 3h$$

The numerical constant must have dimensions MH^{-3} to make the two sides dimensionally equivalent.

This principle has more elaborate implications. Consider a proposed model $\beta_0 + \beta_2 x^2$. This is a restricted form of model. If x is a ratio scale, then this model form is legitimate – it has the same shape whatever scaling is used. However, if x permits translations, then transforming from x to $\alpha + x$ leads to

$$\beta_0 + \beta_2(\alpha + x)^2 = (\beta_0 + \beta_2\alpha^2) + 2\alpha\beta_2 x + \beta_2 x^2$$

thus introducing a term of form $\beta_1 x$. The model shape changes. Similar comments apply to models of the form $\beta_0 + \beta_2 xy$. Further examples of this kind are given in Nelder (2000).

Finney (1977) also gives an example of functions used in immunoassay that give the expected count as a function of dose (Harding *et al.*, 1973):

$$\alpha + \beta/[1 + \gamma \ln\{1 + \exp(x - \delta)\}]$$

where x is a dose (measured perhaps as a weight or a weight per millilitre). Changing the units for dose (e.g from 5 g to 0.005 kg) changes the expected value of the count, so this cannot be correct. There is a coefficient (of units D^{-1}) missing preceding dose, to absorb its units: $\alpha + \beta/[1 + \gamma \ln\{1 + \exp(\vartheta x - \delta)\}]$.

Lindsey (1999) provides a nice example of how dimensional invariance constraints may be breached. This sort of situation can often happen with modern model search computer programs based on the use of genetic algorithms to choose between and combine model terms from a very large set, including all sorts of transformations of the raw variables. After an extensive model search, Lindsey (1999) obtained the following for the mean number of reported AIDS cases:

$$\log(\mu_{ij}) = \phi + \xi_1 \log(t) + \xi_2/t + \theta_1 \log(u)$$

$$+ \theta_2/u + v_1\, t/u + v_2 t \log(u)$$

$$+ v_3 \log(t)/u + v_4 \log(t)\log(u)$$

Here μ_{ij} is the mean number of reported AIDS cases in quarter i with reporting delay j, t is the reporting time, and u is the delay time. Lindsey does say that 'This is certainly in no way the "true" model', but in fact it is a fundamentally misspecified model and cannot possibly be legitimate. In the model, u is delay time (in three-monthly intervals). As such it is a ratio scale and we may express it in different units by multiplying it by some numerical factor. If we do this we see that the term $v_2 t \log(u)$ becomes $v_2 t \log(ku) = v_2 t \log(u) + v_2 t \log(k) = v_2 t \log(u) + v_5 t$. That is, an extra term, $v_5 t$ is introduced into the model. We can make similar transformations for all other terms in the model, but in all other cases the extra term is subsumed into terms which are already in the model: only the parameter values change. This is also true of changes to the measurement scale of time t. The implication of all of this is that the model, as presented, is fundamentally wrong, and depends on an arbitrary choice of units, and not on the empirical system being modelled. This is further brought home if we examine the fit of the model to the data.

Fitting the corrected model (including an extra term $v_5 t$) yields a penalized deviance of 803.8 as compared with the deviance of 800.8 for the incorrect given model. The fact that the given model has a smaller deviance than the corrected model is presumably why Lindsey reported the given model – it provides a better fit to the data. However, if we fit a model using days for the delay time – approximately calculated as 90 times the given delay times (since they are in three monthly intervals) we obtain a penalized deviance for the given model of 814.1. That of the corrected model, of course, remains the same as before, 803.8. That is, if we had happened to use a different set of units, equally legitimate, we would have chosen a different model. Clearly this is absurd: model choice cannot be a function of which of a set of equally legitimate sets of units we happen to use. Taking things in the other direction, if we fit a model using fractions of years for the delay time (approximated by dividing the given times by 4), we obtain a penalized deviance for the given model of 825.9, substantially worse than that of the corrected model, which remains at 803.8.

Lindsey's model cannot be a scientific law because it is neither dimensionally invariant nor dimensionally homogeneous. During the discussion at the meeting at which Lindsey presented his model, I pointed this out, and he replied that his analysis was based on the particular numbers and units observed. That is, he is saying that *these* are the units in which the variables were measured, and the observed data do conform to the equation given above. This is true, but it is then a purely descriptive law, which cannot shed light on the underlying mechanism. Certainly, if I collect more data using the same units, then they will conform to the descriptive law presented. But they will not tell me anything about how the system actually works. Put another way, by adopting a particular set of units, and by not ensuring that one's model is invariant to choice of unit, one is introducing a pragmatic component (unnecessarily, in this case). This is fine if everyone (conventionally) will always use the same units, since then no confusion or misunderstanding can arise. But in this case there is no guarantee of that. In other cases, of course, mere description of the numerical results one will obtain may be sufficient, as, for example, in the use of regression models in credit scoring. These form valuable models for predicting who is likely to default, but they may not reflect any underlying causal relationships. That is, such models are acceptable as empirical models, but not as iconic models. A simple example of this is the equation *height* $\approx$ *weight* for 4-year-old human beings. If measured in units of inches and pounds, then this is a reasonable empirical and predictive model – it can tell you if a child is in the tails of the distribution. However, it cannot be a sensible iconic model – it cannot be taken as a law of nature.

As a final example, consider the relationship $d = 67.39^h - 0.33$, with d the least distance, in feet, from which a road sign can be read and h the height of the letters, in inches. This holds for measurements in those units, but to be contemplated as a description of nature, it would need to be modified to, for example, $d = \alpha\{(67.39)^{\beta h} - 0.33\}$, where, in the special case of using feet and inches as units, $\alpha = \beta = 1$, and if other units are used other values will result. When feet and inches are used, α has units of feet and β has units of (inches)$^{-1}$.

Sometimes things can be deceptive. Let s be the distance a particle in motion covers from rest, v be the speed, t the time, and a acceleration. Then both $s = at^2/2$ and $v = at$ hold. From this it follows that $s + v = at^2/2 + at$ holds. This last equation involves sums of terms with different dimensionality. However, it is in fact a combination of two independent relationships: $s - at^2/2 = 0$ and $v - at = 0$.

It is interesting to note that, given a variable x of dimension D, measures of location and dispersion on this variable have dimension D (or, of course, D^2 for variance, though that is a trivial modification), whereas measures of skewness and kurtosis are dimensionless. These latter tell us about the intrinsic shape of a distribution, whereas location and spread tell us about the configuration relative to external aspects.

It might seem that, for example, the fact that the expectation of a Poisson variate equals its variance contradicts this. This conundrum is resolved by noting that the variable x in this case is simply a count, a frequency, so that its units are fixed. If we were to change the units, and count in hundreds or tenths instead for example, then the equality of the mean and variance would vanish.

7.2 Units of measurement

7.2.1 Background

The earliest forms of physical measurement appear to have arisen in astronomy, surveying and building, where measurement generally took the form of counting the number of replicas of some standard unit (e.g. units of time such as days in astronomy, and units of length such as yards in surveying and building). These standard units were initially defined in purely conventional terms, originally in terms of naturally occurring objects of the same, or at least similar, size. Such objects often occur in a biological context (e.g. foot, hand, inch, seeds), of which more later. The disadvantage of this method is that biological objects are not in fact identical, but differ from each other. This means that a statistical approach is needed – an averaging approach, to uniquely define the size of the basic unit.

From this it was a natural development to define the fundamental units in terms of a single *etalon*, an object of standard size which was stored and used as a basic comparator; the standard metre, which we will also discuss in more detail later, provides an example. While this approach leads to a well-defined unit, physical standard units are subject to deterioration and wear and tear, and in any case the procedure is essentially arbitrary. To remove risk of change due to wear and tear, definitions were introduced which were based on some more basic physical quality, such as properties of atoms, electromagnetic radiation or subatomic particles, which are identical in a fundamental sense, and which can be replicated everywhere. However, this should not conceal the fact that an arbitrary choice is still being made: we can define length in terms of the radiation which results when electrons move between particular shells in a particular type of atom – but which atom should we choose?

The trend described in the previous two paragraphs is a trend towards universal systems – globalization of measurement, if you like – but this trend has a long and tangled history. That this trend is still playing itself out is exemplified by the article, quoted in Section 1.1, from the British newspaper, the *Independent*, published on 23 March 2000 (p. 18), which noted: 'The official report on the loss of the earlier Mars probe, the $125 m (£78 m) Climate Orbiter concluded it had burnt up in the Mars atmosphere because of an elementary miscalculation: metric and imperial units had been confused.' And a letter from the *Daily Telegraph* of 1 September 2000 reads 'that mission notoriously failed on October 1 1999 precisely because of confusion between Lockheed Martin Corporation's data on pressure in pounds and the Jet Propulsion Laboratory's use of newtons ... Shell Oil spent millions of dollars converting pumps from gallons to litres, suffering a slump in custom, whereupon it spent millions more on reconversion to gallons.'

To take another example, the *Daily Telegraph* of 7 February 2001 had an article entitled 'Imperial units are weighed off' by the Home Affairs Editor, Philip Johnston, which reads:

> Ministers have signed the official death warrant for almost all remaining imperial measurements. A little noticed government regulation that comes into force tomorrow sets a deadline of Dec 31, 2009 for an end to the use of pounds and ounces.
>
> Although it is already illegal for traders to sell goods in non-metric quantities, they are allowed to display imperial measurements alongside as 'supplementary indications'.
>
> Under an EU deal in 1999, Britain was given a further 10 years to make the full switch to metric measurements and had until Feb 9 to incorporate the EU directive into law.
>
> The statutory instrument tables by the Trade and Industry Department states: 'These regulations amend the Weights and Measures Act 1985 by inserting a deadline of 31 Dec 2009 for the end of the authorised use of supplementary indications in conjunction with metric units.'
>
> After 2009 the only imperial measurements that will remain will be the mile, the acre, the pint for draught beer and cider and doorstep milk, fluid ounces and troy ounces for precious metals.
>
> However, a DTI spokesman said the option was always open for the Government to negotiate another 10-year extension 'if imperial measurements continue to be widely used'.
>
> Resistance to metrication in America was the main reason for the first extension.
>
> David Delaney, of the British Weights and Measures Association, said: 'This will put an end forever to the use of pounds and ounces.
>
> 'They were obviously hoping no one would notice this regulation so that when trading standards officers start persecuting traders in 2009 they can say it has been law since 2001 and it's their fault for not noticing.'
>
> © Telegraph Group Limited (2001)

In fact, this story ran on. Over the next year several traders fought against the ruling. The *Daily Telegraph* of 19 February 2002 includes a story headed 'Metric martyrs lose their fight: Government welcomes High Court defeat for rebel traders but campaigners proclaim "death of democracy"'. The article says:

> Five market traders who wanted to continue weighing out goods in pounds and ounces lost their appeals yesterday against European laws requiring them to use metric measures. ...

The five so-called metric martyrs were ordered to pay the full costs, estimated at £100,000, of the local authorities that had prosecuted four of them under weights and measures legislation.

… Although 'our imperial measures, much loved of many, seemed to face extinction', there were exceptions and postponements, [Lord Justice Laws] said. Market traders could still use pounds and ounces as 'supplementary indicators' until the start of 2010.

… The best-known of the five traders is Steven Thorburn, of Sunderland. He was found guilty under the Weights and Measures Act in April last year and given a 12-month conditional discharge for using imperial scales which did not bear an official stamp to sell loose fruit and vegetables by the pound.

… John Dove, who runs a fish shop in Camelford, Cornwall, was convicted of selling mackerel at £1.50 a pound. Julian Harman, also of Camelford, was convicted of selling Brussels sprouts at 39p a pound.

© Telegraph Group Limited (2002)

A companion article says:

The survival of yards, pounds, and pints in British commerce is a 'national disgrace', Lord Howe of Aberavon, the former deputy prime minister, claimed yesterday.

The use of imperial measures alongside metric ones was 'seriously damaging' to the economy and left Britain with 'the worst of both worlds', he said.

… Lord Howe said allowing metric and imperial measurements to run alongside each other imposed a 'major cost' on the economy. A CBI report in 1980 said that dual measurement systems increased British production costs by £5 billion a year.

'In the companies on which the survey was based, their increased production costs were equal to nine per cent of gross profit and 14 per cent of net profit,' the Tory peer said.

'You can't stay prosperous, or even survive, for long when you have to cope with such inefficient working, due to the Government failing to get Britain properly metric.'

'Every time a supermarket has to dual-label a cabbage or a carrot, that increases the cost of food.'

… In Britain, people bought petrol in litres but measured fuel consumption in miles per gallon. Ordinance Survey maps used kilometres, while road signs used miles.

© Telegraph Group Limited (2002)

Even within countries one often finds conflicting systems in simultaneous use: for example, the American use of miles in everyday life and kilometres in scientific work. But things are infinitely better now than they used to be: according to Alder (1995), in eighteenth-century France there were 700–800 different measurement terms 'which expressed a mind-boggling total of some 250,000 local variants!'. Between countries (outside a scientific and engineering context) the same name sometimes continues to be used for subtly different quantities. For example, the British pint is just over 20% greater than the US pint, and the British foot is 30.47997 cm while the US foot is 30.48006 cm. Of course, these differences are part of the cultural heritage of the different countries concerned, and one would not want to see them vanish. Indeed, it will be a very long time before globalization is such that Far Eastern and

Table 7.1 Modern lengths of some ancient cubits (Berriman, 1953; Klein, 1974).

Cubit	Length in metres
Roman	0.444
Egyptian 'short'	0.450
Greek	0.463
Assyrian	0.494
Sumerian	0.502
Egyptian 'royal'	0.524
Talmudist	0.555
Palestinian	0.641

Table 7.2 The different sizes of ancient mina weights and number of shekels in a mina (McGreevy, 1995).

Region	Grams	No. of shekels in a mina
Phoenicia	353	25
Babylonia	502	60
Syria	518	50
Egypt (gold beqa units)	648	50
Egypt	933	100

Western nations adopt the same units for everyday measurement. As a further illustration of the geographical and historical inconsistencies between measurements, Table 7.1 gives the modern lengths of eight ancient cubits as given by Berriman (1953; see Klein, 1974, p. 59), and Table 7.2 illustrates the different sizes of ancient mina weights (McGreevy, 1995, p. 31).

If standard units have been a long time coming, their merits have long been appreciated, and, many attempts to impose or adopt standard systems have been made. In Chapter 1 we quoted Deuteronomy 25: 13–14: 'You shall not have in your bag two kinds of weights, a large and a small. You shall not have in your house two kinds of measures, a large and a small.' In England, as far back as 1197, standard iron rods and weights were created and the 35th clause of *Magna Carta* (signed by King John in 1215) states: 'Throughout the kingdom there shall be standard measures of wine, ale, and corn. Also there shall be a standard width of dyed cloth, russet, and haberject; namely [a width of] two ells within the selvedges. Weights [also] are to be standardised similarly' (Klein, 1974, p. 30). This aspiration was not altogether successfully achieved. Not for the first or the last time, attachment to local measures and familiar systems proved an insuperable obstacle. McGreevy (1995, p. 99) quotes Bishop Fleetwood as writing in 1745:

It was a good law of King Edgar that there should be the same money, the same weight and the same measures, throughout the kingdom, but it was never well observed. What can be more vexatious and unprofitable, both to men of reading and of practice, than to find that when they go out of one county into another, they must learn a new language or cannot buy or sell anything. An acre is not an acre; nor a bushel a bushel if you but travel 10 miles. A pound is not a pound if you go from a goldsmith to a grocer, nor a gallon a gallon if you go from the ale-house to the tavern. What purpose does this variety serve, or what necessity is there, which the difference of price would not better answer and supply?

Many other similar attempts have been made to enforce consistent units, often with only moderate degrees of success. The one with most lasting success has been the introduction of the metric and SI units described below. The success of these units – which is still not complete – is both a consequence of the ease of communication and travel which we have experienced in the last century, and a facilitator of these.

In Chapter 1 I described how counting and the use of numbers developed as an abstraction from a one-to-one matching between two sets of objects to a system in which one of the sets was imaginary and consisted solely of notional objects. This involves an intellectual leap, abstracting the counting process from the things being counted. Exactly the same kind of leap was required for other kinds of measurement. Although our modern view of things regards a volume as a volume, regardless of what it is a volume of, it is not a priori obvious that this reflects the fabric of the universe around us. Given this, it will come as no surprise that often different units were originally used to measure (what we now regard as) the same physical quantity of different substances. For example, a *barrel* of ale, based on the corn gallon, was 269 cubic inches while a *barrel* of beer, based on the wine gallon, was 231 cubic inches. Similarly the *sack* differed in size according to whether apples, coal, wheat, and so on was being measured. Sometimes different units continue to be used according to the context of the measurement, up to the present day: the use of the *hand* to measure the height of horses is an example.

Once a basic unit has been defined, a common way of defining sequences of units is progressive doubling. Thus, at one time in England the smallest unit of volume was the *mouthful*. Two mouthfuls made a *jigger* or *handful*, two handfuls made a *jack* or *jackpot*, two of these made a *gill*, and two of these made a *cup*. Continuing this doubling strategy we have the *pint*, the *quart*, the *pottle*, the *gallon*, the *peck*, the *half-bushel*, the *bushel*, the *cask*, the *barrel*, the *hogshead*, the *pipe* (although the volume of the pipe appears to have varied according to the substance it was being used to measure, in the manner mentioned above), and, with some relief, finally, the *tun*. Readers will recognize that some of these ancient measures often continue to be used in special contexts. They will also detect possible relationships between these and other measures – for example, the similarity in name between the tun and the ton, a measure of weight. This is not a coincidence – the similarities do have historical reasons – but we will not go into the details here (see, Klein, 1974, for further information). Other example of the doubling system are the system used for paper sizes in Europe (A5, A4, A3, etc.), and the system of weights used in Burma (which, beginning with the *yegwale* (about 0.02551 grams) progressively doubles to give the *yegwi*, the *pai*, the *moo*, the *mat*, the *ngamu*, and more).

The use of base 10, in which successive units are defined as ten times the preceding unit, also goes back a long way in history and has been used in many different areas, including a Japanese system of weights (beginning with the *mo*, and multiplying by powers of ten to give the *rin*, *fun* and *manne* – although it then breaks down, with 160 *manne* equalling one *kin*), the old Shi'zhi Chinese system of market weights, and many others. In Europe, one of the earliest usages of a metric notation was in 1586 when Simon Stevinus published a work describing the system. His notation was slightly different from the modern notation, in that he explicitly indicated the negative powers of the base 10, so that, for example, 3.142 was written as 3 (0) 1 (1) 4 (2) 2 (3). (Explicit indication of the powers was dropped when Henry Briggs replaced them by the single decimal point in 1617.) All of these decimal notions presumably derive from counting on the fingers of our hands and the common decimal number system. In fact, alternatives have been proposed. For example, a committee established in France to oversee the foundation of the decimal system for weights and measures (see below) considered base 12, on the grounds that, with more factors than 10, it would be more convenient. Not everyone regarded this as an advantage – Lagrange, who served on this committee, favoured base 11.

Of course, once a basic unit has been defined, one need not simply use a constant multiple to obtain larger units. To quote from Statute 2nd of Henry VI AD 1423 Chapter 11: 'formerly the Ton of Wine contained 252 Gallons, the Pipe 126 Gallons, the Tertian 84 Gallons, the Hogshead 63 Gallons, the Barrel of Herring and of Eels 30 Gallons, full packed, and the Butt of Salmon 84 Gallons' (McGreevy, 1995, pp. 115–116). Imperial length measurements are hardly better.

The number of different units of measurement which have been used over the course of history is vast and tangled. The units have evolved in just the same way as languages and even peoples have evolved, and a comprehensive history would fill a vast encyclopaedia. Slowly, however, the metric system is gaining ground as a universal system.

It is probably impossible to state unequivocally when any development such as the metric system originated, but certainly one early important step was the suggestion by Gabriel Mouton, around 1670, that France's many different units should be replaced by a decimal system, with increasing units defined as ten times as large as previous ones (see below). Mouton also suggested that mass units should be based on the unit of length via a given volume of water, an idea which was later adopted. Clearly the proposal for a decimal system was not taken up because later, around 1791, immediately after the French Revolution, the French Academy of Sciences responded to a proposal from Talleyrand (1754–1838), by appointing a committee to consider how best to reform France's chaotic system of weights and measures. Actually, 'system' may be too kind a term: the units often varied locally as well as generally across the nation – and, as we describe elsewhere in this chapter, the same name was sometimes for different values, according to the substance being measured. Klein (1974, p. 108) puts it thus: 'Metrologically, as well as politically, France was still absurdly divided, confused and complicated. A given unit of length recognized in Paris, for example, was about 4 percent longer than that in Bordeaux, 2 percent longer than that in Marseilles, and 2 percent shorter than that in Lille.' Of course, we should not single out France. McGreevy (1995, p. 143) describes an 1838 publication from Switzerland showing '37 different values of units of length called the foot, 68 different values for the ell, 83 different values of dry measure for grain, 70 different measures for fluids, 63 different kinds of dead weights.' Alexander (1850) refers to 110 separate values for the ell in Europe and 66 different values of the pfund in Germany. The mind boggles.

The committee appointed following Talleyrand's suggestion included such luminaries as Borda, Lagrange, Condorcet, Laplace and Lavoisier. It is interesting to note that Talleyrand invited both English and American representatives to sit on the committee, but that both invitations were declined. Subsequent history shows how difficult it is, even for nations, to admit that they were wrong.

The base units for the system the committee proposed were to be one ten-millionth of a quarter of an arc of meridian (based on a measurement of a meridian between Dunkirk and Barcelona) for length (the metre, from the Greek *metron*, or 'measure', and this Greek root also underlies 'metric' in 'metric system') and the mass of water in a cube of side one hundredth of a metre for mass (the gramme). These proposals were accepted by the Constituent Assembly of France in 1791. A decimalized time unit was also introduced, in 1793, but was opposed, not least by clockmakers, and the initiative collapsed within a year. One of the problems with decimalizing time is that it has several natural units which impinge directly on everyday life and for which one is not a multiple of 10 of the other: the day, the month, and the year. The 18th Germinal of 1795 may be regarded as the official launch date of the metric system. It defined six basic units: the *metre* (length), the *are* (area), the *stere* and the *litre* (volume), the *gramme* (weight), and the *franc* (for money).

In 1799 the original definitions of these units were made more concrete by constructing physical standards: for example, a bar of platinum was used for the metre. This principle – that there existed a 'standard metre' against which others could be calibrated – continued until relatively recently, when it was overtaken by progress in physics which permitted more fundamental definitions, referred to above and illustrated below.

McGreevy (1995, p. 148) refers to the presentation ceremony of the etalon of length to the two Conseils du Corps Legislatif in 17 June 1799, and describes the spokesman for the Commission des Poids et Mesures as describing the metric system 'as certainly an idea, beautiful, grand, sublime, worthy of the brilliant century in which we live!'. However, despite this enthusiasm amongst those driving the innovation, not everyone was so pleased, and public protest led to Napoleon permitting a partial reversion to the old system, partial in that he kept the metric units but authorized the use of old units defined in metric terms (e.g. the *pied*, with a length of a third of a metre). It was not until 1837 that the metric system proper was again adopted. Similarly, the spread to other countries was also hesitant: to the Netherlands in 1816, Greece in 1836, Switzerland in 1877, and so on. Indeed, the process in many countries (the United States being perhaps the most notable) is still not complete.

A significant step was made in 1870, when France established a permanent body, the Metre Convention or *Convention du Mètre*, comprising 30 nations. In 1875, the *Conférence Générale des Poids et Mesures* (General Conference on Weights and Measures), referred to elsewhere in this chapter, was set up by 17 nations (including the Unites States, although the United Kingdom, still fighting its rearguard action, did not join until 1884). The Conference held its first meeting in 1889. The Conference elects members of the *Comité International des Poids et Mesures*, and, under the terms of the Metre Convention, this supervises the *Bureau International des Poids et Mesures* (BIPM: see http://www.bipm.fr/enus/). The Comité brings together representatives of the governments of member states once every 4 years, and also runs several consultative committees of expert advisors on scientific and technical matters. In October 1999, a new category of Associate was established for states and economies not yet members of the Metre Convention, and by April 2000 there were 48 members and one associate (Hong Kong).

The metric system looks set to stay for a very long time, in contrast to the rapid creation and demise of the many other systems which have existed. Some of these are mentioned in the following sections. There are far too many to list all of them – indeed entire books have been devoted to this. The United Nations published a survey in the 1960s, the *World Survey of Weights and Measures*, listing many of the strange units still in use at that time, and Cardarelli (2003) is a very impressive compendium of units from throughout the world and throughout history.

7.2.2 The Système International d'Unités

Various metric systems have been proposed for scientific use, going as far back as the early nineteenth century. These include a millimetre–milligram–second system (Gauss and

Weber), a metre–gram–second system (the British Association for the Advancement of Science), and the centimetre–gram–second system (CGS) (also proposed by the British Association, in 1873). The basic units of these systems are self-evident in their names. The CGS system was used widely in scientific circles, but it does have the disadvantage that its basic units are too small for practical use. Because of this it was eventually superseded, starting at about the beginning of the twentieth century, by the metre–kilogram–second system (MKS), which was formally adopted for scientific work in 1954, at the 10th meeting of the General Conference on Weights and Measures. Six years later, at the 11th meeting, this system was given the name *Système International d'Unités* or International System of Units, with abbreviation SI (not, we see, 'SI system', which is repetitive). The 11th meeting established the base units (the metre, the kilogram, the second, the ampere, the kelvin, the mole and the candela) and also the rules to be adopted for prefixes. Of course, in a refrain which the reader will be finding familiar by now, things have not stopped there. The SI is not static, but has continued to develop in response to more sophisticated requirements. This should only be expected: it is an aspect of scientific progress. Nowadays the system includes the radian as the unit of angular measurement (1 radian is the angle which defines an arc of length r on a circle of radius r) and the steradian as the unit of solid angle (1 steradian is the solid angle which defines an area of r^2 on the surface of a sphere of radius r). It also includes a large number of units derived from the basic units, including the newton for force, the joule for energy, the watt for power, the coulomb for charge, the volt for electromotive force, the ohm for electrical resistance, the farad for electrical capacitance, the weber for magnetic flux, the henry for electrical inductance, the tesla for magnetic flux density, and the hertz for frequency. It will be clear from this that one way to achieve a kind of immortality is to work in a scientific discipline at the start, when the basic units are being defined. Beyond these derived units which have been given special names lie others which are simply named by combining their component elements: kilometres per hour, kilograms per square metre, and so on.

7.2.3 Length

Most readers will be familiar with the phrase 'Man is the measure of all things,' attributed to Protagoras by Plato. A glance at the early history of length measurement might make one feel that this is meant in a rather literal, even anatomical, sense. Thus we have:

- the *cubit*. This was based on the length of a man's forearm, from the elbow to the tip of the extended fingers. The range of different values for this length is illustrated in Table 7.1.
- the *inch*. This was defined by King David I of Scotland around 1150 as being the average of the widths of the thumbs of a small man, a medium man, and a large man, measured at the base of the thumbnail. I draw attention to the averaging implicit in this definition: here is a statistical concept being used centuries before it had been properly defined. However, the use of thumbs as the basis of an inch is not the only suggested origin. Writing in the sixteenth century, Sir Rycharde de Benese (1562) wrote:

The lengthe of an ynche after some mennes opynion, is made by the length of thre barly cornes, ye whiche rule is not at all tymes true. For ye length of a barly corne of some tillage is larger, and some shorter, after the fatnes and leanesse of the lande, where it was sowne up(n). Therefor i(n) makyne of an ynche after thi rule It should be somtymes longer and somtymes shorter after the lengthe and shortnes of the barly cornes: the which should make great dyfference in measuring.

Recognizing the intrinsic variability of these seeds, he then went on to say:

Therefore ye that take the length of any ynche more truly upon an artificers rule made of two foote in legth, after the standarde of London, the whiche rule doth conteyne xxiiii ynches in lengthe.

- the *nail*, 'the distance between the last two joints of the middle finger of a man's hand' (McGreevy, 1995, p. 122), used for measuring cloth. It was also defined as 1/16 of a yard and 1/8 of a cubit.
- the *hand*, originally the crosswise width of the four extended fingers (and later defined as 4 inches). The hand is, of course, an example of a measure of length still used in certain special contexts, namely the measurement of the height of horses.
- the *foot*. As with the cubit, numerous versions of the 'foot' have existed in the past, including Persian versions, three different Greek versions, two different Roman versions, the *che* of mainland China's market system, the *shaku* of Japan, the *chok* of Korea, the *tercia* of Costa Rica, the *Fuss* of Germany, a measure based on 2/3 of a cubit, and many others. Many of these are significantly longer than average human feet, suggesting that in many cases the origin was different. Nonetheless, Klein (1974, p. 66) describes how the length of a foot was determined according to the writings of Master Koebel, written in the sixteenth century:

The surveyor should station himself by a church door on Sunday. When the service ends he should 'bid sixteen men to stop, tall ones and short ones, as they happen to come out …'. The chosen sixteen should be made to stand in line with 'their left feet one behind the other.' The resulting sum of sixteen actual left feet constituted the length of 'the right and lawful rood,' and the sixteenth part of it constituted 'the right and lawful foot.'

Once again this is interesting from a statistical perspective, since it makes use of both of the modern notions of averaging and random sampling.
- the *fathom*, originally the height of a man.
- the *li* in China, the distance a coolie could cover, carrying a known burden, between rest stops (McGreevy, 1995, p. 23).
- in Thailand, as late as 1940, a unit was specified in terms of the 'distance through thick jungle, on a clear night, over which one could hear the bark of a dog' (McGreevy, 1995, p. 23).
- the *yard* in England, which may have been defined by the distance from the nose of King Henry I to the tip of the middle finger on his outstretched arm (McGreevy, 1995, p. 60), but other origins have it as the length of a pace. Alexander Thom (e.g. Thom, 1955, 1962, 1964) has analyzed extensive data and concluded that society in Britain and France, over the period from about 1800 BC to 1200 BC, used a consistent unit of length (the *megalithic yard*) which measured approximately 0.829 m, and which varied by no more than a few millimetres across the region. Those who have read the introductory part of this section will perhaps consider this assertion, a priori, to be unreasonable. That such early groups

could maintain consistency over such a period and area stands in stark contrast to more recent experience. If it were true, it would seem to imply a level of civilization substantially greater than that generally assumed for the period. Other authors (Freeman, 1976, for example) have been less convinced by Thom's analyses, and have presented alternative analyses which do not draw the same conclusion.

Early units for longer lengths were based either on the times to cover distances (e.g. how long one could walk in a day, or the Chinese *li* mentioned above) or on divisions of measurements of the size of the Earth. Taking the former approach first, more accurate measures were based on multiples of smaller units. Again there is a large number of variants. One version of the mile is that it is derived from the phrase *mille passuum* – a thousand paces – where a *passus*, or pace, was the length of two steps of a marching Roman legionnaire. Arnold's *Customs of London*, published in around 1500, gives the following (see Klein, 1974, p. 69) 'The length of a barley corn 3 times make an ynche and 12 ynches make a fote and 3 fote make a yerde and 5 quaters [quarters] of the yerde make an elle. 5 fote make a pace. 125 pace make a furlong and 8 furlong make an English myle.' Hylles (1600) gives (McGreevy, 1995, p. 123):

3 Barley Cornes = 1 inch
6 foote = 1 fadome
4 Barley Cornes in breadth make 1 fingerbreadth
4 fingers make 1 Handbreadth
3 Handbreadths make 1 Spanne
4 Handbreadths make 1 foote
6 Handbreadths make 1 Cubit
2 Foote make 1 Single Pace
125 Paces make 1 Furlong
1000 Paces make 1 Italian mile

Gabriel Mouton's suggestion for the decimalization of length measurement, mentioned above, was based on the length of one minute of arc of a great circle around the Earth. He called this a *milliare*, and divided it by progressively greater powers of ten, to give the *stadium, funiculus, virga, virgula, digitus, granum* and *punctum*. Of course, to be precise, this requires accurate and replicable measurement of a minute of arc of a great circle, and precise measurements were not available when Mouton made his suggestion. There is a tendency for us to look back at the efforts of these early 'scientists' with a sense of kindly understanding, as if they were children doing their best to achieve what we can see as obvious. This is, of course, entirely inappropriate. The early investigators were in fact highly sophisticated and careful, and working within established frameworks of their own. The subtlety with which the problems were tackled is indicated by Mouton's work. Appreciating the lack of accuracy in the measurement of great circles, he suggested instead measuring the virgula (about 18 cm) using a pendulum. At a given point on the Earth, pendulums of any given length swing with a particular frequency, so that the number of swings in a given time period can be used to determine length. In fact, Mouton suggested that the virgula should be the length of a pendulum which makes 2959.2 single swings in half an hour. (Other people also suggested that length should be defined in terms of the frequency of a pendulum swing. For example, in 1685, Richard Cumberland suggested that the *horary yard* should be the length of thread, with a bullet tied to the end, required to swing 60 times a second, and in 1794 William Martin defined a unit of length as 39.2 inches, this being the length required to give a 1-second period.)

We mentioned above that, in 1799, the metre, taken as one ten-millionth of the length of a quarter arc of meridian, was formally defined as the length of a particular bar of platinum stored in France. In fact, there is considerably more behind that statement than at first appears: defining the metre depends on determining the length of a quarter arc of meridian. This was painstakingly determined, over a 7-year period, by two Frenchmen, Jean-Baptiste-Joseph Delambre and Pierre-François-André Méchain, who measured the distance between Dunkirk and Barcelona. Unfortunately, as Alder (2002, p. 8) narrates in a fascinating story of the efforts of these two men, a mistake was made, to the extent that 'according to today's satellite surveys, the length of the meridian from the pole to the equator equals 10,002,290 metres. In other words, the metre calculated by Delambre and Méchain falls roughly 0.2 millimetres short.'

The metre bar of platinum was not the first metal etalon used for length measurement, and neither was it the last. In 1588 in England bronze rods had been used to define the yard, and these lasted until 1742 when they were replaced by brass etalons (still for a yard). The platinum bar continued to be the metre etalon for almost 200 years – until the middle of the twentieth century. McGreevy (1995, p. 151) describes the mètre des Archives as using the 'end standard' approach, in which the length of a metre is defined as the distance between the two ends of the bar. One of the disadvantages of this is that it is susceptible to wear. Moreover, the mètre des Archives had a cross-section which permitted a slight but detectable degree of bending when it was used. Because of these problems, the Metre Convention agreed to replace this bar by a new etalon of improved design. After many tests conducted in the 1870s, a design was adopted which had an cross-section shaped like a winged X, used a line standard instead of an end standard (one metre is the distance between two very fine lines engraved on the bar), and was an alloy of 90% platinum and 10% iridium. 'Copies of this prototype bar were sent to all the major standardising laboratories of the world; copy number 16 was allocated to the UK and copy number 27 to the USA. The national copies of the metre differed from the prototype and from each other by less than 0.01 mm, with a probable error not exceeding 0.2 micrometre' (McGreevy, 1995, p. 151). Chisholm (1877) gives a full history of these bars. At its first meeting, in 1889, the General Conference on Weights and Measures adopted the prototype, a copy of the mètre des Archives, as its standard metre.

Physical bars are all very well, but they are not ideal, and even bars using a line standard have problems. They are subject to wear and tear, accidents (for example, in 1834 a fire destroyed the Houses of Parliament, seriously damaging the imperial yard etalons which were kept there), distortion when moved or used, and are susceptible to the conditions under which they are measured, as well as natural relaxation processes in their material (for example, it was observed in 1947 'that the imperial standard [yard] bar was decreasing in length by about 10^{-6} inch every 23 years' (McGreevy, 1995, p. 154)). Because of these properties of physical objects it had been suggested, as far back as the first half of the nineteenth century, that a more consistent definition of length could be based on phenomena not subject to wear and tear. The lengths of pendulums are an example of this idea, but it was the wavelengths of other kinds of vibration which were eventually adopted. In particular, the wavelength of electromagnetic radiation was chosen, although it was not until the first half of the twentieth century that instrumental techniques made this feasible. Thus, in 1927 the Seventh meeting of the General Conference on Weights and Measures tentatively adopted an alternative definition of the standard metre, defining it as equal to 1,533,164.13 wavelengths of the red spectral line of cadmium. In 1960 the Conference formally replaced the definition based on the physical metre bar by the following: 'The metre is the length equal to 1,650,763.73 wavelengths in a vacuum of the radiation corresponding to the transition between levels $2p_{10}$ and $5d_5$ of the krypton-86 atom.' This definition in turn gave way, in 1983, to the current definition, which is the length of the path travelled by light in a vacuum during a time interval of 1/299,792,458 of a second.

Although the metre has been defined in terms of fundamental properties of matter, in a sense it is not, itself, fundamental. More fundamental measures can be defined in terms of universal constants. One example is the *Bohr radius*, defined as $a_0 = h^2/4\pi^2 m_e e^2 = 5.29177 \times 10^{-11}$ metres, where h is Planck's constant, m_e is the rest mass of the electron, and e is the charge of the electron. Klein (1974, p. 195) gives others.

Units other than decimal power multiples of metres are sometimes used for extreme lengths in scientific applications. Examples are the *light year* (the distance that light travels in a vacuum in a year – about 9.463×10^{15} metres) and the *parsec* (the distance at which a star would have an apparent displacement ('parallax') of one second of arc as the earth orbits the sun, equal to 3.2616 light years), although, especially at the opposite extreme, decimal powers are used: for example, the *micron* (10^{-6} metres) the *angstrom* (10^{-10} metres), and the *fermi* (10^{-15} metres). We see that the notion of using the time to travel a distance as the basis for measuring that distance has been used throughout history, from the Chinese *li* mentioned above, to today's light years.

Although it was a long time before other countries adopted metric measurements, in fact both the US and the UK defined their *yard* in terms of the metre: unfortunately, however, they adopted slightly different lengths, with the US yard being 0.91440183 metres and the UK yard being 0.9143992 metres (later changed to 0.9144; and still later to 1,420,212 times the wavelength of the red line in the spectrum of cadmium – adopting a system based on fundamental atomic properties). As we have mentioned above, this is not the only way in which the US and the UK have adopted slightly different sizes for their basic units.

Length and angle measurement play a fundamental role in measuring many other properties. Prior to the advent of direct electronic readouts, many, perhaps most numerical values were read from length scales, sometimes mapped to circles, as in a clock or gauge, to give an angle. Modern familiarity with gauges and indicator scales makes their interpretation immediate and intuitive, but it was not always so (and one wonders if it will remain so for much longer: one of the author's graduate students confided that they found it difficult to tell time from an analogue clock dial, having been raised on digital clocks). In particular, the early measurement of electrical units experienced a controversy over whether 'direct reading' instruments were a legitimate way to measure – see Section 7.2.10.

7.2.4 Area

Biological, if not anatomical, origins also seem to underlie many early units of area. An early form of area measurement arose in agriculture, where an area of land was defined in terms of the weight of grain which was needed to sow the area. Thus, for example, in Germany a volume of a *scheffel* of grain would seed an area of land also known as a *scheffel*. Following a similar sort of principle, similar also to the notion of determining long distances by how long it took to walk them, areas have been defined by how long it would take to walk or ride around them (a modern observer will note the effect of shape on this measure). Yet other units of agricultural area were defined in terms of how much work was required to cultivate them: for example, how long it would take to plough a given field. From a human perspective, this is by no means a silly idea. It is taken further in the 'knyghtes fee', based on an amount of land awarded to a knight in return for his support, and which would feed him and several peasant households.

The word 'rood', which appeared in the previous section as a unit of length, was also used as a unit of area. One rood was the area of a *furrow*, which was a strip of land one *rod* wide (see the previous subsection, equal to 16½ imperial feet) and 40 rods long. Some subtleties sometimes appear: McGreevy (1995, p. 21) describes the *double remen*, which is the length of the hypotenuse of a square of side one royal cubit. This is useful, since a square of side one double remen will have twice the area of a square of side one royal cubit.

The relationship between length and area measures, while familiar and apparently obvious to us, is not really so. After all, they relate to different concepts (which might reasonably be measured in entirely different units – as are, for example, weight and height), and it takes some degree of sophistication to conceptualize one as based on multiplying pairs of the other. Indeed, although in most modern systems of area measurement the unit of area is defined as the area of a square with side unit length, this is not the only possibility. For example, for electrical conductors in cables, electrical engineers in the US have defined the unit of (circular) cross-sectional area, the *circular mil*, to be the area of a circle of diameter 1 mil (which is a thousandth of an inch). It follows that the area of a circle of diameter d mils is d^2 circular mils. This is fine, and very convenient, provided one only works with conductors with circular cross-sections, but requires the introduction of adjustments involving π when other shapes are involved: an object of area A in traditional units has area $A4/\pi$ in 'circular' units.

7.2.5 Volume and density

Volume measurements appear usually to have originated in containers for liquid or grain (and, in these sorts of applications, the word *capacity* is often used instead of *volume*). For obvious reasons care had to be taken: solid contents such as grain can be heaped in a container, and the extra quantity in the heap depends on a nonlinear way on the radius of the rim of the container. An example of the care which has been taken to overcome this nonlinearity is given in the *Fleta*. This is a Latin document, thought to have been written about 1290 by someone imprisoned in the Fleet Prison in London, which says: 'let the bailiff or his deputy keep the barns and deliver the corn to the reeve by stricken measure [that is, flat to the container's rim] and not by heaped … stricken measure [that is, heaped], I say, because fraud can very quickly be effected with a heaped measure, since 4 heaped measures will contain 5 stricken, more or less, if the measure is a broad one, and, if it is narrower, 6 will be contained in 5, and if still narrower, 7 will be contained in 6' (McGreevy, 1995, p. 82). This is clearly an attempt to explain a nonlinear relationship. Compounding this is the danger of settling – a phenomenon which helps explain why cartons of granular products are often not quite full.

Since there are many natural (biological or otherwise) objects of roughly uniform size for measuring volume or capacity (analogous to seeds for measuring weight) there has been a great variety of volume units. Examples of this still occur in cooking, where we use the units of *cup*, *teaspoon* and *tablespoon* to measure liquid volume. A cup is approximately 0.23 l, a teaspoon is approximately 4.9 ml, and a tablespoon three times this. In fact, these measures are defined exactly in terms of *drops*: a drop is approximately 0.0648524 ml, a teaspoon is exactly 76 drops, and a tablespoon is exactly 228 drops. Of course, it would be remiss not to note that a *British teaspoon* is not quite the same, being 95 drops. There are also less familiar cooking units used occasionally, each of which have proper definitions: a *coffeespoon* is exactly 19 drops. These and other exotic definitions are given in Johnstone (1998).

Of course, once again, having defined a basic unit, it is easy to define multiples and submultiples. Thus the Old Testament (Exodus 16: 36) tells us that 'An omer is the tenth part of an ephah' (McGreevy, 1995, p. 25), where both the omer and the ephah are volume measurements.

Often different units were developed for solid and liquid volume measures, and even for different types of liquid or solid. This was mentioned above, and McGreevy (1995, p. 119) gives a nice example:

> In the reign of Elizabeth I (1558–1603) there was a dispute concerning the size of the barrel used in London for the sale of herrings. Legislation was proposed in 1571 to remedy a certain malpractice which had grown up over the years. Fishermen had been packing their fish in barrels of 32 gallons, based on the wine gallon or Exchequer standard gallon. But the correct gallon for herrings was distinctly larger; the 32 gallons used for this purpose were only equal to about 28 gallons. The latter, larger gallons should have been used for herrings. The practice gave the sellers an extra 14% profit, based on a fixed price per barrel. It was proposed in the new legislation that there should be 'at least 1000 herrings' in a barrel, since this was estimated to conform to the old standards which had in practice given way to illegal measures.

If barrels of herrings have been sorted out to the satisfaction of modern herring purveyors, other confusions remain: within the United States a dry quart is 0.038889 cubic feet while a liquid quart is 0.033421 cubic feet.

The litre is the most common modern measure of volume used nowadays in metric countries. This is was originally defined as a cubic decimetre, but creating an accurate etalon to match this definition is difficult, and for the period between 1901 and 1964 it was defined as the volume occupied by 1 kilogramme of pure water at 4°C. This only lasted until certain small inconsistencies were noted between the definitions of units of mass, volume, and density based on this, and the definition of a cubic decimetre was then readopted. Hence millilitres and cubic centimetres are equivalent.

Collier's Encyclopedia of both 1987 and 1992 states that 'The liter, a metric unit of volume or capacity, is named for a founding father of the metric system, Claude Emile Jean-Baptiste Litre, whose family sold wine in the first litre (liter) bottles to bear the name.' The journal *Chemistry International*, no. 1, 1980, p. 32, contains a summary of Litre's life. Both this article and the encyclopedia definition appear to be based on an article by Ken Woolner of the Department of Physics at the University of Waterloo in Canada, which appeared in *CHEM 13 News*, April 1978, a publication from the University of Waterloo aimed at high school teachers, which outlined Claude Litre's life story. The one page article noted that 'In celebration of the 200th anniversary of the death of this great investigator, the Conférence Générale des Poids et Mesures has decided to use his name for the SI unit of volume. The official abbreviation will be L, following the standard prescription of using capital letters for units for named individuals.' The body of the article gave considerable details about Litre's life:

> Claude Émile Jean-Baptiste Litre was born on February 12, 1716, in the village of Margaux in the heart of the Médoc region of France. His father was a manufacturer of wine-bottles, as had been his grandfather and great grandfather. Indeed, Litre bottles had been a vital adjunct of the Bordeaux wine industry since the 1620s. This family tradition of concern for the problem of liquid containment, and knowledge of the properties of glass, was undoubtedly a major influence upon Litre's later work on the measurement of volume.
>
> By the age of 16, Litre had demonstrated a budding mathematical talent, and he was sent to Paris to study with Pierre de Maupertuis (1698–1759), who became his scientific mentor.

The article goes on to describe how Litre joined an expedition to Swedish Lapland to measure the curvature of the Earth, how he met Celsius, developed a very successful business making chemical glassware, and in particular how he became renowned for the accuracy of the apparatus he made:

> Before Litre, no one had ever made an accurate cylinder of clear glass, and yet his cylinders varied in internal diameter by less than 0.1% over their whole length. And no one, before Litre, had so precisely graduated a cylinder of glass – into tenths, hundredths, and sometimes even thousandths! His graduated cylinders, and his burettes (he invented the device, and its name) were coveted by chemists all over Europe.

It describes how his book, *Études Volumétriques*, was translated into English (by none other than Joseph Priestley) as well as into German, and how he was awarded a special gold medal by the Royal Society. He died in the 1778 cholera epidemic on 5 August.

> In his *Études Volumétriques*, Litre had chosen, for his standard volume, a measure very close to the old *flaçon royal* of Henri IV, introduced in 1595 to standardize the taxation of wine. However, he recognized the arbitrariness of this unit, and suggested that in any rationalized system of units, volume could be specified in terms of a standard mass of a standard liquid. He suggested mercury. But Litre's dream of a rationalized system of units did not start to materialize until 15 years after his death, when the mathematician Lagrange (1736–1813) was appointed to head a commission to draw up such a system. And in 1795 the metric system was born.
>
> Litre's method of specifying volume was adopted, although the commission decided to use distilled water rather than mercury as the standard liquid.

Unfortunately, Claude Litre never existed. His genesis

originated with a technical quibble: Some chemists wanted to make the abbreviation of the litre measurement to be 'L' instead of 'l', to avoid confusion with the figure 1. 'But the rules of nomenclature of the Conférence Générale des Poids et Mesures require that upper-case letters be reserved for units which are named for individual scientists', Woolner says. 'Since no such person existed for the litre measurement, it seemed reasonable one should be invented. (University of Waterloo Gazette, May 17, 1999, p. 1).

Woolner's article in CHEM 13 News was the result.

'I realized that drama, revolution and romance were all very good, but the joke would be better if the article came across as a piece of sober historical research' Woolner wrote ten years later. 'The idea was to give a straightforward account with correct dates and accurate historical details, but with one person added to the great stream of history. So that's how I wrote it.' ... And over the years other scientists filled in missing sections to Litre's biography.

These included the creation of a daughter named Millicent.

Woolner later writes (*CHEM 13 News*, September, 1988, Number 178, pp. 45–46):

I also found out I had been 'scooped' by Bruce Dodd of the Canadian Government Specifications Board, who had published a 'Research Note' on Marco Guiseppe Litroni in the June 1977 issue of *Standards Engineering*. The link between Litroni and Litre was provided in 1979 by Steve Marriott of the British Standards Institution, who speculated that the name 'Litre' was adopted by Litroni when he settled in France after a mafia-induced flight from his native Tuscany.

This later note by Woolner includes references to other contributions to the history.

Woolner also noted (*CHEM 13 News*, September 1988, Number 178, pp. 45–46):

There is a special corner of Hell reserved for failed humorists, and the first indication that a place was being prepared for me came as early as November 1978, when a college librarian wrote from California requesting '... any information and sources you have used in the preparation of this most interesting article.'

This is probably not the end of the story: 'A local high school teacher recently wrote to *CHEM 13 News* complaining that until he read Woolner's explanation, he had been teaching his students about Litre' (*Winnipeg Free Press*, 28 Feburary 1989, p. 1).

This story makes one wonder about all those other improbable sounding people whose names have become attached to other units: farad, coulomb, ohm, tesla, watt, poise, and so on – and as for 'henry', well! On the other hand, perhaps the 'hand' used in measuring the heights of horses is not based on the span of four fingers at all ...

Density is a derived variable, defined as mass per unit volume. Density measurements of liquids can be made using a *hydrometer*, an instrument which floats in the liquid, with a stem projecting above the liquid's surface. The lower it floats, the greater is the volume of liquid displaced by the hydrometer, and the less dense is the liquid. Graduations on the stem are used to read off the density directly – though things are slightly complicated by the fact that the density will be greater at lower depths. The history of measuring the densities of liquids is a fascinating one in its own right, owing more to issues associated with alcoholic beverages and adulteration, and the taxation of such liquids, than to science. A large number of different measurement scales through which to record the density have been created at different times and in different places, including the Bates, Brix, Twaddell, Tralles, Cartier, and Baumé scales. Just one example is given by the alcoholometer invented by Gay-Lussac. This covered a range of densities for water–alcohol mixtures, with 0 corresponding to pure water and 100 to pure alcohol (measured at 15°C), and is based on the proportion of alcohol content by volume. The word 'proof', used in measuring the alcoholic content of drinks in Britain, has gone through various meanings. Nowadays it indicates the percentage change in volume that would occur were the liquid to be diluted or concentrated sufficiently to yield a 100 degree proof, a density of 12/13 that of pure distilled water at 60°F. The American usage of the same term is slightly different (as the reader is doubtless beginning to expect), being about 12.4% weaker than the British meaning.

7.2.6 Weight

The earliest attempts to construct weight measures were probably based on seeds. For example, one *pennyweight*, equal to 5% of a troy ounce, was based on the weight of 32 wheat corns. Similarly, (McGreevy, 1995, p. 2): 'An ancient Hindu unit of weight was the berry of the shrub now known as *Abrus precatorius*. In modern measure, 80 of these berries would be equal to about 6.8 grams and these would together form the unit.' Other weight units were based on black and white mustard seeds, barleycorns and rice grains.

Defining weights using biological specimens is all very well, but something more enduring, convenient, and recognizably consistent is required for practical purposes (e.g. in the marketplace). Thus objects of known weight were constructed. Limestone weights were used in Egypt around 3800 BC, for example. Rocks are all very well, but they are not very dense, so that heavy weights are relatively large and inconvenient. Because of this, metals, notably bronze and brass, replaced them in Egypt around 1500 BC – although, according to McGreevy (1995, p. 29), they reverted to stone after the death of Tutankhamun in 1352 BC.

There is an interesting interplay between coins and weights, with coins often serving as basic units of weight. This explains the dual use of the word 'pound' in Britain and some other countries right up to the present day. Unfortunately, the manufactured weights of early coins were not very consistent, and they became less consistent with use. Furthermore, those made of precious metal, such as silver, had a tendency to diminish in size as their edges were clipped. Irregularity in the individual weights of coins was often tackled by making payments in terms of a given weight of coins, rather than a given number (indeed, at times, payment in terms of the number of coins – 'by tale' – was illegal). This even more strongly intertwines notions of weight and value.

Early monetary inflation is also recorded in terms of the relationship between the coinage and weight. As Robert Recorde wrote in 1534 in *The Groüd of artes teachyng the worke and practice of Arithmetike,* 'the leaste portion of weyghte is commonly a grayne, meanynge a grayne or corne of whete, drye, and gathered out of the myddell of ye eare. Of these graynes in tymes passed 32 wayed juste 1 penny of troye, and then was but 20 pennes in an once, but nowe are there 46 pennes in an once, so yet there are not fully 14 graynes in 1 penny. But now of onces (after troye rate, which is the standarde of Englande) 12 do make 1 pound' (McGreevy, 1995, p. 104).

In Section 7.2.1 I mentioned the strategy of doubling to define successively larger units. In fact this strategy also formed the basis of some concrete physical systems. For example, weights used for scales are often in the form of a nested sequence of (for example) 1 oz, 2 oz, 4 oz, 8 oz, etc. The merit of such a system is that combinations of these can be used to make up any weight from 1 oz up to 1 oz less than twice the largest: the binary system did not begin with computers. With an extra 1 oz weight, each weight will be the sum of all smaller weights.

As in other physical measurement domains, some special disciplines and areas retain their own specialized units. Klein (1974, p. 82) gives the origin of the word *carat*, a unit equal to 0.2 grams used for measuring the weight of diamonds, as the Arabic word *qirat*, meaning the seed of the coral tree. McGreevy (1995, p. 2) gives alternative possible origins, including, *keration*, from the seeds of Greek carob trees, and the beans from Abyssinian *kuara* trees. (The word 'carat' also has another quite distinct meaning, measuring the purity of gold articles in terms of 1/24 fractions. Thus an 18-carat item would be 3/4 gold and 1/4 some other metal.) Although, at the time of writing, metrication is slowly progressing, in Britain the weight of people is still commonly expressed in stones and pounds, where a stone is 14 pounds. This can be confusing to American ears, since in the US it is common simply to use pounds. Indeed, after the publication of Hand *et al.* (1994), the author was castigated by an American colleague for using the 'archaic' units of stones in a table of data describing the weights of the crews of the annual boat race along the Thames between Oxford and Cambridge Universities (Table 414 in Hand *et al.*, 1994). In fact, the boat race itself is something of a historical anomaly, and I doubt that it will ever adopt metric weights.

The oldest instrument for measuring weight appears to be the two-pan weighing scale, certainly in use as early as 3500 BC. The simplest such instrument has equal arms, so that it balances if the weights on each side are the same. The object to be weighed will be placed on one side, and a standard weight (or set of standard weights) on the other. A more sophisticated version (called a *steelyard*) has arms of unequal length and permits the position of the known weight to be moved along the longer arm. This allows a heavy object to be weighed by balancing it against a lighter standard weight on the long side. These are sometimes used in scales for weighing people (one occasionally sees them on seafronts), so that, for example, a weight of a few pounds on one side balances a 12 stone man on the other. Yet another variant (called a *bismar* by the Anglo-Saxons or *auncel* by the Normans) has a fixed weight at one end, and achieves a balance by moving the fulcrum.

The bismar is, in fact, less accurate than the steelyard. Suppose we are weighing an object of weight w_1 using a counterweight w_2, so that $w_1 d_1 = w_2 d_2$, where d_i is the distance between the fulcrum and weight w_i. Now replace the object by one of weight $w_1 + \delta w$, with $\delta w > 0$. If a steelyard is used for the weighing, we will move the counterweight a distance $\delta d > 0$ to increase the distance from the fulcrum to the counterweight, so that

$$(w_1 + \delta w)d_1 = w_2(d_2 + \delta d) \tag{7.1}$$

On the other hand, if a bismar is used, we will move the fulcrum by a distance $\Delta d > 0$ so that

$$(w_1 + \delta w)(d_1 - \Delta d) = w_2(d_2 + \Delta d). \tag{7.2}$$

From (7.1), $\delta d = (w_1 + \delta w)d_1/w_2 - d_2$, and, using this in (7.2), we find

$$\Delta d = \frac{(w_1 + \delta w)d_1 - d_2 w_2}{(w_1 + w_2 + \delta w)}$$

$$= \frac{\delta d \cdot w_2}{(w_1 + w_2 + \delta w)}$$

Now, since $w_2/(w_1 + w_2 + \delta w) < 1$, it follows that $\Delta d < \delta d$. That is, the bismar achieves the same rebalancing effect by moving the fulcrum through a distance shorter than the distance through which the steelyard's counterweight is moved. This means that the bismar's measurements will be less accurate. Moreover, we see from (7.1) that $\delta w \times d_1 = w_2 \times \delta d$, so that the steelyard adjustment δd is proportional to the change in weight δw, whereas the bismar adjustment Δd is nonlinearly related to δw. The consequence is that the latter's accuracy changes as the weight being measured changes.

This inaccuracy had been appreciated earlier, but presumably based on practical experience, rather than analysis. Thus we find a statute of Henry VI requiring (McGreevy, 1995, pp. 124–125): 'whereas it hath been of old times accustomed in all the counties of England, that all cheeses which ought to be sold by the Wey should be weighed by the Auncel, and because that at the last Parliament holden at Westminster it was ordained that the said Auncels, in respect of the great Deceit of the same, should be destroyed …'. McGreevy (1995, p. 125) also quotes a 1461 document as saying: 'Aunsell wyght is forboden by the Parlement; and also holy Chyrche hath cursyd all theym that by oor sell by that weyght, for itt is false' (they didn't mince words in those days).

By some extraordinary coincidence, the day after writing the above paragraphs, I received, from the UK's Royal Mail, a small cardboard device which would permit me to weigh letters so that I could determine the value of the stamp to use. This device is a cardboard cylinder with equilateral triangular cross-section. Continuing along the axis of the cylinder, but projecting beyond its end, is another triangular cylinder with smaller cross-section. This smaller cylinder has a series of slots cut in its length. Sufficiently heavy envelopes placed in a slot will cause the device to pivot about the end of the larger cylinder. The slots are marked with weights of 100 g, 60 g, 40 g, and 20 g. So this device uses a similar mechanism to those outlined above, but the fulcrum is fixed, and so is the 'standard weight' (this being the weight of the bulk of the larger cylinder). What is changed is the distance, from the fulcrum, of the object to be weighed.

Hooke's law (Robert Hooke, 1635–1703) states that the extension of a spring is proportional to the applied load, a law which makes possible the *spring balance.* In fact, he first published this law as a postscript in his book *Description of Helioscopes*, in 1675, writing 'the true Theory of Elasticity or Springiness, and a particular Explication thereof in several Subjects in which it is to be found: And the way of computing the velocity of Bodies moved by them. ceiiinosssttuu'. The last word is not where I accidentally leant on my keyboard, but is in fact an anagram, with Hooke revealing the translation to be *Ut tensio sic vis* in 1677: 'That is, the Power of any Spring is in the same proportion with the Tension therof: That is, if one power stretch or bend it one space, two will bend it two, and three will bend it three, and so forward. Now as the Theory is very short, so the way of trying it is very easie' (Hooke, 1678).

Other forms of weighing device have been developed for specialized scientific applications. The development of particularly sensitive devices has led to a deeper understanding of the limits to what can be attained. *Flexure-strip* balances, using flexures of an alloy of copper and beryllium, can measure to a few parts in 10^{11} in air, and even better *in vacuo.*

In most everyday situations, the distinction between weight and mass is irrelevant, and the same units are used for both. In physics, however, the distinction is important, and different units must be used. Weight is a force – that which will accelerate the body in question at a rate of g metres per second per second (where g is the acceleration due to gravity at the surface of the Earth – about $9.8 \, \mathrm{m \, s^{-2}}$). The names for these units depends on what one regards as the basic unit. For example, the Metric Technical System defines the mass unit, the *metric slug*, to be the mass which is accelerated at a rate of $1 \, \mathrm{m \, s^{-2}}$ when a force of 1 kg is applied. Conversely, in the SI, the newton (Isaac Newton, 1642–1727) is defined as the force which will accelerate a mass of 1 kg by $1 \, \mathrm{m \, s^{-2}}$. The original kilogram, defined in 1795, was one thousand times the weight of a cube of water of side 1 cm at 0°C at standard atmospheric pressure. There were problems in using this definition. One is that water freezes at 0°C, and it was found very difficult to cool water to this temperature without producing some ice, which in turn meant that the volume was distorted. Because of this, the definition was changed, in 1799, to refer to water at 4°C, the temperature at which it has its greatest density. A metal block was made with this weight, and it became the standard kilogram (although it was initially called the *kilograve*). This standard object was replaced, 90 years later, by a cylinder of 90% platinum and 10% iridium weighing the same, stored in Sèvres in France. As with any physical etalon, this cylinder is subject to natural physical degradation processes. In particular, it has been discovered that it absorbs tiny amounts of mercury vapour from the air, with a consequent increase in weight. Although in principle it would be possible to define a unit of mass in terms of basic units (e.g. a certain number of atoms of a particular substance), analogously to the modern definition of metres in terms of the speed of light, in practice this presents difficulty.

In chemistry, the mole or gram mole is used as a unit of mass, being the mass, in grams, equal to the molecular mass of the material, using atomic mass units.

As with other properties, to avoid the complication of multiple powers of ten littering the pages, extreme values have been given their own names, and they have also been defined in terms of basic atomic values. Thus, for example, the international *atomic mass unit*, or *dalton* (John Dalton, 1766–1844), is one-twelfth of the mass of the carbon 12 isotope. There are also other variants of the atomic mass unit, defined in terms of other atoms. They differ in less than 10^{-30} kg. Atomic mass units are not to be confused with the *atomic unit of mass*, which is the rest mass of the electron, m_e, described below.

For rotational force or torque a new unit is needed, and this is provided by the newton-metre, defined as the torque due to a force of 1 newton acting at right angles at the end of a metre-long lever.

7.2.7 Pressure

Concepts such as pressure, which are relatively recent developments, do not have the colourful history of more basic measurement concepts, although sometimes there has been a struggle to arrive at a sensible system. If a force of 1 newton acts uniformly over an area of 1 square metre, a pressure of 1 *pascal* (after Blaise Pascal, 1623–1662) is said to be exerted. This same unit was previously called a *tor* (after Evangelista Torricelli, 1608–1647), not to be confused with the *torr*, which is 133.322 pascals. Such definitions are based on various physical phenomena, including the length of a column of mercury or water (see temperature measurement, below) and relationships between gravitational weight and area (hence the common *psi* – pounds per square inch – used in pressure gauges). So, for example, the torr is the pressure that will support a 1 mm high column of mercury at 0°C, under a standard gravitational field of $g = 9.80665$ m/s^2. At high pressures the *bar* is used ($= 10^5$ pascals). A measure called the *standard atmosphere* is also sometimes used, equal to 1.01325 bar.

7.2.8 Energy

The SI unit of energy is the joule (after James Prescott Joule, 1819–1889), being 1 newton acting through a distance of 1 metre. The CGS unit is the *erg*, equal to 1 dyne acting through a distance of 1 centimetre. This unit is in fact not named after Professor Erg, but is derived from the Greek *ergon*, meaning 'work'.

Energy (work done) is force times the distance over which it acts, and force times the velocity at which it acts is power (the rate at which energy is consumed). Perhaps the earliest unit of power was the *horsepower*, derived from the use of such animals in mines and textile mills. According to Klein (1974, pp. 237–238), James Watt (1736–1819) calculated that an average horse had a power of just under 33,000 foot pounds per minute, and called this unit the horsepower. Later on, in response to the need for a unit of power linked to electrical concepts, Watt was honoured by having the (absolute) *watt* named after him – equal to 1 joule per second. (As with so many other units, however, slightly different versions have been defined: between 1908 and 1948 the *international watt* was used, and the absolute watt is 1.00019 times as large as the international watt.) Other units of power are also in standard usages in different contexts. For example, in cars, *brake horsepower* is the quantity of power that an engine provides.

Einstein's famous equation, $E = mc^2$, shows us how mass and energy are related. The fact that the velocity of light is so large means that small quantities of mass are equivalent to large amounts of energy. (Actually, that statement is not strictly true. It is only true in the context of amounts of energy and mass which humans deal with in everyday life. It is easy enough to define new units so that $c = 1$.) This means that small masses can be used as useful energy units. For example, the rest mass of the electron, m_e, is equal to 9.109558×10^{-31} kg which is equal to 8.2×10^{-14} joule.

The *electron volt* is equal to 1.6×10^{-19} joule, and is defined as the energy that an electron gains when it moves up an electrical potential difference of 1 volt.

7.2.9 Sound

The sounds that we can hear have an enormous range of intensities – with the most intense being up to 10^{12} times as great as those we can only just detect. 'Intensity' here is defined as equal to P^2/DV, where P is the pressure driving the sound, D is the density of the medium it is travelling through, and V is the velocity with which it is moving. This is power per unit area. Because the range is so large, ratios – or, in fact, a logarithmic scale – are used for reporting sound intensities.

The *bel* (Alexander Graham Bell, 1847–1922) is simply the logarithm, to base 10, of a ratio between two intensities, and the more familiar *decibel* (db) is simple a tenth of a bel. These are thus comparative measures between two sounds: a change in 1 decibel means that one sound has 10 times the intensity of the other.

To determine the 0 db level, a panel of people with 'normal' hearing was exposed to a 1 kHz sound, initially below the limit of perception and then of gradually increasing intensity: 0 db was set at that level which could just be detected by a high proportion of the group. 'Level' here can refer to either the pressure or the intensity of the sound. Using this process, if pressure is used in the definition, 0 db has been set at 2×10^{-5} pascal, and if intensity is used, 0 db is 10^{-12} watt per square metre.

Decibels are measures of physical – that is, objective – intensity of sounds, but we humans rate sounds in terms of how *loud* they are. This is a subjective expression of how the sound is perceived. Perceived loudness is not merely a simple function of intensity – other factors are also relevant, some subjective. Measures of loudness and of loudness level have been developed. The *phon*, for example, is the loudness level of a sound with physical intensity of 1 db, and the *sone* is the loudness of the sound of a pure 1 kHz tone with an intensity of 40 db. Of course, one of the difficulties with measuring sound is that it seldom arrives in pure tones, and will generally be a mixture of tones of different frequencies.

Sounds may seem louder if they are intrusive or unwanted. A dripping tap or car theft alarm can come to assume gargantuan proportions at night. Unwanted sound may also be called 'noise', and efforts have been made to define and measure this. For example, the *noy* is defined using perceived noise levels in a 180 Hz frequency band centred on 1 kHz.

Complications such as the circumstances in which the sound arises have led to special types of measurement. For example, in studies of aircraft noise, a unit call the *noise and number index* (NNI) is used, which takes account of both the sound and the number of aircraft. The difficulties here are reminiscent of those of measuring quality of life, and such measurements have a large pragmatic component.

7.2.10 Electrical and magnetic units

The *ampere* (after André Marie Ampère, 1775–1836) is a unit of electrical current, defined as follows. If a constant current is maintained in two parallel conductors of infinite length, with negligible cross-section and 1 metre apart in a vacuum, then that current which will produce a force between the conductors of 2×10^{-7} newton per metre of length of the conductors is 1 ampere. Since this is defined in terms of length and force, we see that the unit of current is equal to the product of the square root of a length unit, the square root of a mass unit, and the reciprocal of a time unit.

One *coulomb* (after Charles Augustin de Coulomb, 1736–1806) is the charge which passes when 1 ampere flows for 1 second. The coulomb is a large unit relative to the charges encountered in most scientific work, and much smaller subdivisions (10^{-6}, 10^{-9}, 10^{-12} parts) are often met in electrostatic work. Coulomb invented the torsion measuring instrument, in which the charge exerts a twisting force on some tiny wires, and permit a pointer to change direction.

The *volt* (after Count Alessandro Volta, 1745–1827) is a unit of electrical potential difference: 1 coulomb of charge moved through a potential difference of 1 volt will gain (or lose) 1 joule of energy. This means that if a current of 1 amp flows in response to a potential difference of 1 volt, it will be delivering 1 watt of power.

Of course, as with other units, electrical units have evolved over the course of time. Take units of resistance as an example. The modern unit, the *ohm* (or 'absolute ohm', after Georg Simon Ohm, 1787–1854), is defined as the electrical resistance between two points if a constant potential difference of 1 volt between these two points produces a current of 1 ampere. Earlier units included the *statohm* (equal to 8.98776×10^{11} times the size of the present ohm), the *international ohm* (which was replaced by the absolute ohm in 1948 and which was 1.000495 times the absolute ohm, and was defined as the 'resistance of a uniform column of mercury containing just 14.4521 gram of that liquid in a column 1.063 meter long, and maintained at 0°C' (Klein, 1974, p. 448)), the French version of the international ohm (used before 1911 and equal to 1.003395 absolute ohms), an ohm used in England in 1884 (0.99767 absolute ohms), the siemens (William Siemens, 1823–1883, 0.9412 absolute ohms), and the *abohm* (10^{-9} absolute ohms).

Resistance properties of substances are typically given using *resistivity*, defined as the resistance, in ohms, of a block of the material 1 square metre in cross-section and 1 metre long. Of course, measuring resistances of such blocks would pose practical difficulties, and the most common way to measure resistivities is via a wire of more convenient fixed cross-section and length, using known physical relationships, of the kind described in Section 7.1, to deduce the resistivity.

Other electrical and magnetic units named after their discoverers (or, at least, recalling Stigler's law of eponymity (Stigler, 2000), which states that 'no scientific discovery is named after its original discoverer', after someone who was involved in the development and promotion of the units) include the *farad* (Michael Faraday, 1791–1867, for capacitance), the *henry* (Joseph Henry, 1797–1878, for inductance), the *weber* (Wilhem Weber, 1804–1890, for magnetic flux: 1 weber equals 1 ampere times 1 henry), the *maxwell* (James Clerk Maxwell, 1831–1879, equal to 10^{-8} weber), the *tesla* (Nikola Tesla, 1857–1943, for magnetic flux density, equal to the number of webers per square metre), the *gauss* (Karl Friedrich Gauss, 1777–1855, 10^{-4} tesla), and the *oersted* (Hans Christian Oersted, 1777–1851, a unit of magnetic field strength used in the CGS system of units). Of course, more unit names have been introduced than are nowadays used and, as we have seen elsewhere, the same names are sometimes used for different units, even for different physical properties, and different names are sometimes used for the same units.

Sometimes it is convenient to work with complementary or inverse properties: wavelength and frequency will be a familiar example, and the use of the dioptre (the reciprocal of focal length in metres, with converging lenses given positive signs and diverging ones negative signs) to measure the power of lenses is another. In electrical units, things are sometimes simplified if we think in terms of conductance, the reciprocal of resistance. In such cases, the units are sometimes named by the simple expedient of inverting the original names. Thus we have the *mho* for conductance (not to be confused with the *Mohs* scale for hardness – see below) and the *daraf* for elastance (reciprocal of capacitance).

As the above tour of just some of the zoo of names for electrical units may hint, the concepts underlying them did not spring cleanly into being, but were the result of a process of gradual development and evolution – as with other units and properties. Furthermore, not only the units but also the measurement procedures were developed gradually. In particular, the end of the nineteenth century saw a sometimes heated debate about whether it was appropriate to use 'direct reading' instruments – those calibrated directly in amps, volts, and ohms – rather than requiring the measurer to convert these manually from tables. For example, in a fascinating exploration of this controversy, Gooday (1995, pp. 265–266), who attributes the controversy to the different needs of the electrical engineering industry and the physics laboratory, says:

Professor George Carey Foster of University College, London, pointedly objected to students' use of direct-reading instruments, alleging that these devices yielded only 'unscientific information.' He had 'never' given his students direct-reading instruments to use, preferring that they should use 'old-fashioned instruments' such as a tangent galvanometer, or a potentiometer and Clark Cell to measure current, insisting always that each of them "reduce their observations to absolute measure for themselves.'

However, Professor Silvanus P. Thompson, inheritor of Ayrton's methods of training in Technical Physics at the Finsbury Technical College, came to Ayrton's defense. He challenged the arbitrariness of Foster's injunction against the laboratory use of direct-reading ammeters and voltmeters while placing no similar prohibition on the deployment of such nonelectrical direct-reading instruments as stopwatches and thermometers. Most specifically he 'differed' with Foster over the use of direct-reading electrical instruments. Flatly denying that they were 'harmful' he demanded to know why it was that Foster did not instead insist upon laboratories being equipped with such 'old-fashioned instruments' as water clocks for the absolute measurement of time or why the student making absolute determinations of mass should be allowed to luxuriate in a precalibrated set of weights? 'Where,' Thompson enquired, 'should the line be drawn?'

Gooday later (p. 271) remarks that 'For Foster, then, self-reliance in experimentation was more important as a *moral* guarantor of metrological authenticity than the degree of *precision* achieved through inappropriately assisted measurement techniques'. With the advent of computer aids, we have seen many similar examples of the argument that the 'older' methods lead to superior depth in some sense: mental versus calculator arithmetic, and knowledge of Latin as an aid to producing good English spring to mind.

7.2.11 Time

In this section we are concerned with measuring *time intervals*, the gaps between events. We are not concerned with constructing or positioning events on some conventional time continuum – that is, with determining calendars and calendar dates.

Time is one of the most puzzling of phenomena. Despite this, time, or at least duration, is perhaps that physical property which can be measured most accurately. Think of such units as nanoseconds and picoseconds – tiny, but in everyday use in contexts such as computing and physics. Time intervals are also remarkably uniform in the units of measurement used around the world, achieving this uniformity long before units of weight and length achieved what uniformity they have via the metric system.

It is amusing to note that we measure small units of time using powers of 10 (seconds, milliseconds, microseconds, etc.) and large units of time using powers of 10 (years, decades, centuries, millennia, etc.) but connect these two systems together using a system that dates back five or six thousand years to ancient Mesopotamia (60 seconds equal one minute, 60 minutes equal one hour, 24 hours equal one day, and 365 days equal one year). The reason for this lack of decimal relationships in the units is that there is more than one naturally occurring unit (days, months, and years) to be catered for in the system. Or perhaps we could look at this from the other perspective and say that the reason for the use of decimal relationships for very large and very small durations is that there was no need to measure or even talk about such durations until recently, by which time scientific units, with their metric system, had become established.

Of course, attempts have inevitably been made to produce simpler systems, for example, a base 100 system, with 100 seconds in a minute and 100 minutes in an hour. In post-Revolutionary France in 1793, the source of so much sensible innovation in the area of practical measurement, a proposal was made based on an ancient Egyptian system: each year was to comprise 12 months, each of 3 periods of 10 days, with 5 extra days added each year (6 in a leap year). Each day had 10 hours, each hour 100 minutes, and each minute 100 seconds. As history shows, however, the system did not last.

Units such as days, months, and years, of course, derive from astronomical observations, and these have a monstrously tangled history of their own. Things are complicated by the fact that astronomical 'periods' are generally not nice factors of each other (hence the leap year). The time unit a *second* was derived from links to angular rotation as the stars crossed the skies, which explains the use of the same noun for small time intervals and small angular differences. According to Klein (1974, pp. 103–104), 'Originally 1/60 hour had been called (in Latin) *minuta primam*, or "first small one." In the late 16th century, around the time when Galileo was measuring how long it took balls to roll down inclined planes, the *minuta secundam*, or "second small one," came into use also. Its name was later shortened, fortunately, to the one word: *second*.'

Things are further complicated by the fact that astronomical 'periods' are often not really periodic at all, but have subtle variations: days, months and years are not constant in length – they evolve over time. If a day of daylight is divided into an equal number of hours, then the length of those hours will change over the course of the year (as indeed they did in medieval Europe, being indicated by the sounding of church bells: the word *clock* itself derives from the medieval Latin word *clocca*, meaning 'bell'). Perhaps because of this, the idea that time could be divided into equal intervals is not an obvious one. This is no problem for most situations in everyday life, but can cause difficulties in situations when more accurate measurements are needed.

Later clocks were based on other periodic macroscopic artefacts, and permitted smaller units of time to be measured. Examples of these are water clocks (in which the Chinese developed particular expertise) and pendulum clocks. At the height of the technology, around AD 1300, Chinese water clocks achieved accuracies of about one part in 3000. A simple form of water clock consists of a large vessel full of water, with a small hole in the bottom. The unit of time is the length of time it takes for the water level to fall from one mark to another – the same principle as the hourglass or eggtimer, but using liquid instead of sand. A more complicated form had a small vessel with a hole in it floating in a larger container of water. Eventually the small vessel, initially empty, filled with water and sank – and the unit of time was the time until it sank. Mechanical clocks in the West only began to achieve comparable accuracy in the sixteenth century, but the introduction of the pendulum led to accuracies of around one part in 9000.

Macroscopic periodic systems are all very well, but only limited accuracy can be achieved with them. The next step was achieved by harnessing naturally occurring resonant frequencies on a smaller scale: in particular in quartz crystals and later (in the 1940s) in ammonia molecules and in caesium and other atoms. Molecules and atoms have the advantage over macroscopic systems that they are identical and do not degrade over time. In 1956, the second was formally defined as 1/31,556,926 part of the year 1900, but this did not last long and was soon replaced by measures based on atomic vibrations.

Quartz clocks are based on the piezo-electric effect. If pressure is exerted on such a crystal, an electrical potential difference is set up. Conversely, if an electrical potential difference is applied, the crystal slightly changes its shape. Such crystals have a natural frequency of physical oscillation, and by matching this with an oscillating applied electrical potential a stable oscillation results. Using electronic counting instruments to count the cycles yields a clock.

Atomic clocks make use of the fundamental quantum nature of matter: the fact that the electromagnetic radiation absorbed or emitted when an electron in an atom moves from one state to another is a constant (which depends on the atom and the levels moved to and from). Caesium clocks are accurate to about one part in 10^{12}. Caesium clocks were adopted as the basis for measuring time intervals in 1967, at the 13th General Conference on Weights and Measures, when the second was defined as 9,192,631,770 periods of the radiation corresponding to the transition between the two hyperfine levels of the ground state of the caesium-133 atom.

Even though the second and the smaller and larger units derived from it are ubiquitous, as when measuring other properties, other units are used in special contexts. For example, in the context of radioactive substances, we speak of the *half-life* of substances, this being the time it takes for radioactivity to fall to half of its initial value. Since the rate of decay is exponential, the half-life is the same whatever the initial radiation level. The range of half-lives of different radioactive materials is huge, from, for example, 3×10^{-7} second for polonium-212 to 4.5×10^9 years for uranium-238.

Although we have not discussed the creation of conventional descriptions of time continua, such as calendars, they have their own complications and history. Sobel (1995) describes the struggles by John Harrison to develop clocks sufficiently accurate and robust for mariners to take on sea voyages so that they could determine their longitude. Bryson (1995, pp. 78–79) illustrates that not all of these stories refer to the ancient past:

> Until as late as 1883, there were no fixed times in America. When it was midnight in New York, it was 11.47 in Washington, and 11.55 in Philadelphia. In 1869, when Leland Stanford struck the golden spike that marked the completion of America's first transcontinental railway ... the news was instantly telegraphed to a breathlessly waiting nation. In Promontory, Utah, the great event happened at 12.45, but in nearby Virginia City it was deemed to be 12.30. In San Francisco it was 11.46 or 11.44, depending on which authority you believed in, and in Pittsburgh the information was simultaneously received at six places and logged in at six different official times.
>
> In an age when most information arrived by horseback, a few minutes here and there hardly mattered. But as the world became more technologically sophisticated, the problem of variable timekeeping did begin to matter. It was a particular headache for the railways and those who travelled on them. In an effort to arrive at some measure of conformity, most railway companies synchronized the clocks along their own lines, but often these bore no relationship to the times used either locally or by competing railways. Stations would often have a multiplicity of clocks – one showing the station time, another the local time and the rest showing the times on each of the lines serving that station. Passengers unfamiliar with local discrepancies would often arrive to catch a train only to find that it had recently departed. Making connections in a place like Chicago, where fifteen lines met, required the careful study of fat books of algorithms showing all the possible permutations.

Velocity is an indirect measure, defined as the distance covered per unit time. As such its units are generally combination units: miles per hour, feet per second, and so on. In some special situations, however, new units have been coined. For example, the *Mach number* describes the ratio of the speed of an aircraft relative to the air it is travelling in to the speed of sound in that air. So, at Mach 1 the aircraft is moving at the speed of sound, at Mach 2 at twice the speed of sound, etc. Note that, since the speed of sound varies with both the pressure and volumic mass of a medium, the meaning of Mach 1 varies: in air at 20°C at sea level, Mach 1 is about $340\,\mathrm{m\,s^{-1}}$, while at 11,000 m it is about $290\,\mathrm{m\,s^{-1}}$.

An analogous term appears in *Star Trek*, where *warp factor* describes the ratio of the speed of the *Enterprise* to the speed of light.

7.2.12 Heat, temperature and entropy

The history of measuring temperature is one of the most interesting episodes in the history of measurement. The struggle of deciding how to measure it was interwoven with the struggle to define exactly what it was, and the struggle to understand heat in general. Indeed, it is perhaps not clear a priori that temperature will permit a precise quantification without a substantial pragmatic component: we are aware of differences in temperature, but there is clearly a subjective aspect to this and it is not unreasonable to suppose that this would render precise quantification as difficult as measuring, say, depression or affection. This might explain why the earliest instruments for measuring temperature were called *thermoscopes* rather than *thermometers*.

Temperature measurement is indirect, in that it necessarily involves using another physical system. The length of a mercury column, expanding in response to changes in temperature, will be a familiar example, as will the change in electrical resistance as temperature changes. One can, of course, simply take the length of the mercury column as a pragmatic measurement: scratch numbered graduations on the glass containing it, and one has a temperature scale. On the other hand, physical theories relating the expansion of mercury to temperature are relatively simple (at a basic level, anyway), in that they do not involve large numbers of variables. This means that temperature measurement is fairly direct.

A large number of different physical systems, which change in response to temperature changes, have been used as the basis of thermometers. They include the familiar ones of the change in volume of given quantities of mercury and alcohol, but at more extreme temperatures they also include such physical phenomena as magnetic properties of certain salts, the resonant frequency of quartz crystals, the time it takes high-frequency sound to traverse a substance, the change in pressure of fixed volumes of gas, the light radiated from a substance at high temperatures (such measuring instruments are called pyrometers), spectroscopic methods, electrical resistance, and electric currents generated by thermocouples. Indeed, no other indirectly measured physical quantity appears to be measured using so many different physical systems.

It is known that Newton suggested a temperature scale (and doubtless many others did as well) based on specifying the temperatures of two natural phenomena and using these to define the scale. Newton used the freezing point of water and the temperature of soft coal fanned by a bellows. Carlo Rinaldi, an Italian, suggested the freezing and boiling points of water. Ole Roemer, a Dane, took these and two other points, the temperature of the coldest mix which could be obtained by mixing water with ice and salt, and the temperature of the human body, as four points defining his scale. He then divided the complete range (from the water–ice–salt mix to the boiling point of water) into 60 degrees (familiar from time and angle). This made each of his degrees about 1.9 times the size of a modern kelvin or degree Celsius and 3.4 times the size of a degree Fahrenheit.

Like Roemer, Daniel Gabriel Fahrenheit (1686–1736) initially used alcohol, in glass tubes, as the expanding substance in his thermometers. In fact, the two met in 1708. Alcohol expands about eight times as much as water for a given rise in temperature, so yielding more accurate measurement. Furthermore, since it has a lower specific heat than water, a given rise in temperature can be achieved with less energy – so that it responds more quickly. Its disadvantage is that it boils at a much lower temperature than water, which limits its range of usefulness. Furthermore, both water and alcohol have a concave meniscus, which makes it difficult to assess precisely where the 'top' is, limiting accuracy. Fahrenheit therefore turned to mercury – which had also been used as the expanding medium in a thermometer made by Ismael Boulliau. Mercury, being opaque, is more easily seen than either water or alcohol, its meniscus is convex, and it does not wet the container, as both water and alcohol do. Furthermore, it freezes at $-27°$F and boils at over $338°$C. One disadvantage of mercury is that it expands even less than water for a given rise in temperature. Fahrenheit overcame this by the ingenious (though now very familiar) solution of constraining the expansion of a relatively large volume of mercury to a single dimension by attaching a reservoir of mercury to a narrow tube. Thus, in 1714 was born the mercury thermometer, which continues to be used right up to the present day (although, as with other measuring instruments, in many applications electronic displays have taken over).

Fahrenheit's scale was initially based on Roemer's by dividing each of Roemer's degrees into four and basing it on a slightly lower lowest temperature achieved by adding ammonium chloride to the water–salt–ice mixture. This yielded a scale which had $32°$ for the freezing point of water and $240°$ for the boiling point. He later changed the latter to $212°$, after deciding that $240°$ appeared to be too high.

The Réaumur scale was developed by René Antoine Ferchault de Réaumur a decade or so after Fahrenheit had begun to produce his mercury thermometers. Réaumur's thermometers were based on mixtures of water and alcohol, and his scale was based on assigning extra degrees to each extra 1% expansion of the original volume of the mixture. This led to a scale in which the freezing point of water was at $0°$ and the boiling point of water was at $80°$ under normal atmospheric pressure. Thus he produced a scale which was defined purely in terms of changes in state of water, and did not involve such things as mixtures of water and salt, or human body temperatures. Once defined, the scale was adopted for the more easily used mercury thermometers developed by Fahrenheit. The reader may have spotted that there are some subtleties concealed behind this account. The boiling point of alcohol is lower than that of water, so it is not possible to heat a water–alcohol mixture to the boiling point of water – the alcohol begins to boil off, changing the mixture. Klein (1974) has concluded that Réaumur extrapolated from a temperature halfway between the freezing points and boiling points of water.

Anders Celsius (1701–1744), a Swede, also used the freezing and boiling points of water at normal atmospheric pressure as the fixed points for his temperature scale. However, his 1741 scale differed in a crucial way from those described above: he set the freezing point of water to be $100°$ and the boiling point to be $0°$. This was reversed in 1750 by Martin Strömer. The C in $°$C originally referred to Centigrade – the 100 divisions of the range, and it was not until 1948 that the Ninth General Conference on Weights and Measures made the symbol officially represent Celsius.

The Fahrenheit scale is still widely used, especially in the United States. The Réaumur scale was used for many decades in France and Germany. The Celsius scale continues to be the most widely used of these scales, in part because the size of its unit degree is equal to the size of a kelvin.

Using the macroscopic properties of physical substances to define temperature scales is all very well, but it will be apparent that there is much scope for subjectivity, arbitrariness and disagreement. Réaumur made a step in the right direction by defining temperature in terms of percentage volume increase, but even this is not a complete solution. Different substances do not have the same rates of expansion, so that one could build two thermometers, both having $0°$ at the freezing point of water and both having $100°$ at the boiling point of alcohol, but one based on mercury and the other on alcohol. If these were then both plunged into a liquid which showed a temperature of 20° on the mercury thermometer, it would not show 20° on the alcohol thermometer.

The temperature scales described above are defined by dividing the range of expansion of a substance between the temperatures of two specified conditions into equal intervals. These two conditions were chosen arbitrarily – though in such a way that (the researchers hoped) they could be replicated elsewhere and by other experimenters. In fact, however, it is possible to choose one of these conditions in a non-arbitrary way. Early work had shown that the pressure of a fixed volume of gas decreased as the temperature was reduced. By extrapolation to zero pressure, it seemed that there was a coldest possible temperature. This could be adopted as the zero point of a temperature scale – an 'absolute zero'. In particular, we can redefine the scales described above in this way, using the size of the degrees of each system as the basic counting unit, but beginning them at this absolute zero point. By this means, if we take the Fahrenheit degree, we obtain the Rankine thermodynamic scale of temperature, and if we take the Celsius degree we obtain the kelvin thermodynamic scale. The freezing point of pure water under normal atmospheric pressure is at 489.67 degrees Rankine and at 273.15 kelvin (K – note, nowadays *not* the 'degree Kelvin'). In fact, instead of the freezing point of water, more rigorous determinations were based on the 'triple point' of water: the temperature at which ice, water and water vapour co-exist in a state of equilibrium.

This triple point occurs at 0.01°C. Using this, the kelvin, the SI's unit of temperature was defined at the Thirteenth General Conference on Weights and Measures, in 1967, as 1/273.16 times the thermodynamic temperature of the triple point of water.

To facilitate practical temperature measurement in regions which are not close to the triple point of water, determinations of the temperatures of other important physical changes of state have been made. These include the melting points of tungsten (at 3380°C), gold, silver, zinc, tin, and other metals, the boiling points of oxygen, neon, and hydrogen (in fact, the boiling point of hydrogen at two different pressures), and the triple points of oxygen and hydrogen. At temperatures away from the triple point of water, however, the relationship between the kelvin scale and the scale defined by these other changes of state is not always clear-cut: in particular, the size of the unit degree defined in terms of these additional points is not uniform, and varies over the scale.

While on the subject of heat and temperature, perhaps we should also mention the *tog* – the thermal resistance of textiles such as bedclothes. A substance has a thermal resistance of 1 tog if a temperature difference of 1 K applied to an area of 1 square metre induces a heat flow of 1 watt. Duvets used as bedclothes have ratings of around 9–14 togs. It is not only the physicists who have to measure temperature.

We have already seen that the joule is a unit of energy. The *calorie* (or more correctly, the *gram calorie*) is the amount of energy, in the form of heat, required to raise the temperature of 1 gram of water from 14.5°C to 15.5°C under a 1 atmosphere pressure. A 1000 calorie unit, a kilocalorie, is sometimes called a *Calorie*, with a capital initial letter. The 'calories' referred to in diet books are these kilocalories. 1 kilocalorie is equivalent to 4,185.5 joules. In general, the *heat capacity* of a body is the amount of heat required to raise its temperature by 1°C at a given temperature (heat capacities generally change, often quite dramatically, with temperature). When the quantity of matter is taken into account, so that it becomes the amount of heat required to raise a given quantity of the substance by 1°C, it is termed the *specific heat* of the substance. When the amount of heat is measured in calories, we see from the definition of the calorie above that this means that the substance is implicitly being compared with water: the heat capacity of water is being used as the baseline standard.

Specific heat is related to the deep concept of *entropy*: a measure of the randomness or disorder in a system, or, more formally, the amount of energy in a system which cannot be extracted to do work. The change in entropy of a system is given by the change in heat content divided by the temperature at which the change occurs. By integrating this, we obtain a definition of entropy – and we also see why interest is generally in differences in entropy. One unit of entropy is simply the calorie/kelvin, often written as eu for 'entropy unit'. Entropy is an example of something we can (indirectly) measure but which we cannot experience directly (unlike temperature, weight, etc.)

7.2.13 Light and other kinds of radiation

Light is electromagnetic radiation which has a frequency between 4.3×10^{14} Hz and 7.5×10^{14} Hz. Just as we can distinguish between objective and subjective measurements in sound (intensity, loudness, and so on) so we can distinguish between objective physical measurements of light and measurements of physiological responses to light. *Radiometry* is the general science of measuring electromagnetic radiation, but *photometry* is the measurement of human response to light.

Originally units of light intensity were defined in terms of the illumination given by candles of specified shape and made of specified substance, but these left many sources of variation unspecified and were not very accurate. Later versions used lamps which burnt oil. Once again, however, the evolution of measurement technology towards definitions based on deeper theoretical concepts led to more fundamental definitions, and for a while the *candela* was defined in terms of the theoretical concept of black body radiation: the intensity of radiation from such a body at a temperature of 1,772°C. Actually realizing this intensity of radiation was no trivial task. Platinum was placed at the bottom of a well made of the heat-resistant substance thorium dioxide and heated by microwaves until it reached its melting point of 1,772°C. The resulting radiation was matched against that from an electric lamp, whose intensity could be controlled. When the two match, the electric lamp is able to serve as a standard 1 candela lamp. Later this was replaced by the luminous intensity in a given direction of a source that emits monochromatic radiation of frequency 540×10^{12} Hz and has a radiant intensity in that same direction of 1/683 watt per steradian (unit solid angle).

Luminance measures luminous intensity per unit area, and has units of the *nit* (1 candela per square metre) or the *stilb* (10,000 nit). These units are not named after Mr Nit and Mr Stilb, but from the Latin *nitere* and the Greek *stilbein* (both of which mean 'to shine'). Given the candela as a unit of intensity, one can define a unit of luminous flux. The SI unit, the *lumen*, is the luminous flux emitted by a source of intensity 1 candela in a solid angle of 1 steradian. This definition assumes a point source of light, something which does not exist in reality. Some subtly different units are the *lux*, a measure not of luminance, but of *illuminance*, the illumination obtained when 1 lumen of luminous flux is uniformly distributed over 1 square metre, and the *phot*, equal to 10,000 lux.

A *bolometer*, a device for measuring radiation, works by absorbing incident radiation on a blackened surface, and measuring the consequent rise in temperature of the surface (using the change in electrical resistance). The indirect nature of such instruments is apparent – converting radiation into a temperature rise and then into a resistance change.

Further illustration of the indirectness of many physical measurements is given by Klein (1974, p. 404) in the context of radiation. He writes: 'Most so-called "measurements" of radiation emission are actually long distance deductions or extrapolations. They are based on the distribution and relative magnitude of radiations absorbed at the receiving end … There is no way to apply a thermometer to the surface of the sun or other shining star. Every thermometric substance we use on Earth – mercury, platinum, or any other – would be vaporized before it could be placed at that 6000°K solar surface.'

Light, of course, is simply radiation occurring in one small interval of the spectrum of electromagnetic radiation. Elsewhere in this spectrum, in the X-ray and gamma ray intervals, for example, and when the other types of radiation, such as alpha or beta radiation, are concerned, other measurement tools and units are needed, which match the physical properties of the radiation. Thus, for example, for X-rays, the *roentgen* (Wilhem Conrad Roentgen, 1845–1923) is the amount of radiation required to produce 2.082×10^9 ion pairs per cubic centimetre of dry air. These ion pairs would be detected by electrostatic measurements. The roentgen is now used as a measure of emitted radiation. It is also sometimes useful to have measures of absorbed radiation. One such is the *rad*, defined as the absorption of 11 erg per gram of energy in the formation of ion pairs, and another (now obsolete) is the *rem* (for *Roentgen Equivalent Man*) intended to represent a dosage equivalent to that arising from irradiation by a certain flux of X-rays. In 1974 the General Conference on Weights and Measures defined the *becquerel* (Antoine Henri Becquerel, 1852–1908) as one nuclear disintegration per second and the *gray* (Louis Harold Gray, 1905–1965) as the unit of absorbed radiation dose. The gray thus replaced the rad, and is equal to 100 rad. In 1980 the *sievert* (Rolf Maximilian Sievert, 1896–1966) was introduced as a measure of absorption which takes account of the nature of the source of the radiation. It is defined as the number of grays multiplied by 20 for alpha radiation and 1 for beta and gamma radiation.

Perhaps, in this context, we can also quote the following: 'Another little known technical term is the boggle. This is a measure of the time a writer sits staring blankly at his typewriter trying to think up some way of explaining terms like "roentgen." It is equal to 1.5 ounces of Irish whiskey. A roentgen, by the way, is 10^{14} boggles' (Novick, 1970).

Other units have been defined for other kinds of radiation. Thus, for example, the *curie* (originally after Pierre Curie, 1859–1906, but now regarded as also honouring Marie Curie, 1867–1934) is the amount of any radioactive substance yielding 3.7×10^{10} radioactive disintegrations per second, though the *rutherford* (Ernest Rutherford, 1871–1937) has also been proposed, this being 10^6 radioactive disintegrations per second.

7.2.14 Other measurable physical properties

The preceding sections have described some of the history and technology behind physical measurement units and methods. The evolution from a tremendous variety of systems, with different systems being used in different places and at different times, towards a unified system will be apparent for just about every measurable property. Likewise, the evolution from units based on simple physical or biological systems to those based on deep theoretical, and unchanging, physical properties of matter will be apparent. The same evolution is clear for physical properties we have not discussed above. It is obviously not possible, in a single book, to cover all physical attributes (though books such as Cardarelli, 2003, and Johnstone, 1998, make a sterling effort). Some not described above will be mentioned briefly in this section.

Viscosity describes the tendency of fluids to resist flow, and is measured by timing such flows. For example, the Höppler viscometer consists of a slanted glass tube, filled with the fluid to be measured, and down which a small sphere is dropped. The dimensions of viscosity are those of force times time per unit area. In SI units, this is newton-seconds per square metre. One-tenth of one newton-second per square metre is called the *poise* (Jean Poiselle, 1799–1869) although Klein (1974, p. 620) remarks that 'almost every imaginable unit combination of force times duration divided by area has been employed' to measure viscosity.

Measures of concentration are typically given in terms of so-many parts per given volume: as in, for example, 10 micrograms of lead per 100 ml, or 80 mg of alcohol per 100 ml of blood. The abbreviation *ppm*, for parts per million, is common in this regard (although there can be ambiguity, since it may refer to volume or mass). The term *pH*, used to measure acidity and alkalinity, is defined as the negative logarithm, to base 10, of the concentration of hydrogen (H^+) ions. Pure water has a neutral pH level (of 7), acidic liquids have a lower level, and alkaline liquids have a higher level.

The *Beaufort scale* (Francis Beaufort, 1774–1857), very much a pragmatic scale, was adopted in 1838 by the British Royal Navy and in 1874 by the International Meteorological Committee to provide a scale for wind force at sea, and a land equivalent has been developed. The land version has 13 qualitatively defined points, including force 0 (calm: smoke rises vertically, wind speed from 0 to 1.5 km/h), force 1 (light air: direction of wind shown by smoke drift but not by wind vanes, wind speed from 1.5 to 5 km/h), 2 (light breeze: wind felt on face, leaves rustle, ordinary vane moved by wind, wind speed 6 to 12 km/h), up to 11 (violent storm: very rarely experienced, accompanied by widespread damage, wind speed 103–116 km/h), and 12 (hurricane: wind speed greater than 116 km/h).

The *Richter scale* (Charles Francis Richter, 1900–1985) is a numerical scale, with integral values from 1 to 10, which attempts to give the order of magnitude of the energy released by an earthquake at its source. The Mercalli scale (Giuseppe Mercalli, 1850–1914) is an alternative measure of earthquake intensity, which indicates the intensity at a given point on the earth's surface. The pragmatic nature of this measurement is made clear from the definitions of the numerical values of this 12-point scale. For example, a value of 2 is defined as 'Felt by a few persons who are at rest, especially on upper storeys of buildings. Some suspended objects may be noticed to swing somewhat' (Klein, 1974, p. 705).

In Chapter 2 we referred to the Mohs scale of hardness. This was a purely ordinal scale, described in 1822 by Friedrich Mohs (1772–1839) based on the scratch resistance of a set of ten minerals: diamond (scale point 10) will scratch corundum (scale point 9), which will scratch topaz (scale point 8), and so on through quartz, feldspar, apatite, fluorspar, calcite, gypsum, and talc (scale point 1). This was not the first hardness scale based on this principle – an earlier one had been developed for metals by Richard Kirwan (1733–1812). Other ways of measuring hardness have also been devised, in which hardness is mapped to a property which is itself a ratio scale. For example, some approaches are based on indentation tests. The Brinell scale (Johann Auguste Brinell, 1849–1925) measures the size of the depression made by a steel ball pressed into the surface whose hardness is to be measured. The ten minerals above, in the same order, have Brinell hardness measures of 667 (diamond), 304, 178, 147, 137, 64, 53, 12 and 3 (talc). Another approach uses a *scleroscope* to determine the height to which a steel ball dropped onto the surface will rebound. The fact that the property used to measure hardness is itself a ratio scale (size of depression, height of bounce) does not mean that hardness itself has a ratio scale – unless one pragmatically defines hardness as the result of the measurement procedure.

This chapter, perhaps more than any of the others, has been littered with units named after the men (and in one case woman, though even that one case was by the back door) who developed them. Occasionally names of units have arisen from other sources. Some arise as acronyms, and some from Latin or Greek sources. Others have more whimsical origins. In nuclear physics, the *barn* is an area of $10^{-28}\,\mathrm{m}^2$, and is used in determining how likely a particle is to interact with a nucleus (this area being the typical area of a nucleus). The name has its origin in the phrase 'you couldn't hit a barn door', and was used as a code word when it was coined in 1942 as part of the US nuclear programme during the Second World War. Other such strange units have come and gone: the *gillette* described the power of a laser, in terms of how many stacked razor blades the beam would burn through.

7.2.15 Conclusion

This section owes much to the extensive discussion of the evolution of physical units of measurement up to the early 1970s given by Klein (1974) and the discussion of the historical background given in McGreevy (1995). The more philosophical discussion of the meaning of measurement in physics in Section 7.1 is based on discussions in Roche (1998), Krantz *et al.* (1971), and the other cited sources.

Although this chapter has illustrated how measurement systems have gradually become rationalized and unified as the world has shrunk, older systems still leave their mark, both in particular specialized usages, some examples of which we have seen, and on language. Thus we have the Swedish engineer who described the British as approaching metrication 'inch by inch' and the British sports commentator, describing a high jump event, who referred to 'just inching the bar up centimetre by centimetre'. Perhaps a fitting way to conclude this chapter is with a letter to *The Times* on 28 November 2002:

> Sir, Your graphic ('Tanker disaster', November 21) on the loss of the *Prestige* shows that the depth to which it has sunk is 11,500 ft, or 3,500 m, or '9.2 times the height of the Empire State Building'.
>
> The correct Imperial unit for the discussion of height and depth is surely Nelson's Column. That being so, the *Prestige* rests 62.16 NCs down. Furthermore, it weighed 81,564 tonnes deadweight, as much as some 11,652 African elephants each of seven tons, and was, at about 800 ft, as long as 36.67 London buses (30 ft).

Yours faithfully,

ANDREW BERNARD

8

Measurement in economics and the social sciences

Measures are more than a creation of society, they create society.
Alder (2002, p. 342)

8.1 Introduction

In social and economic measurement problems we are often, perhaps typically, concerned with aggregate objects. This notion, of an object which is an aggregate of individual objects, with its own properties and attributes, is one which is not always easy to accept, and it is certainly one which developed gradually. In physics, of course, things developed in the opposite direction, with macroscopic objects having an evident and irrefutable existence, and atoms being conjectured as possible underlying building blocks. Indeed, the general acceptance of the existence of atoms as real entities only occurred in the early twentieth century.

In recent decades the idea of objects being composed of large numbers of small and simple building blocks, with possibly complex interactions, has taken off in a more elaborate way, with the general recognition and exploration of the fact that the aggregate object may possess 'emergent' properties, not possessed by the individual elements (see, for example, Haken, 1983; Prigogine and Stengers, 1985). Desrosières (1998, p. 70) has a wonderful expression for this creation of a new level of reality from that of the subsidiary elements via statistical aggregation: he describes it as 'the magical transmutation of statistical work'. He also traces the debate about the existence of these higher-level objects back further in time. He illustrates with the case of the Franciscan order. The success of this order meant that they managed many monasteries, with much agricultural land, but, because of their vow of poverty, they could not actually own these. Technically, they were owned by the Pope. However, in order to avoid the administrative load, he wished to hand over the ownership to the Franciscans. Of course, if they were handed back to individual Franciscans it would rupture the Franciscans' poverty vow, but not if they were handed over to an aggregate entity, the Franciscan Order. William of Occam supported the Francisans, arguing 'that it was impossible to return these possessions to the order as a whole, since the Franciscan Order was only a *name* designating *individual* Franciscans'. One is reminded of the fact that modern corporations serve as if they were individuals in the eye of the law.

Porter (1995, p. 37) has also described this way in which statistics creates higher-level realities:

> The concept of society was itself in part a statistical construct. The regularities of crime and suicide announced in early investigations of 'moral statistics' could evidently not be attributed to the individual. So they became properties instead of 'society,' and from 1830 until the end of the century they were widely considered to be the best evidence for its real existence.

He later says:

> A more commonplace entity, to us, is the crime rate. There were, of course, crimes before the statistician occupied this territory, but it may be doubted whether there were crime rates. Similarly, people sometimes found themselves or people they met to be out of work before this had become a statistical phenomenon. The invention of crime rates in the 1830s and of unemployment rates around 1900 hinted at a different sort of phenomenon, a condition of society involving collective responsibility rather than an unfortunate or reprehensible condition of individual persons.

The nineteenth century witnessed extended debates about the role which could be played by descriptors of these aggregate objects – for example, about the role of a population average. Certainly, it is a summary statistic for that population, but is it then appropriate to regard it as a descriptive measure of the population as an object? If so, surely that interpretation means that the aggregate object, the population, has some kind of existence of its own, separate from its component elements. The most familiar debate of this kind hinged around Adolphe Quetelet (1796–1874) and his notion of the 'average man' (*l'homme moyen*).

The social sciences are concerned with how people interact and function in groups and organizations. Economics is based on concepts such as inflation, productivity, creditworthiness, consumption, capital growth, and so on. Likewise, the behavioural sciences speak of attitudes, ability, enthusiasm, determination, intelligence, accident proneness, and so on. All these are things which cannot be measured directly, but, if proper understanding of the concepts is to be achieved, and if we are to be able to manipulate them, then measure them we must. Put another way, if the power of science – and, in particular, of statistics – is to be brought to bear on the phenomena represented by such concepts then it is necessary to represent them numerically. Fortunately, although none of the above can be measured directly themselves, they are all related to things which can be measured directly. Ability is related to how well people perform in certain tasks, attitude is reflected in how people react or would react to different circumstances, and determination is indicated by how persistently people will try before abandoning a task. If we understand, or are prepared to assume, something about the relationship between the concepts of interest and the variables we *can* measure, then we can try to extract the information about the concepts of interest from observations on the other variables, via indirect measurement procedures.

Note that even the higher-level attributes which might appear straightforward to measure can pose difficulties. Rusnock (1995) describes some of the issues associated with early attempts to measure population sizes (which might, at first glance, seem susceptible to simple counting):

> The concept of population gained saliency in early modern Europe, in large part due to the consolidation of centralized monarchies during the sixteenth and seventeenth

centuries. Historians locate the origins of the modern nation-state precisely during this period, when definitions of the royal state came to include population as well as territorial extent. The emphasis on population by the monarchies of the sixteenth century had a clear basis in taxation: monarchs such as Francis I in France and Henry VIII in England sought to devise new ways to finance increasingly costly and sizable wars.

(Rusnock, 1995, p. 17)

In the absence of regular censuses, estimates of the population of eighteenth-century France were based on various methods:

The two most common methods of estimating population involved a universal multiplier. In the first, the number of hearths taken from tax records was multiplied by the number of people per hearth – usually 4 or 5. In the second method, the number of annual births taken from parish registers was multiplied by a number, anywhere between 25 to 28, which yielded a figure for the total population. This number was typically attained by taking a census of several small parishes and establishing the proportion of births to total population.

(Rusnock, 1995, p. 25)

However, taking figures from tax records was criticized on the grounds that it underestimated population sizes, as people found ways to avoid being mentioned. Similar suspicions related to censuses, which were generally taken for tax reasons. The most accurate approach seemed to be based on birth records, scaled up by overall population figures based on censuses of small regions. Measurement, it appears, is never straightforward.

One of the key roles of measurement in economics and the social sciences is to monitor change over time. This might be for the purposes of intervention (is inflation too high?) or simply congratulation (more youngsters than ever before are entering higher education). In any case, measurement here, as elsewhere, serves to objectify the phenomenon being observed. Of course, there is always the possible objection that the measures are not measuring 'the right thing'. However, once the construction of a measure has been made explicit, it can always be adjusted and altered so that it is closer to what one wants (though changing measures can invite its own criticisms, as any government will know when it attempts to refine the way in which economic indicators such as the unemployment rate are calculated): in our terminology, the pragmatic aspects can be varied.

Bennett (1999) has examined cultural changes in this way. In the original version of his book, published five years before, he concluded (Bennett, 1999, p. 1) that 'we had indeed experienced substantial cultural decline. To be specific, virtually every important indicator not only got worse, it got much worse. I wrote then that unless these exploding social pathologies were reversed, they would lead to the decline – and perhaps even to the fall – of the American republic.' However, the 1999 revised edition shows 'many significant positive developments. The decade of the nineties has seen progress in some key social indicators: reductions in welfare, violent crime, abortion, AIDS, divorce, and suicide; upswings in SAT scores and charitable giving.' But the picture is mixed: 'During these same 1990s, we also experienced social regression in several important areas.' Of course, much has happened in the world since then, and I look forward to seeing the next revision of the book. However, from our point of view the key thing is that we can only see these changes because quantitative measurements have been used. Without these, any such observations would be purely subjective. Without the statistically constructed entities which are being measured, we would have no objective idea of whether things were improving or deteriorating. The choice of measures Bennett makes is interesting and an enlightening perspective on what he considers important. He lists total crimes, violent crimes, juvenile violent crime, imprisonment, punishment, drug use, out-of-wedlock births, single-parent families, marriage and cohabitation, divorce, child well-being, welfare, abortion, three measures of educational performance, spending in schools, public schools, out-of-wedlock teenage births, teenage abortion, teenage suicide, teenage drug and alcohol abuse, television, movies, recreation and leisure, church membership, depression, voter turnout, trust and cynicism, charitable giving, immigration, and military service. It is very clear from this list that others might have chosen different indicators, that one man's deterioration is another's improvement, and that change need not mean either deterioration or improvement, but merely change.

There is no limit to the measurements which may be taken when studying economic and social phenomena, and most of them involve substantial pragmatic aspects. This means that a comprehensive overview is out of the question. Instead, therefore, I have merely attempted to sample the space of such measures, illustrating the sorts of problems encountered when measuring in such domains and showing the solutions which have been found.

8.2 Performance indicators

One of the most striking developments in the application of measurement technology in the UK in the past couple of decades has been the dramatic increase in the use of *performance indicators* in the public sector. In part this has been driven by the politico-economic process of devolution and privatization, whereby responsibility is handed over to the private sector, but it has also been partly driven simply by the need for improved management practices and the need for clear measures of those practices and their effectiveness. In a word, the need for *accountability*. Often extra regulatory bodies are set up to monitor the performance of organizations within a sector. For example, Oftel appointed auditors to verify that the monitoring of telecommunications systems (when measuring things such as call duration for the purposes of charging) achieves a certain level of accuracy.

Performance indicators are perhaps most commonly used in a comparative way (e.g. the UK government's aim to reduce the number of working days lost from work-related injury and ill health by 30% by 2010), although sometimes more absolute measures are used (e.g. the UK government's one-time intention that 50% of the population should experience higher education).

Examples of performance indicators used by local government authorities are net cost per 1000 population, pupil–teacher ratio, gross cost per pupil, population per police officer, serious offences per 1000 population, highway maintenance cost per kilometre, and rent arrears as a percentage of rent collectable. As this list illustrates, in assessing performance of a complex organization with multiple roles, tasks and objectives, it is necessary to use multiple measures. Indeed, in introducing such measures in the National Health Service, the UK government described 123 indicators (Department of Health and Social Security, 1983), later increased to 418. These included measures of such things as low birthweight rates, nurses per bed, provision of land in relation to resident population, costs in relation to inpatient cases, admission rates and average length of stay. Some of these describe the processes carried out by the service, others relate to costs and efficiency, and others relate to the resources and environment. Clearly there is no limit to the number of performance indicators which could be constructed – there are no effective limits to the number of attributes social organizations can possess. This means that a comprehensive description could involve a vast number of indicators. Some sort of balance has to be struck between the number of indicators and the effectiveness of the indicator profile in describing the organization and the service it provides. It has been suggested that 10–20 indicators provide a reasonable number. Sometimes aggregate indicators are constructed, often by forming weighted sums of component indicators. If the weights are determined subjectively, this will provide a pragmatic definition of an attribute deemed to be of interest. Examples are given in Audit Commission (2000b), including a composite indicator of potentially avoidable deaths from ten different conditions which uses equal weights, and 'the average cost of handling a Housing Benefit or Council Tax Benefit claim, taking into account differences in the types of claim received'. League tables of educational establishments (see below) provide other examples.

In his helpful review of performance indicators in the public sector, Smith (1990, p. 69) points out that 'to suggest that managers should be striving for median behaviour on all indicators is clearly absurd and may be asking them to adopt inappropriate practices for their locality. However, the performance indicator culture is likely to push managers increasingly towards a superficial view of the service, in which indicators are cited when convenient and ignored otherwise.' He also cautions that 'The performance indicator initiatives are likely to encourage management to move further from the use of a modelling or systems approach. Data will increasingly be viewed as a convenience for supporting or refuting rhetoric, and not as an intrinsic part of the manager's creative armoury.' I do not agree with this pessimistic view, believing that the use of a multiplicity of measures can protect against this danger. But it is clearly a danger which must be kept in mind.

Various financial ratios are often used as measures of performance for companies in the private sector, including such things as price–earnings ratio and the ratio of assets to liabilities (see, for example, Barnes, 1987). More recently, however, there has been a trend towards including other indicators, such as measures of quality of goods, of customer satisfaction, of impact on the local community, and of environmental impact.

Two papers published by the Audit Commission (2000a, 2000b) describe how to construct and use performance measures in local and central government and the UK's National Health Service but, of course, the lessons apply much more widely. Audit Commission (2000a, p. 6) describes two objectives for performance measurement: improved public services and improved accountability. This paper also notes that 'the choice of performance indicators will have a major impact on the behaviour of the organization. It is therefore necessary to understand the processes that will have a beneficial impact on performance and to choose indicators that reflect these.' It stresses six key principles:

- *Clarity of purpose*: who will use the information and how it will be used.
- *Focus*: this, in the first instance, should be on the priorities of the organization. Audit Commission (2000a) also remarks that 'organizations should learn how indicators affect behaviour, and build this into the choice and development of their performance indicators.' Of course, it is a precondition for this that organizations are clear about their priorities and strategic objectives, as well as that these are expressed sufficiently clearly that they can be couched in explicitly measurable terms. Focus also means that one should avoid the temptation to include every indicator one can think of, and Audit

Commission (2000a, p. 12) suggests applying the criterion 'What action could the recipient of the information take on the basis of this information?' such that 'indicators where the answer to this question is that the recipient would never take action on the basis of the information should not be used for measuring performance.'

- *Alignment*: the indicators should fit into the general objectives and processes of the organization.
- *Balance*: the profile of measures should give a balanced view of the organization's performance. If emphasis is placed on only certain aspects of a system, then performance is likely to be distorted to favour those aspects.
- *Regular refinement*: the indicators should be adapted to changing circumstances. Perhaps this is one of the biggest differences between measurement in behavioural, social and economic domains on the one hand, and measurement in the physical sciences and engineering on the other: generally, 'performance indicators' in the latter area will not evolve over time. The discussions on test equating (Section 5.10) and Goodhart's law (Section 4.5) are also relevant here.
- *Robustness*: careful and detailed definition is essential, so that the measures are sufficiently robust and comprehensible for their use.

Others have stressed similar attributes, such as relevance, accuracy/reliability, validity, timeliness, accessibility, comparability over time (and perhaps space), and coherence. In a report on performance indicators in the public services which was published just as this book was completed, Bird *et al.* (2003), on behalf of the Royal Statistical Society, called for:

- *performance monitoring protocols* – to ensure that statistical standards are met. Protocol is an orderly record not only of decisions made (from design to analysis and reporting) but also of the reasoning and calculations that led to those decisions;
- *independent scrutiny* – to safeguard the wider-than-government public interest, the individuals and institutions being monitored, and methodological rigour;
- *the reporting of measures of uncertainty* whenever performance data are published, including as league tables or star ratings – to avoid over-interpretation and the false labelling of performance;
- *research on different strategies than 'name and shame'* for the public release of performance data, *and better designs* (including randomization) for evaluating policy initiatives – the first to allay ethical and effectiveness concerns, the second for robust evidence about 'what works';
- *much wider consideration of the ethics and cost-efficiency of performance monitoring.*

The fact that the choice of performance indicators will guide the direction of the organization is a feature of the pragmatic aspect of the measurement procedure. The indicators are being defined so that the organization can monitor particular types of performance, but this will inevitably lead to the tendency to optimize performance in those directions, as measured by the performance indicators. The organization will thus construct the measures with an eye on the direction in which they would like to move. But, by virtue of this, great care has to be taken. Mullen (1985) gives the apocryphal example of a nail factory achieving its target, which was measured in terms of weight of nails, by producing just one giant nail. If the percentage of students achieving grades A–C in British A levels is to be used as a key indicator, as it was, then an easy way to improve apparent performance is by refusing to allow those predicted to obtain lower grades to sit the examinations. Moreover, one should also be wary of the *ratchet effect*. This is the encouragement of only small improvements, since large ones will be impossible to match up to in future years.

Related to this is the Macnamara fallacy, quoted by Charles Handy (1995), which says: 'The first step is to measure what can be easily measured. This is OK as far as it goes. The second step is to disregard that which can't be easily measured or to give it an arbitrary quantitative value. This is artificial and misleading. The third step is to presume that what can't be measured easily really isn't important. This is blindness. The fourth step is to say that what can't easily be measured really doesn't exist. This is suicide.' Handy summarizes this as 'What does not get counted does not count', and we might go one step further and say 'What gets measured gets done'. Audit Commission (2000a, p. 6) says 'The emphasis on performance measurement in the public sector carries with it the risk that the process becomes an end in itself. It is important that organizations do not lose sight of the fundamental objectives of performance measurement.'

Measuring performance incurs some cost – the 'measurement burden' referred to in Chapter 1. In general, the cost of collecting performance data should be substantially outweighed by the benefits of that collection. This probably means that one should tend to favour indicators which are part of the everyday running of the organization, rather than indicators which are specially measured for monitoring purposes. Measures used for the running of the organization will certainly be more sympathetically received by those being monitored using them. It also means that some assessment of the measurement cost should be taken.

Audit Commission (2000b) notes the difference between outcome measures and process measures. *Outcome measures* may be the ideal thing to measure, but outcomes may occur a long time in the future. In the case of educational programmes, for example, the outcome may be many years down the line. In such cases an alternative is to look at *process measures*, such as the proportion of the population receiving the new educational programme. The validity of such measures depends on the strength of the theory linking process and outcome. Audit Commission (2000b, p. 25) also cautions that performance indicators 'that measure activity rather than performance will provide less useful data and information overload'. While on this topic, we should also add a third type of measure: *input measure*. In many cases, the outcome should be measured relative to the input, since what really counts is the 'value added' by the process: clever students will generally produce clever graduates even if the education they receive is not as good as it might be; a surgeon who operates only on less severe cases will have a high success rate. Naive reporting of outcome independent of the quality of input has stimulated some fierce debates, perhaps especially in the health services domain, with waiting times for operations being a familiar example. On the other hand, in many other cases, the input measure will involve units quite different from those of the outcome measure, so that assessment of the added value is difficult. Because of this, Goldstein and Spiegelhalter (1996) argue for relative comparisons between institutions (they are particularly concerned with education and health), adjusted for baseline measurements, and prefer the phrase 'adjusted comparison' to 'value added'.

Of course, if outcomes are to be measured relative to, or adjusted for, input measures, then this gives additional opportunities for manipulating the results – not by changing the outcomes (or the components of the outcome calculations), but by changing the input baselines. Green and Wintfield (1995) describe an example of this, in which reported congestive heart failure increased from 1.7% to 7.6% between 1989 and 1991, so that the expected mortality was increased: if observed mortality remained roughly constant, the apparent result would be a very significant improvement.

Apart from their role in assessing change, performance indicators are also widely used to compare institutions or organizations. In both of these roles it is important to recognize that the indicators are not exact: they are measured with uncertainty. Inevitably, the popular media, in the interests of a good punchy story, often overlook this fundamental fact. Goldstein and Spiegelhalter (1996, p. 397) point out that the main constraint on fine comparisons between institutions arises from sample size limitations, but also draw attention to input data which is opportunistic ('they happen to be available') rather than chosen by design, selectivity bias due to students with particular characteristics changing schools, imprecision in the definitions of the outcome (e.g. the example given by McKee and Hunter, 1995, of how a 'finished consultant episode' is defined in health systems assessment) and, of course, measurement error in the predictor variables. Furthermore, models which compare outcome with input, or use input characteristics as covariates, are inevitably out of date by the length of time between the input and outcome measurements. In the case of education, this may amount to several years. This raises questions about the relevance of the models to the current institutions, which may have changed substantially over such a period.

Perhaps it is important to conclude with the reminder that one should expect performance indicators to be improved over time, as well as evolve over time to reflect the changing circumstances in which they find themselves. Since these indicators will both guide and influence the direction an organization takes, constructing, monitoring and assessing them should be regarded as a crucial aspect of any organization. I repeat: what gets measured gets done.

8.2.1 League tables: an example

Higher education in the UK has undergone dramatic changes in the last two decades. These changes have stimulated interest in the comparative merits of different universities. In response to this, in the early 1990s, *The Times* began publishing what became an annual league table of universities. Initially, these were not well received: 'We believe the tables are wrong in principle, flawed in execution and constructed upon data which are not uniform, are ill-defined, and in places demonstrably false' (The Committee of Vice Chancellors and Principals, quoted in *The Times*, T2, 7 May 2003, p. 6). Despite this, these tables or similar alternative listings have proved popular with potential students in helping them select a place to study.

The evolution of the tables, and how social indicators improve through a process of criticism and refinement, is nicely encapsulated by the following:

> Only the most virulent critics any longer claim that the data used to compile *The Times* tables are ill-defined or demonstrably false, even if they take issue with the combination of measures used to rank universities. This is because the figures come from the universities, mainly via the Higher Education Statistics Agency, and each exercise is followed by close analysis with representatives of the university sector. This has led to improvements almost every year, to the point where only half the original indicators survive. (John O'Leary, *The Times*, T2, 7 May 2003, p. 6)

Universities are multi-faceted institutions: they have several objectives. *The Times* tables present nine measures, which are pragmatically combined in a weighted sum to produce an overall ranking. The nine measures are: teaching (itself a product of a complicated and lengthy teaching quality assessment procedure), research (taken from the Research Assessment Exercise – see Section 8.2.2), entry standards (the average A-level point score of new students), student–staff ratio, library/computing spend per student, facilities spending per student, the proportion of students awarded first and upper-second class honours degrees, the percentage of graduates in graduate-level employment, and the proportion of students completing their courses in the expected time. The final score is adjusted so that the top ranking university obtains 1000.

Several things are immediately obvious from this. For example, the component variables are non-commensurate, so that they must either be rescaled or the weights chosen to take this into account. And note that there is considerable scope to 'play to the measure' (sometimes called 'gaming'): a university could (to take an extreme example) award all students first class degrees (though one might hope that the external examiner system, which seeks to maintain broad comparability between universities, would prevent too extreme such abuse). Furthermore, the mere fact of aggregating component variables does not render the result supremely accurate if the component variables are doubtful. Grade inflation, discussed in Section 5.9.2, may lead to suspicions about the value of A-level results, one of the component variables, and the assessment of research quality from the Research Assessment Exercise (Section 8.2.2), another of the component variables, is not without its weaknesses. Above all, the aggregate university measure used in the ranking is clearly strikingly pragmatic: the overall measure is *how the compilers chose to define 'quality' of a university*. It is not at all far-fetched to imagine that others would choose to use different variables (how about adding something about the extent to which the lecturers stimulate the students?) or to combine them in different ways.

For the curious, the top ten in the 2003 ranking has, in decreasing order, Oxford, Cambridge, Imperial College London, The London School of Economics and Political Science, Bath, University College London, York, Warwick, Bristol, and Durham (www.timesonline. co.uk/uniguide). There is generally some slight shifting around from year to year.

8.2.2 The Research Assessment Exercise

In this exercise, research at all UK universities (which make a submission) is graded, at a great deal of effort and expense. The definition of research used in the 2001 Research Assessment Exercise was related to the Frascati definition:

> Research and experimental development (R&D) comprise creative work undertaken on a systematic basis in order to increase the stock of knowledge, including knowledge of man, culture and society and the use of this stock of knowledge to devise new applications.
>
> R&D is a term covering three activities: basic research, applied research and experimental development.
>
> Basic research is experimental or theoretical work undertaken primarily to acquire new knowledge of the underlying foundation of phenomena and observable facts, without any particular application or use in view.

Applied research is also original investigation undertaken to acquire new knowledge. It is, however, directed primarily towards a specific practical aim or objective.

Experimental development is systematic work, drawing on existing knowledge gained from research and/or practical experience, that is directed to producing new materials, products or devices, to installing new processes, systems and services, or to improving substantially those already produced or installed. (OECD, 1993)

This is all very well, but the categories used were defined as:

5* Quality that equates to attainable levels of international excellence in more than half of the research activity submitted and attainable levels of national excellence in the remainder.

5 Quality that equates to attainable levels of international excellence in up to half of the research activity submitted and to attainable levels of national excellence in virtually all of the remainder.

4 Quality that equates to attainable levels of national excellence in virtually all of the research activity submitted, showing some evidence of international excellence.

3a Quality that equates to attainable levels of national excellence in over two-thirds of the research activity submitted, possibly showing evidence of international excellence.

3b Quality that equates to attainable levels of national excellence in more than half of the research activity submitted.

2 Quality that equates to attainable levels of national excellence in up to half of the research activity submitted.

1 Quality that equates to attainable levels of national excellence in none, or virtually none, of the research activity submitted. (HEFCE, 2001)

Apart from the peculiarity of a scale in which the categories are labelled, in order, as 1, 2, 3b, 3a, 4, 5, 5*, it is clear that the distinctions between some of the levels show remarkable degrees of sophisticated interpretation. (For the record, at the time of writing, a new top level, 6*, has recently been added to this curious scale.)

8.3 Economic indicators

The discussion of dimensional analysis in Chapter 7 showed that the variables used in physics and engineering could be described in terms of a space with relatively few dimensions. In part this was a consequence of the largely representational measurement of physical concepts. Unfortunately, the same is not true of economics. Concepts in economics are much more complex than those in physics, so that measuring them necessarily includes important pragmatic aspects: there are many different, but related, measures for very similar concepts. A few of these measures are described in this section, in an attempt to convey the complexity. The next section examines price indices, and in particular the UK's Retail Prices Index (RPI), in some detail.

The gross national product (GNP) of a country is the output produced by the country's residents, while gross domestic product (GDP) is the output produced within the country, regardless of whether, for example, it is produced by foreign-owned organizations with some of the profit being sent overseas. GDP can be calculated in various ways. The *output method* simply takes the added value (the difference between the value of the raw materials, labour costs, etc., and the sold products) of all goods and services produced over the period in question. The *income method* works from the other end, summing all the incomes earned by workers, land (e.g. rent), capital (e.g. interest), and so on. The *expenditure method* adds consumer spending, government spending, investment and exports, and subtracts imports to arrive at a total. Obviously, each of these components requires a complex methodology in its own right to arrive at a figure. There are also other subtleties. For example, in Britain spending by charities is separated out.

Perhaps not surprisingly, there is considerable continuing debate about how GDP (and GNP) should be measured. This includes argument about whether non-paid but crucial activities, such as housework, should be included in the calculations. This would be difficult because there is no explicit rate for such work and no record is kept of the hours spent. Clearly pragmatic notions play a very large role in the operational definitions of GDP and GNP. It also seems likely that the GDP figures are becoming less reliable over time, as activity shifts from traditional manufacturing industries towards service and information industries such as telecommunications and finance.

The concept of *money supply* is crucial in economics, and is believed to play a fundamental role in demand, prices, and so on. Unfortunately, opinions differ about exactly what this fundamental role is. Monetarists believe that changes in money supply influence demand, which in turn influences prices rather than unemployment and output. Others believe that the relationship between money supply and demand is less clear-cut, and that demand influences unemployment and output more than prices. At this point, I suppose, I am obliged to insert some comment such as 'if all the economists in the world were laid end to end, they would not reach a conclusion' (attributed to George Bernard Shaw, 1856–1950). As with other macroeconomic concepts, there are various definitions of money. *Narrow money* includes cash and money which can be spent via cheque or credit card. *Broad money* also includes bank and building society deposit accounts. Particular definitions are denoted Mx, where x is an integer. Thus, in the UK, M0 is cash and M4 is M0 plus sterling deposits of the UK non-bank private sector. As the reader will no doubt have feared, different countries use slightly differing definitions, but the differences are not too important when interest focuses on change. In general, M0 and M1 are measures of narrow money, and those with integers above 1 measure broad money.

Price indices, and in particular the UK RPI, are described in detail below. The RPI is a measure of how the prices of goods and services we consumers buy change, and so is used to measure inflation. At the opposite end of the scale lie *producer price indices*. As one might guess, these measure the changes in prices charged by producers. These indices are calculated as a weighted sum of components, analogously to the RPI, with the weights depending on the size of the industrial sector, and with prices collected from surveys. Producer price indices do not include services, unlike retail price indices.

Consumer confidence is also important in economics. This describes the attitude of consumers to current economic conditions and to what the future is likely to bring. Clearly the issue of measuring a concept such as confidence overlaps substantially with the psychological measurement issues discussed in Chapter 5, but, of course, here we are interested in an aggregate measure, describing some population, rather than the confidence of individuals. In general, the measures are based on surveys of consumers.

In addition to the large number of well-established indices measuring different aspects of the economy, many rather unusual ones have been suggested based on observation of people's behaviour. They include the *Big Mac* index, which compares the cost of hamburgers between countries, and the *Hemline* index, which is based on the length of women's skirts. Peter Temple (2003, p. 189) suggests that men's business clothing becomes smarter in more difficult economic conditions (when jobs are not so certain) and that less formal dress-down styles are more acceptable in benign economic conditions.

Recently, interest has turned to developing indicators of sustainable economic development. Custance and Hillier (1998) describe sustainable development as 'achieving a balance between three broad objectives – maintenance of economic growth, protection of the environment and prudent use of natural resources, and social progress which recognizes the needs of everyone'. Clearly the input components to such indicators include things not included in earlier indicators of economic growth.

8.4 Price indices

8.4.1 General

Index numbers are statistical summaries which can be used to compare different situations, times, conditions or objects. They are widely used, but perhaps the most familiar examples are stock market price indices such as the FTSE 100, the Dow Jones, and the Nikkei index, and price indices such as the RPI in the UK and the Consumer Price Index in the USA. On the subject of price indices, Mitchell (1915, p. 10) comments: 'It is a curious fact that men did not attempt to measure changes in the level of prices until after they had learned to measure such subtle things as the weight of the atmosphere, the velocity of sound, fluctuations of temperature, and the precession of the equinoxes. Their tardiness in attacking that problem is the more strange because price changes had frequently been a subject of acrimonious debate among publicists and a cause of popular agitation'.

Prices fluctuate over the course of time for a variety of reasons: other suppliers enter the marketplace, raw materials become scarcer or the supply is threatened (e.g. by war), new technology is developed, or perhaps fewer people want to purchase a given type of item because of changing fashions, and so on. The quantity of an item or service (which, for convenience, here we shall jointly refer to as a 'commodity') which can be purchased for £1 or $1 will decrease if the price of that commodity rises and will increase if the price falls. One can put this the other way round, and say that the *value of money* (relative to a commodity) has decreased if £1 will now purchase a smaller amount of the commodity. However, there is a complication: the prices of different commodities will rise and fall in different ways – the supply of them and the demand for them will differ between commodities. Worse still, the prices of different commodities may differ from supplier to supplier. If we want a single overall measure of the changing purchasing power of £1, we will need somehow to aggregate the purchasing power over different commodities and suppliers. Such aggregates or averages are called 'price index numbers'. Examples are the RPI in the UK, the Harmonised Index of Consumer Prices (HICP) in the EU, and the Consumer Price Index in the USA.

Index numbers have a long history. Kendall (1969) refers to work published by Bishop Fleetwood in 1707, and Mitchell (1915), referring to work published in 1764 by

G.R. Carli, says: 'In an investigation into the effect of the discovery of America upon the purchasing power of money, he reduced the prices paid for grain, wine, and oil in 1750 to percentages of change from their prices in 1500, added the percentages together, and divided the sum by three, thus making an exceedingly simple index number.' This example illustrates the simplest use of price index numbers: to compare prices in two situations. Typically these different situations are different times, though they need not be – one could compare different geographical areas (e.g. urban and rural) or different types of purchasing pattern. In this discussion, however, we will refer to times for convenience. One of these times will be the 'base' or 'reference' time, and the other will be the 'current' time. In the UK, the RPI may be dated from 1914, when the Board of Trade began publishing the percentage changes of the prices of the main items of family expenditure. A crude weighting scheme, based on the expenditure data from a 1904 survey of working-class households, was used to provide weights to combine the price changes into a single overall index. This index was used for over 20 years, until criticism that the weights (e.g. the underweighting given to electricity relative to candles) were no longer appropriate forced a change. The Second World War interfered with the introduction of a new index, and an Interim Index of Retail Prices was introduced in 1947, being replaced by the first Retail Prices Index in 1956. The weights for this index were based on the Household Budget Inquiry, later to be replaced by weights regularly updated from information collected in the Family Expenditure Survey. The modern version of the RPI is described briefly below.

If one was concerned with a single product, then things would be simple. Suppose, for example, one wanted to know how the price of a sack of flour (of given weight and quality) changed over time. If the price was £1.20 at the end of 2000 and £0.90 at the end of 2001, then the price index for 2001 relative to 2000 would be $100 \times 90/120 = 75\%$. Here, 2000 has been taken as the base year.

This example is simple because there is a unique price at each time, and this can be measured directly. Things are not so straightforward, however, when multiple commodities are involved. Each of these may change by different percentages over the same time, and some way has to be found to aggregate these changes.

We can think of the problem as being the standard indirect pragmatic measurement problem of both defining and estimating an underlying general price level; this necessarily involves pragmatic choices. Fluctuations in this general price level cause fluctuations in the price levels of each of the commodities, but these commodities also have their own random components. Somehow we want to extract the common aspect from these multiple fluctuating time series. We attempt to do this by defining a 'basket' of commodities, and summarizing how their overall price differs between situations or over time. Issues arise in how the basket of commodities is chosen, and what summary measure to use. There are also various additional complications. The different commodities are not commensurate: sugar may be measured in kilograms (or perhaps even pounds), petrol may be measured in litres (or perhaps gallons), tomato ketchup may be measured in number of bottles (which may come in different sizes), and so on. If the different commodities are to be combined in a meaningful way, we need to decide what units to use to measure each of them.

In practice, there may be a huge number of items which could be included in the index, perhaps running into the tens of thousands (contrast Carli's three: grain, wine, and oil – but then he did not have a computer). Think of the number of distinct items on a supermarket's shelves, let alone all the items and services which do not appear there. Furthermore, different people, with different lifestyles, will buy different sorts of goods: they will have different baskets. For example, meat will not figure in a vegetarian's purchases, so that a rise in the price of meat will not affect the value of money for that person. This can lead to apparently paradoxical conclusions. Such conclusions are overcome by defining and using a common basket, typically defined on the basis of data collected from surveys of consumer retail outlets.

Suppose that $p_i(t)$ is the price of some unit of commodity i at time t, and $q_i(t)$ is the quantity of commodity i at time t. From this it follows that the value of the ith commodity purchased at time t is $p_i(t)q_i(t)$. Relative to the total basket of items purchased at time t, the value of item i is $p_i(t)q_i(t)/\sum_{j=1}^{n} p_j(t)/q_j(t)$. The ratio of the price of the ith item at time t to its price at time s, $p_i(t)/p_i(s)$, is called the *price relative*, and, similarly, the ratio of quantities, $q_i(t)/q_i(s)$, is called the *quantity relative*.

The key theoretical question is how to combine the prices and quantities of the different commodities to produce the measure (and definition) of the index. An early suggestion was the ratio of the means of the prices of the n commodities at times t and 0:

$$I_1 = \frac{\sum_{i=1}^{n} p_i(t)}{\sum_{i=1}^{n} p_i(0)} \tag{8.1}$$

where time $t = 0$ has been taken as the base time and n is the number of commodities included in the index. Now, price is measured on a ratio scale: rescaling simply converts between currencies. If the different commodities in (8.1) may be measured in different currencies then statements about the relative size of I_1 at different times are meaningless. For example, suppose that $n = 2$, with the prices in certain currencies at the three times 0, t, and s, being $\{p_1(0) = 1, p_2(0) = 1\}$, $\{p_1(t) = 3, p_2(t) = 1\}$, and $\{p_1(s) = 1, p_2(s) = 2\}$. Then the value of I_1 at t is $I_1(t) = (3 + 1)/(1 + 1) = 2$, which is greater than the value of I_1 at s, which is $I_1(s) = (1 + 2)/(1 + 1) = 1.5$. However, if we change the currency in which the second commodity is measured in such a way that the price of this commodity is multiplied by a factor of 4, we obtain

$$I_1(t) = \frac{3 + 4}{1 + 4} = \frac{7}{5} < I_1(s) = \frac{1 + 8}{1 + 4} = \frac{9}{5} \tag{8.2}$$

In general, for this index we can have

$$\frac{\sum_{i=1}^{n} p_i(t)}{\sum_{i=1}^{n} p_i(0)} > \frac{\sum_{i=1}^{n} p_i(s)}{\sum_{i=1}^{n} p_i(0)} \tag{8.3}$$

while

$$\frac{\sum_{i=1}^{n} \alpha_i p_i(t)}{\sum_{i=1}^{n} \alpha_i p_i(0)} < \frac{\sum_{i=1}^{n} \alpha_i p_i(s)}{\sum_{i=1}^{n} \alpha_i p_i(s)} \tag{8.4}$$

where α_i is the rescaling factor for the ith currency, so that one's conclusions about how aggregate prices were changing would depend on the choice of currencies. Note, however, that if all the commodities are measured in the same currency – that is, $\alpha_i = \alpha$ for all i – then the two expressions in (8.4) reduce to the corresponding expressions in (8.3): the index is invariant to a common choice of currency so that we can then compare prices at different times.

The lack of invariance of I_1 to different changes in the price units is overcome in another early suggestion, the mean of the ratio of the n prices:

$$I_2 = \sum_{i=1}^{n} \frac{p_i(t)}{p_i(0)} \qquad (8.5)$$

This measure is still sometimes used in some circumstances.

Another alternative to I_1 which is also invariant to different changes in the price units of different commodities is the geometric mean of the price relatives:

$$I_3 = \left[\frac{\prod_{i=1}^{n} P_i(t)}{\prod_{i=1}^{n} P_i(0)} \right]^{1/n} \qquad (8.6)$$

On the other hand, the question of whether I_3 is a useful index of consumer prices is a different matter. It is not enough that the index should have attractive *theoretical* properties: it must also have relevant and attractive *practical* properties. Another index with a form very similar to this will occur later.

Neither I_1 nor I_2 takes account of the quantity of the items purchased. This means that they may be useful for some purposes, but not for others. In particular, they will not indicate how 'an average shopping basket' has increased in price: if the price of item k has increased astronomically, but it is only purchased in minute quantities then it will affect I_1 substantially even though it might be irrelevant to almost everyone in the population.

In fact, of course, the elements in index I_1 need to be weighted, to take account of the amount of each commodity purchased in the 'average shopping basket'. The idea of weighting the prices appears to have been introduced by Lowe (1822), who used the general form:

$$I_4 = \frac{\sum_{i=1}^{n} w_i p_i(t)}{\sum_{i=1}^{n} w_i p_i(0)} \qquad (8.7)$$

The index will represent an 'average basket' if the weights are measures of quantities purchased (e.g. obtained from surveys). If w_i is set to be the quantity of commodity i in the 'average basket' at the base year ($t = 0$) then $100 I_4$ is called the *Laspeyres* index (Laspeyres, 1871):

$$I_5 = \frac{\sum_{i=1}^{n} q_i(0) p_i(t)}{\sum_{i=1}^{n} q_i(0) p_i(0)} \qquad (8.8)$$

The effect of this weighting by $q_i(0)$ is to yield a measure of what the basket at time 0 would cost at time t, relative to what that basket cost at time 0. That is, the measure is *standardized* to the time 0 basket, so that only changes of price contribute to changes in the index.

If, on the other hand, w_i is set to be the quantity of commodity i in the 'average basket' at the current year t then $100 I_4$ is called the *Paasche* index (Paasche, 1874):

$$I_6 = \frac{\sum_{i=1}^{n} q_i(t) p_i(t)}{\sum_{i=1}^{n} q_i(t) p_i(0)} \qquad (8.9)$$

There is an obvious symmetry about these formulae in terms of price and quantity, and one can similarly define *quantity index numbers*:

$$Q_5 = \frac{\sum_{i=1}^{n} p_i(0)q_i(t)}{\sum_{i=1}^{n} p_i(0)q_i(0)} \qquad (8.10)$$

$$Q_6 = \frac{\sum_{i=1}^{n} p_i(t)q_i(t)}{\sum_{i=1}^{n} p_i(t)q_i(0)} \qquad (8.11)$$

Of course, the weighting schemes above are not the only possible weighting schemes, and many others have been proposed, including weighting the prices using the average of the quantities at times 0 and t. Note that the weighting schemes in I_5 and I_6, by using the same quantities as weights in the numerator and denominator ($q_i(0)$ in I_5 and $q_i(t)$ in I_6) mean that an overall change in quantities will not affect the index. In contrast, if $q_i(t)$ was used in the numerator and $q_i(0)$ was used in the denominator, then purchasing double the quantities at time t relative to the quantities at time 0 would lead to a doubling of the index, even if the prices remained constant.

The property of invariance to different units of currency for the different commodities, illustrated above, is an example of a property of an index which one might regard as attractive. In a seminal book on the subject of index numbers, Fisher (1922) described such criteria (calling them 'tests') and examined a large number of indices using these criteria. The following are examples of such criteria:

- The index should be invariant to changes in currency provided that this is common to all prices (the $\alpha_i = \alpha$ case mentioned above). This is called the *commensurability* test.
- The *time-reversal test*, which requires that the index for time t relative to time s should be the reciprocal of the index for time s relative to time t.
- The *circular test*, which requires that the product of the index for time s relative to time r and the index for time t relative to time s should equal the value of the index for time t relative to time r.

In fact, it is not possible to construct an index which satisfies all of Fisher's criteria simultaneously (Eichhorn and Voeller, 1976), but Fisher settled on the geometric mean of the Laspeyres and Paasche indices as the ideal form, and this is now called the *Fisher ideal index*. A set of tests can be considered as a set of axioms describing index numbers, and Eichhorn and Voeller (1976) and Balk (1995), in his excellent review of axiomatic price index theory, distinguish between axioms and tests: to them axioms are 'more or less self-evident' whereas 'more debate is possible' about tests.

Index values can be calculated for each subperiod over a period of time. For example, if we are now at year t, we could calculate index values for each of the intermediate years, using s as a base year and $s + 1$ as the current year for $s = 0, \dots, t - 1$. We could also calculate the index for year t using year 0 as the base year. Only if the index satisfies the circular test will the product of the intermediate index values equal the overall value. That, is denoting an index calculated for current year p and base year q by $I(q, p)$, only if I satisfies the circular test will it be true that

$$I(0,t) = \prod_{s=0}^{t-1} I(s, s+1) \qquad (8.12)$$

In general, a *chain index* is one which satisfies (8.12), $I(r, r) = 1$, and $I(q, p) = I(p, q)^{-1}$ for $p \neq q$. Clearly the chain property is an attractive one, as it enforces an internal consistency between values of the index. It means, in particular, that the index between two times some distance apart is invariant to the way the index has changed between those two times. Note that any index could be used as the base form, $I(s, s + 1)$, in (8.12).

One very important form of index is the *Divisia* index, based on the following argument. The instantaneous rate of change of the price of the ith commodity at time s is $dp_i(s)/ds$. To make this invariant to changes in the currency units, we standardize it to

$$\frac{1}{p_i(s)} \frac{dp_i(s)}{ds} \tag{8.13}$$

which is equal to

$$\frac{d}{ds} \log p_i(s) \tag{8.14}$$

Now the instantaneous standardized rate of change of an entire basket of commodities can be defined as a weighted sum of terms such as that given in (8.14):

$$\sum_{i=1}^{n} w_i(s) \frac{d}{ds} \log p_i(s) \tag{8.15}$$

A sensible choice of weights, $w_i(s)$, is to weight the ith commodity by the relative amount spent on that commodity in the basket:

$$w_i(s) = \frac{q_i(s) \, p_i(s)}{\sum_{j=1}^{n} q_j(s) \, p_j(s)} \tag{8.16}$$

Since (8.15) is the instantaneous standardized rate of change of the entire basket, by analogy with (8.13) we can define the price of the overall basket at time s as P, with P satisfying

$$\sum_{i=1}^{n} w_i(s) \frac{d}{ds} \log p_i(s) = \frac{1}{P(s)} \frac{dP(s)}{ds} \tag{8.17}$$

It follows that

$$\sum_{i=1}^{n} w_i(s) \frac{d}{ds} \log p_i(s) = \frac{d}{ds} \log P(s) \tag{8.18}$$

Integrating this from 0 to t gives the log price at t relative to that at time 0:

$$\int_{0}^{t} \sum_{i=1}^{n} w_i(s) \frac{d}{ds} \log p_i(s) ds = \log P(t) - \log P(0) \tag{8.19}$$

from which the change in price of the overall basket between times 0 and t is

$$I_7 = \frac{P(t)}{P(0)} = \exp \int_{0}^{t} \sum_{i=1}^{n} w_i(s) \frac{d}{ds} \log p_i(s) ds \tag{8.20}$$

This, with the weights given in (8.16), is the Divisia index, originally described by Divisia (1925). Note that, by virtue of the integral in (8.20), this index is path-dependent: its value depends not only on the prices at the end points, but also on the values the prices assume between the end points.

Under some circumstances it may be reasonable to assume that the weights in (8.16) are approximately constant. For example, this may be the case when a rise in price for the ith commodity leads to a reduction in the quantity of the ith commodity purchased. Of course, as with any model, this should not be taken too far. If one is prepared to assume that the w_i are constant over time, $w_i(t) = w_i$, then (8.20) simplifies. We have

$$\frac{P(t)}{P(0)} = \exp \sum_{i=1}^{n} w_i \int_0^t \frac{d}{ds} \log p_i(s) ds$$

$$= \exp \sum_{i=1}^{n} w_i \left[\log p_i(t) - \log p_i(0) \right]$$

$$= \exp \sum_{i=1}^{n} w_i \left[\log \frac{p_i(t)}{p_i(0)} \right]$$

$$= \prod_{i=1}^{n} \left[\frac{p_i(t)}{p_i(0)} \right]^{w_i}$$

The index

$$I_8 = \prod_{i=1}^{n} \left[\frac{p_i(t)}{p_i(0)} \right]^{w_i} \tag{8.21}$$

which is obviously similar to the geometric mean of the price relatives in (8.6), is sometimes called the *Walsh index*.

The axiomatic approach to index number construction might be described as a mathematician's approach. In this, one identifies desirable properties of index numbers, formalizes these as functional equations, and then finds numerical constructs satisfying these equations. It also seems appropriate to describe this strategy as predominantly pragmatic, in that the properties to be satisfied by the index are chosen by its creator, rather than reflecting the behaviour of an underlying empirical system. However, this mathematical/pragmatic approach is not the only possible approach. One alternative is a statistical approach in which one tries to formulate and model an underlying process. It seems appropriate to describe this strategy as predominantly representational (in the sense of reflecting the behaviour of an empirical system). In this approach one will conceptualize some underlying variable, measure manifest variables related to this latent variable, and from these measurements try to deduce the value of the underlying variable. The Divisia index, with the price level of the overall basket $P(s)$ given in equation (8.17), is a step in this direction. We can think of this approach as supposing that there is an underlying level of 'value' of money, and that this fluctuates over the course of time. As it fluctuates, so the prices of commodities will change, increasing if purchasing power drops and decreasing if it increases (though unfortunately, as we noted above, generally not in phase or by the same proportion – real life is more complicated than that). Because there is a smallest unit of currency, these changes will generally be step functions, even though the underlying value of money is thought of as being a smoothly varying function of time. The price curve is then the reciprocal of this 'value of money', and this, or at least changes in this, is the latent variable we wish to measure.

Theil (1960) has taken this modelling approach further. He assumes there is an underlying price value $P(s)$ and that the observed prices of the individual commodities are related to it by a simple model of the form

$$p_i(s) = \beta_i P(s) + e_i(s) \tag{8.22}$$

In a standard factor analysis model (see Chapter 3), this is replicated across subjects, with $P(s)$ being randomly drawn for each subject from some common distribution, but in our case we cannot assume that $P(s)$ is sampled from some distribution. In factor analysis, we estimate the covariance matrix Σ between the several variables by a matrix of the form $\Sigma = \Lambda\Lambda^T + \Psi$, where Λ is the 'factor loading matrix' and Ψ is diagonal. One approach to fitting a factor model is to choose the elements of Λ and Ψ to minimize some measure of the discrepancy between Σ and $\Lambda\Lambda^T + \Psi$ (though nowadays maximum likelihood is a more common approach). We can adopt a similar strategy to estimate the $P(s)$, $s = 0, \ldots, t$, in (8.22). To achieve this, multiply (8.22) by the quantity of the ith commodity at time r, $q_i(r)$, and sum over the n commodities, to give

$$\sum_{i=1}^{n} q_i(r)\, p_i(s) = P(s) \sum_{i=1}^{n} \beta_i q_i(r) + \sum_{i=1}^{n} q_i(r)\, e_i(s) \tag{8.23}$$

In fact, we can put this in matrix form. Define the $(t + 1) \times n$ matrix $\mathbf{P} = \{P_{si}\} = \{p_i(s)\}$, the $(t + 1) \times n$ matrix $\mathbf{Q} = \{Q_{ri}\} = \{q_i(r)\}$, the $(t + 1)$-vector $\mathbf{p} = (P(0), \ldots, P(t))$, the $(t + 1)$-vector

$$\mathbf{q} = \left(\sum_{i=1}^{n} \beta_i q_i(0), \ldots, \sum_{i=1}^{n} \beta_i q_i(t) \right)$$

and the $(t + 1) \times (t + 1)$ matrix

$$\mathbf{R} = \{R_{sr}\} = \left\{ \sum_{i=1}^{n} q_i(r)\, e_i(s) \right\}$$

Then equation (8.23) can be written as

$$\mathbf{PQ}^T = \mathbf{pq}^T + \mathbf{R} \tag{8.24}$$

Our aim is to estimate $\mathbf{p}$, and this can be done by finding the elements of the vectors $\mathbf{p}$ and $\mathbf{q}$ which minimize some summary of the elements of $\mathbf{R} = \mathbf{PQ}^T - \mathbf{pq}^T$. An obvious such summary would be the sum of squared elements of $\mathbf{R}$. This particular summary yields $\mathbf{p}$ equal to the first principal eigenvector of the matrix $\mathbf{PQ}^T\mathbf{QP}^T$. Of course, although this leads to a straightforward solution, it is not the only summary, and arguments have been made for alternatives – see, for example, Fisk (1977). Indeed, more generally still, the simple additive model of (8.22) is not the only possible form, and others have been suggested.

Further discussion of index numbers is given in Stuvel (1989), Diewert and Nakamura (1993) and Selvanathan and Prasada Rao (1994).

8.4.2 An example: the UK Retail Prices Index

The RPI 'is defined as an *average measure* of *change* in the prices of *goods and services* bought for the purpose of *consumption* by the *vast majority* of *households* in the *UK*' (Baxter, 1998; emphasis in original). It is published monthly. Around 650 price indicators of goods and services, including information on food and drink, housing, transport and clothing, are used to construct the index. The level of detail is indicated by the fact that, for bread, large sliced white loaves, large unsliced white loaves, small brown loaves, large wholemeal loaves, bread rolls, pitta bread and baguettes are included, but in other cases representative items are used. For example, Baxter (1998, Section 2.1) says 'for home killed lamb, prices are collected for "loin chops with bone" and "shoulder with bone". Other joints, and loin chops and shoulders without bones, are not priced; it is assumed that their price movements are close to those of the joints of lamb that are priced'. The items for inclusion are reviewed annually, to reflect changes in purchase patterns and the introduction of new products.

Prices are collected by market research organizations, which collect over 130,000 prices per month, and a further 10,000 prices collected centrally each month by the Office for National Statistics (ONS). The market research organizations use a random sampling design to collect information from about 150 shopping locations each year, with about 20% of these locations being changed each year. The prices are weighted by quantity information obtained chiefly from the Expenditure and Food Survey, but spending patterns of pensioner households deriving at least three-quarters of their income from state pensions and benefits, and those with household income in the top 4% are excluded. The latter is perhaps a hangover from the initial aims of the 1914 index to produce an index for 'working class families'. (I cannot help but remark that, at the other end of the continuum, Forbes.com produces a Cost of Living Extremely Well Index, for 'tracking of the price fluctuations of items that are affordable only to those with very substantial means [which] provides a useful barometer of economic forces at the top end of the market'. The index includes 42 goods and services, such as one year's tuition, room, board and insurance at Harvard, a kilogram of beluga malossol caviar at Petrossian, Los Angeles, and a face lift at the American Academy of Facial, Plastic and Reconstructive Surgery in New York.)

Since the RPI is a measure of how things change over time, it is important to be precise about the date to which each value refers, and the RPI refers to prices on the second or third Tuesday of each month. In a similar vein, one must decide whether the date of receipt of a good or service, the date at which the good or service is used, or the date of payment is the relevant date. The distinction matters if lump sums are paid for goods or services which are then consumed or manifest over an extended period of time.

The use to which an index will be put will be the determining factor in its construction. Inevitably, however, an index such as the RPI will be put to many uses. In particular, the RPI is used as:

(i) a macro-economic indicator, for example, measuring inflation for monitoring economic policies and comparisons with other countries. However, as with many other things European, different countries have calculated their price indices in different ways. The HICP is a standard approach to be used by all EU countries, designed to facilitate international comparisons and not to replace the national indices. (In fact, just as this book was going to press, the UK Chancellor of the Exchequer took some City commentators by surprise. Instead of a gradual transition from the use of the

RPI to the HICP as the Bank of England's official inflation target, he simply made the switch on 10 December 2003. It is also now called simply the Consumer Price Index, and abbreviated to CPI.)

(ii)   a comparator for adjusting tax thresholds, index-linked salaries and pensions, etc.

(iii)   a price regulation indicator for index-linked contracts, etc. There are also various specialized versions of the RPI, including the RPIX (which excludes mortgage interest payments) and the RPIY (which excludes mortgage interest rates, value added tax, council tax, and other indirect taxes).

The RPI has a hierarchical structure, with indices at lower levels aggregated to form higher-level indices. Prices collected at the lowest level may be categorized according to the stratum of geographic region and type of shop in which they were collected, and prices for identical items are averaged (either by using the average of price relatives or the ratio of price averages) to yield an overall index for that item. These are merged (using weighted averages) to yield section indices, which are further merged (again using weighted averages) to yield group indices (e.g. bread, footwear, and rail fares are sections; food and housing are groups). The weights are based on aggregate expenditure on items (from various sources, including the Family Expenditure Survey mentioned above), which means that households spending more have greater influence – the so-called *plutocratic* weighting system. An alternative would be a *democratic* weighting system, in which each household had equal weight. According to Baxter (1998, Section 6.2) democratic weights have never been used, but the choice depends on the use to which the index is to be put: 'Plutocratic weights are more appropriate for an index used as a general measure of inflation, for current cost accounting and for deflating expenditure estimates. Democratic weights might be more appropriate for an index used for uprating purposes, e.g. for pensions or social security benefits.'

Within each year, the RPI is similar to a Laspeyres index (see equation (8.8)), but with the quantities calculated using data for the most recent 12 months, and then the index values for successive years are chained together.

A current topic of discussion is how to take account of the advent of the internet and electronic ordering and purchasing of goods. At the time of writing, books purchased from internet sources are explicitly included in the index, but more and more goods are being ordered by this route. Many questions need to be answered, including whether or not to include delivery charges, how to distinguish business and retail prices, how to handle purchases from overseas websites, how to construct a sampling frame of outlets, and how to deal with the volatility of the internet marketplace.

8.5 The American Customer Satisfaction Index

In measuring economic performance, there has been a growing trend to tap into other aspects of life. An example of such an instrument is the American *Customer Satisfaction Index*. The model for this indicator 'is a set of causal equations that link customer expectations, perceived quality, and perceived value to customer satisfaction. In turn, satisfaction is linked to consequences as defined by customer complaints and customer loyalty – measured by price tolerance and customer retention' (http://www.theacsi.org/model.htm). The model is thus an indirect and largely pragmatic model. Some of the components are measured in a simple and fairly direct way; for example, customer complaints are simply measured as the percentage of respondents who reported a problem with the product. Others are necessarily more complicated: perceived quality is tapped via three questions relating to overall quality, reliability, and the extent to which a customer's needs are met.

8.6 Crime

Pragmatic measurement plays a prominent role in the measurement of crime. Not only are there questions about what should be included in the statistics but, even more fundamentally, there is the fact that legislation, and hence what constitutes crime, changes over time and space. It is because legal systems evolve that comparisons over time are so difficult, and because legal systems are different in different parts of the world that meaningful international comparisons are so difficult.

In practical terms, further complications are caused by the social interactions necessarily involved. For example, the reporting of crime is influenced by perceptions of what might be achieved by reporting it (perhaps the victim believes the police will regard it as trivial, or have no hope of catching the perpetrator), as well as by such things as social attitudes to victims. By no means all crime is reported. Naturally, of course, as in other social areas, in which people or organizations adjust the processes to optimize measured indicators, 'playing to the measure' or 'gaming' occurs: if police 'productivity' is based on a particular measure, inevitably recording practices will evolve to maximize apparent performance on that measure. Furthermore, feedback mechanisms can come into play in a way more subtle than this. For example, before capital punishment was abolished in the UK, juries were hesitant to reach a verdict of 'guilty' to murder, since it was then mandatory for judges to sentence the perpetrator to death. Yet further complications arise from crimes involving multiple victims: are they single or multiple crimes, and how often should they be counted in the overall measure?

In Britain, information on the extent of crime now comes mainly from two complementary sources. The annual *Crime in England and Wales* reports, collated from police statistics, and the *British Crime Survey* (BCS), a survey of the public. The definitions used in these sources differ, and influences of the kind mentioned above mean that the trends apparent from these two sources may not be in the same direction, which naturally confuses the public. For example, 'violent crime appears to be rising according to police statistics, but when one takes into account recording changes, this increase appears to be much smaller, and for those violent crimes reported to the BCS the trend over the past half decade has been down, and significantly so' (Simmons *et al.*, 2002). This should be in the context of slightly under 50% of BCS incidents being reported to the police, and of those which are reported only about 60% are recorded. Of course, these are aggregate statistics, and they differ between different types of crime. Since 1995 the BCS has reported a fall in crime year on year, with an overall reduction of 22% between 1997 and 2002. This appears to contradict (to the great confusion of the public) with the apparent 7% increase in recorded crime in police statistics. In fact this increase is largely attributable to changes in recording practice – that is, to changes in the (pragmatic aspects of) the measurement process.

Perception of crime is even more difficult to measure than the actual crime rate. In part this is because, even with the difficulties of defining 'crime', it is still a relatively well-defined concept, with an attempted formal definition, but this may not perfectly match what is perceived as 'morally wrong'. In the UK we have an offence of 'criminal damage' which may not coincide with what is perceived as damage by any individual. To be a 'crime' an action has to be so classified by the legal system. For example, causing someone to die may or may not be murder, depending on the circumstances, but a relative of the victim may not be inclined to make much of a distinction. In the context of perception of crime, Simmons *et al.* (2002) remark that 'Readers of the national tabloid papers are much more likely to consider the national crime rate to have increased over this period, compared to broadsheet readers (42 percent versus 26 percent)'. Perception of crime can have direct and immediate behavioural consequences, not merely for the freedom of law-abiding citizens to walk the streets at night, but also for less law-abiding citizens, as is elegantly illustrated by the success of the policy of zero tolerance. If some minor social infringement is ignored and allowed to pass, then the perpetrator is encouraged to commit larger insults. And then where does one draw the line? The zero-tolerance policy spells out that line explicitly, shifting the onus of deciding where to draw the line from the individual. It spells out the social morality, so that the individual does not have to make their own choice.

8.7 Workload measurement

The measurement of workload is a clear example of a predominantly pragmatic measurement procedure. The standard approach is to break jobs down into tasks, along with counts of the frequencies with which those tasks are undertaken, and to assign relative weights (often in terms of time units) to each of those tasks, so that an overall measure of load is obtained.

An example is given by Hurst (1999), who describes the events which occur in a system for assessing the case workload in the juvenile justice system of the USA. He says:

> In delinquency cases, such events will include, at a minimum, an arraignment and/or a detention hearing, an adjudication hearing, and a disposition hearing. If a juvenile is detained, some States require a detention hearing every 7 days for the duration of the pretrial detention. In dependency cases, there may be as many as 35 discrete case events.
> (Steelman, Rubin, and Arnold, 1993; Halemba, Hurst, and Gable, 1997)

> Event frequency for particular types of cases is then established by reviewing case records. The time that a judge spends on each event in each case, including bench and nonbench time, is recorded in minutes. The sum of all these times becomes the case weight. The case weights are an average – derived by examining a sample of cases of a particular type – but are thought to represent the amount of time needed to process a typical case.

University staff workloads are becoming the focus of increased attention as more and more auditing and monitoring systems are put in place. University staff have various types of duty, often classified as teaching, research (including research supervision) and administration. As far as the teaching aspect is concerned, the delivery of the material is only one link in a chain of other events which include preparing the material to be presented, creating assessment material and marking schemes, and marking coursework and examinations. Typically preparation takes much longer than the actual presentation. Furthermore, class size is an important factor since it clearly impacts on marking time. Presentation itself may be divided up into different categories, such as lectures, seminars, tutorials and classes. Preparation will typically be divided into preparation for new courses and reviewing material for courses which have been presented before. Preparation for new courses can require very substantial time. For example, it will often require searching for and reading potential new course texts, extracting and restructuring material, creating notes, slides, computer materials, handouts, problem sheets, answer sheets, etc. Furthermore, the different levels at which the material is to be taught may require different weights: first year undergraduate courses may be less demanding than advanced postgraduate courses, but may be much larger. Some disciplines place much emphasis on project work, with associated reports, dissertations and theses.

Research is a category which can easily become squeezed if not monitored closely, because it is often a solo activity – perhaps simply requiring thinking and writing. An emergency committee meeting may take priority just because the research can, in principle, be done at another time whereas the committee meeting cannot. There is also the question of balance between time spent seeking research funding (applying for grants, visiting corporations) and time spent actually working on the research.

Administration will include running sections, departments, or divisions, but will also include service on and chairing committees, recruitment and selection of students, coordinating the preparation of examination papers, interfacing with the external examiners, maintaining the student records, dealing with pastoral issues, timetabling, overseeing the maintenance of the university's infrastructure, and so on.

A fourth category of 'duty', or at least 'obligation', also exists, and it may not be clear how to include this in the calculations. This is the category associated with promoting the university externally: service for learned societies, conference organization, journal editing, serving on national commissions, serving on advisory boards or boards of directors of major corporations, fund raising, consultancy work, etc.

The question of what units to use in such assessments arises. At different universities I have seen currency units, time units and FTEs used. 'FTEs' are full-time equivalents of students, a unit which is used in other contexts in higher education, including funding decisions. This, then, is a sensible unit, not least because it permits an indication to be gained of whether a department is, overall, desperately overstretched.

Of course, once a measurement system is in place, effective management can begin. This is beyond the scope of this book, but will include issues of equity across staff, how to reward achievement in certain areas (the weighting system can be manipulated to effect this, so that the measurement becomes a control process), and how to handle anomalies such as a consistent overload (it being unlikely that everyone will have exactly the same load).

Measuring the workloads of medical staff involves issues similar to those of university staff, even if the component attributes leading to the total are different.

8.8 Conclusion

The examples described above illustrate the heavy, even dominant, role that the pragmatic aspects take in much of social and economic measurement. The objects and systems being described in such contexts are so complex that pure representational measurement is rarely possible. Definition of the attribute being measured necessarily occurs at the same time as the measurement instrument is constructed.

9

Measurement in other areas

Theory, concepts, and methods of measurement are born into a world, by a single creative act, in inseparable union.
John Wheeler, in foreword to Barrow and Tipler (1988)

9.1 Introduction

The chapters of this book have described the theoretical principles underlying measurement and the construction of measurement systems and tools in certain areas. In those chapters, I have made a conscious effort to span the range of different types of measurement situations and types. However, there is an infinite number of other situations in which measurement is used. No book on this topic could ever hope to be comprehensive, not least because the development of measuring instruments is a continuing process. This chapter continues the examination of measurement technologies by providing brief examinations of measurement in several haphazardly selected additional domains.

9.2 Measuring probability

There is a long-standing debate about the meaning of probability. That is, not about the mathematics (the probability 'calculus', one might say), which is generally based on a system of axioms proposed by Kolmogorov in 1933, but about the interpretation (the probability 'theory', as one might put it). This debate reached its apex in the latter half of the twentieth century, but has since rather died down, though not because the debate has really been resolved. From a representational measurement perspective, we will be interested in conditions under which events may be mapped by an order-preserving function satisfying Kolmogorov's axioms. The empirical relationship being represented by this mapping would then be 'at least as probable as'. Krantz *et al.* (1971, pp. 200–201) say: 'It is not evident why the measurement of probability should have been the focus of more philosophic controversy than the measurement of mass, length, or of any other scientifically significant attribute; but it has been.' They then note:

> Theories about the representation of a qualitative ordering of events have generally been classed as subjective (de Finetti, 1937), intuitive (Koopman, 1940a, b; 1941), or personal (Savage, 1954), with the intent of emphasising that the ordering relation ≥ *may* be peculiar to an individual and that he *may* determine it by any means at his disposal, including his personal judgment. But these 'mays' in no way preclude orderings that are determined by well-defined, public, and scientifically agreed upon procedures ... Presumably, as science progresses, objective procedures will come to be developed in domains for which we now have little alternative but to accept the considered judgments of informed and experienced individuals.

Krantz *et al.* (1971) then go on to quote Ellis (1966, p. 172):

> [the development of a probability ordering is] analogous to that of finding a thermometric property, which ... was the first step towards devising a temperature scale. ... The comparison between probability and temperature may be illuminating in other ways. The first thermometers were useful mainly for comparing atmospheric temperatures. The air thermometers of the seventeenth century, for example, were not adaptable for comparing or measuring the temperatures of small solid objects. Consequently, in the early history of thermometry, there were many things which possessed temperature which could not be fitted into an objective temperature order. Similarly, then, we should not necessarily expect to find any single objective procedure capable of ordering all propositions in respect of probability, even if we assume that all propositions possess probability. Rather, we should expect there to be certain kinds of propositions that are much easier to fit into an objective probability order than others.

Krantz *et al.* (1971, p. 202) nicely illustrate the difficulties in other areas (which we may not notice now, since they have become part of the background conceptual fabric against which we think):

> In any case, the difficulties in measuring probability do not appear to be inherently different from those that arise when we apply extensive and other measurement methods elsewhere. In measuring lengths, for example, rods are practical only for relatively short distances; certainly other methods must be used in astronomical research, and considerable disagreement exists among astronomers about which of the alternatives is appropriate (for a general discussion, see Chapter 15 of North, 1965).

One basic objective tool for measuring probabilities in certain situations is the actuarial approach. Here a collective of objects (typically people, in actuarial situations) as similar as possible is investigated, and the probability of a property (event, etc.) is taken as the proportion of objects in the collective which have the property (event, etc.). A classic example is the probability that a person will die at a certain age, where the population of similar people can constitute the collective.

This approach is clearly directly based on a frequentist notion of probability. In many situations, however, the notion of a collective is not so clear-cut, and an objective measure is not so readily found. For example, it would be difficult to define an appropriate collective for the probability that the current Prime Minister of the UK will be assassinated. This seems to be a one-off situation, so that constructing an appropriate collective would be difficult. Doubt about the collective would translate into inaccuracy and doubt about the value of any estimated probability. (Sometimes the word 'uncertainty' is used in such situations, rather than probability, but increasingly many statisticians take the view that probability and uncertainty are the same – after all, the individual objects in a collective are not actually entirely identical, so some subjective decision has been made about how to define the collective.) Many statisticians now accept the viewpoint that probability has no external existence (hence the statement, in capitals, in the preface of Bruno de Finetti's seminal work (de Finetti, 1974): '*PROBABILITY DOES NOT EXIST*'), and instead adopt the subjective view of probability as the degree of belief the observer has that an event will occur. Although this is likely to vary from observer to observer, one can still seek to measure each observer's degree of belief. One can do this directly, by asking people what their probabilities are (and using calibration methods to refine these values – see, for example, Aczél and Pfanzagl, 1966; Murphy and Epstein, 1967; Savage, 1971) or one can attempt indirect approaches, such as via notions of betting.

Bets are often stated in terms of odds, rather than probabilities, though these are equivalent. To say that there are odds of '4 to 1 against' means there is a probability of $1/(1 + 4) = 0.2$ of the event occurring. If the probability of an event E occurring is p, the odds in favour of this event are $p/(1 - p)$. The size of bet you are prepared to take indicates the extent to which you expect certain outcomes to occur. Suppose you wish to bet a fixed amount, \$1, say, that a certain event will occur, and suppose that a bookmaker offers you odds of r. This means that, if the event does not happen, you pay the bookmaker \$1, but if the event does happen the bookmaker pays you \$$r$. If you thought the event was very likely, then you would be prepared to accept a small value of r – less than \$1 even. After all, if you think the event is very likely, you expect to have little chance of losing your \$1, but a good chance of gaining the bookmaker's \$$r$. On the other hand, if you thought the event was very unlikely, then to compensate for the high probability of losing your \$1 you would want a large value of r. For example, you might consider it worth risking losing the \$1 if you thought there was a small chance of winning \$1000. Somewhere in the middle of this there is a value of r which precisely balances your belief in the probability that the event will occur, and we can use this notion to measure our personal probability of the event occurring.

Unfortunately, there are certain complications with this procedure. We took, as the 'stake' in the above, the amount \$1. If, instead, we had used a stake of \$100,000 (with a chance of winning \$100,000r from the bookmaker), then you may have looked at the bets rather differently. For example, you may not have been so willing to risk losing the \$100,000 for the chance of winning \$100,000r. This suggests that we should base our estimates on low stakes. However, if the stakes are too low, then you probably won't care very much either way. If the initial stake was \$0.01 then you probably would not mind risking it at all, and so might take the bet whatever odds were offered. A second complication is that people have a tendency to be 'risk averse' – to reject a bet even if the odds are favourable. It is as if one requires some allowance to be made for the very risk of taking the bet. This phenomenon manifests itself in insurance premiums, which cost more than would be required for the insurance company to break even in the long term (after all, the company is there to make a profit). People are prepared to pay the extra in order to avoid the greater risk they are insuring against.

We can sidestep these complications by avoiding directly asking you to balance the chance of winning and losing, and instead ask you to compare two gambles. So, consider the following two choices:

Choice *A*. If event *E* occurs you receive $1, but if event *E* does not occur you receive nothing.
Choice *B*. An urn contains n_b black balls and n_w white balls and a ball is picked at random. If a black ball is drawn you receive $1, but if a white ball is drawn you receive nothing.

Initially, the urn in *B* has $n_b = n_w$ (equal to some moderately large number, say 100). You are now invited to choose between *A* and *B*. If you choose *A*, n_b is increased by 1 and n_w is reduced by 1. If you choose *B*, n_b is reduced by 1 and n_w is increased by 1. This exercise is repeated until you are indifferent to the choice between *A* and *B*. Having reached this stage, it means (or, at least, is taken to mean) that you believe that the probability of event *E* occurring in choice *A* is the same as the probability of drawing a black ball in choice *B*. From our frequentist approach, based on collectives, we can estimate the probability of the latter as $n_b/(n_b + n_w)$. This is thus the estimate of your belief that *E* will occur.

The assessment of subjective probability has attracted a great deal of research attention, from psychologists as well as statisticians. Important references are Chapter 5 of de Finetti (1974, Volume 1) and Wright and Ayton (1994).

9.3 Software metrics

Writing in 1991, Norman Fenton argued for the importance of measurement in software engineering (Fenton, 1991, p. ix): 'A primary reason why software engineering remains more an ideology than a discipline is that measurement has been almost totally ignored by some of the leading personalities who have shaped its direction.' He went on to argue why measurement was fundamental to software engineering and listed some of the activities which needed properly defined metrics, including cost and effort, productivity, quality, data collection, reliability, performance, algorithmic complexity, and structural complexity.

Fenton (1991, p. 42) listed three classes of entities which need to be measured in software engineering: processes, products, and resources. Likewise, Kitchenham (1992) distinguished between 'metrics which assist in the *control* or *management* of the development process' and 'metrics which are *predictors* (or *indicators*) of either product qualities or control metrics'. An important distinction is also made between internal attributes ('those which can be measured purely in terms of the product, process, or resource itself') and external attributes ('those which can only be measured with respect to how the product, process, or resource relates to its environment'). Thus, a thrown rock will have a mass (internal attribute) and a speed (external attribute). In general, external attributes such as cost-effectiveness, productivity and portability are the ones of most interest to managers and users, but these are often difficult to measure. Fenton says: 'Moreover, there are normally no agreed definitions; attributes like *quality* are so general that they are almost meaningless. Thus we are forced to make contrived definitions of the attributes in terms of some other attributes which *are* measurable. For example some people define quality of a software system (an external product variable) in terms of *the number of bugs found during formal testing* (an internal process attribute).' In the language used in this book, this is presenting a pessimistic take on the fact that measurement processes involve pragmatic aspects, which are often indirect. This can be regarded as a merit (one can precisely define what one is measuring) rather than a demerit ('we are forced to make contrived definitions').

In order to construct effective indirect pragmatic measurement definitions of external attributes it is necessary to identify the constituent features which should be combined to yield the definition. For processes, these will include time, effort, and cost-effectiveness; for products they will include size, modularity, functionality, complexity, understandability, portability and maintainability; for resources they will include price, age, size, etc. Specific processes, products, and resources will have associated specific relevant component attributes.

For example, the Jelinski–Moranda model of software reliability (Jelinski and Moranda, 1972) assumes that failures occur according to a Poisson process with a rate decreasing as more faults are discovered. In particular, if ϕ is the true failure rate per fault and N is the number of faults initially present, then the failure rate for the ith fault (after $i - 1$ faults have already been detected and fixed, not introducing new faults in doing so) is

$$\lambda_i = (N - i + 1)\phi$$

The indirect nature of the measurement procedure, the fact that the measurement is defined in terms of a model, is very apparent in this example.

To take another example, cost models are often couched in terms of the effort required. The COCOMO model (Boehm, 1981) is one such, taking the form

$$\textit{effort} = k(\textit{size})^c$$

with appropriate parameters. Effort might be measured in person-months and size in thousands of delivered source instructions.

Although one might expect internal attributes to permit more straightforward definitions and hence measurement procedures than external attributes, even internal attributes are often far from simple. Take the length of a program, for example. One might decide to measure this using the number of lines of code. But this leaves open questions such as whether comment lines are to be counted and, perhaps more importantly, how lines with multiple commands are to be counted. Software complexity is clearly more difficult to define than program length, and this is reflected in the diversity of attempts to define and measure it. Many measures of software complexity are based on the flow of control – the possible routes through the code. A natural representation for such structures is as graphs, with statements corresponding to nodes, and edges corresponding to flow of control; for example, McCabe's *cyclomatic number* (McCabe, 1976) is defined as $e - n + 2p$, where e is the number of edges in the control graph, n is the number of nodes, and p is the number of connected components. Zuse (1990), in particular, gives a detailed discussion of the measurement of complexity of software systems.

9.4 Database management systems

Various measures of performance are relevant to databases. The most important might be precision and recall, defined in terms of relevance. *Relevance* is the extent to which the retrieved documents match the needs of the user. *Precision* (P) is the ratio of the number of relevant documents which are retrieved to the total number of retrieved documents. Finally, *recall* (R) is the ratio of the number of relevant documents which are retrieved to the number of relevant documents in the database. Broadly speaking, precision and recall are inversely related: precision is high when few irrelevant documents are retrieved, and this is most easily achieved by retrieving few documents (those most relevant), which leads to low recall. Of course, this is only an informal relationship.

Precision and recall have been combined in a pragmatic manner to yield overall measures of the effectiveness of a retrieval system. For example, the following two measures compute kinds of average of precision and recall (averaged over many retrieval operations). *Meadow's measure* is based on the complements of precision and recall:

$$1 - \sqrt{\frac{(1-P)^2 + (1-R)^2}{2}}$$

Vickery's measure is:

$$1 - (2P^{-1} + 2R^{-1} - 3)^{-1}$$

A third measure, *Heine's measure,* is intended to apply to individual retrieval operations (indexed by i):

$$1 - (P_i^{-1} + R_i^{-1} - 1)^{-1}$$

Van Rijsbergen (1979) has experimented with varying the relative importance accorded to precision and recall in the above measures.

Yet other measures are based on the notions of sensitivity and specificity discussed in Chapter 7, and, of course, other measures will be defined in terms of cost, speed of accessing material, etc.

Effective measures of relevance are important in the context of search engines on the Internet, which can easily throw up many tens of thousands of matches. One would like to start one's examination of these with those most likely to be useful!

9.5 Informetrics

Boyce *et al.* (1994, p. 99) define 'bibliometrics as the search for measurable regularities in the production and utilization of recorded information, and informetrics as the search for measurable regularities in all communication processes'. Since these things are intended to be measurable, there must be ways to measure them. A wide variety of 'regularities' are of interest in these areas, and have been the subject of considerable research in the data retrieval literature. More recently, this has been stimulated by the advent of the Internet and the development of search engines. This section illustrates just a few of these measures.

9.5.1 Characterizing documents

Documents are characterized by the words they contain, and, in general, one might regard a word that occurs more frequently in a document as more important in characterizing the meaning of that document. Of course, if a word occurs frequently in many documents, it will be less important in singling out the characteristics of an individual document. These two points lead to a definition of *inverse document frequency weight* as

$$w(i,j) = f_{ij}\left(\log_2\left(\frac{n}{n_j}\right) + 1 \right)$$

where f_{ij} is the number of times word j occurs in document i, n is the total number of documents in the collection, and n_j is the number of documents in which word j occurs. This definition clearly has a heavy pragmatic aspect.

A more elaborate approach is based on discriminative concepts, where one explores the extent to which a word allows one to distinguish between documents. If we represent each document by a binary vector, with each position in the vector corresponding to a different word, then we can use the similarity measures described in Chapter 3 to decide how similar two documents are. In a similar way, we could use the binary vectors describing which documents a particular word occurs in to measure the similarity of two words, in the context of the available document collection.

Natural language texts (in English, at least) have been found approximately to conform to the relationship

$$f \times r = c$$

where f is the frequency of occurrence of a word in a document, r is the rank order of frequency of occurrence, and c is a constant. This relationship is known as *Zipf's law* (e.g. Zipf, 1949). c is thus given by the frequency of the word which ranks first. Since the minimum frequency is $f = 1$, the corresponding value of r is c, so c approximately indicates the number of distinct words – it is a rough measure of richness of vocabulary in the document collection. In fact Zipf's law only provides an approximate fit, and various modifications have been suggested.

9.5.2 Impact factors

Scientists and scholars publish their work in journals in order to convey their ideas to other people. It is natural, therefore, that they should wish their work to appear in those journals which are most widely read, or, at least, which have the most influence on others. One measure of this influence is *impact factor* developed by the Institute for Scientific Information (Garfield, 1979, 1994). The impact factor for a given journal in year i is defined as the ratio of the total number of citations in year i of articles published in that journal in years $i - 1$ and $i - 2$ to the total number of articles published in that journal in those two years. The denominator removes the effect of journal size (number of articles published each issue, number of issues per year), and the restriction to the two most recent preceding years eliminates the fact that an old-established journal could have citations referring to its older body of articles as well as removing the effect of a single very highly cited article which appeared in the journal years before. Of course, it does not resolve issues of specialism: a highly regarded journal in a small niche area will not receive as many citations as a general journal, all else being equal.

Modifications of the above impact factor include versions based on a longer time period (e.g. a 5-year impact factor), and versions which exclude citations to the same journal.

9.6 And beyond

This book has, of necessity, focused on measurement in just a few areas, but there is, of course, no limit to the things that can be measured or the ways in which those things may be measured. There are various compendia of measurement units. Indeed some people seem to collect them, in much the same manner as others collect stamps or coins. Such compendia include *For Good Measure* (Johnstone, 1998), the *Oxford Dictionary of Weights, Measures and Units* (Fenna, 2002), and the *Encyclopaedia of Scientific Units, Weights and Measures* (Cardarelli, 2003). The last of these, in particular, properly deserves the name 'encyclopaedia', providing a comprehensive description of modern and ancient systems from throughout the world. Nowadays, of course, there are also websites devoted to such collections; for example, http://www.ex.ac.uk/cimt/dictunit/dictunit.htm. These collections are mainly concerned with units which have a strong representational basis, and hence are oriented towards the physical sciences and engineering. Books which describe the more pragmatic measurements used in, for example, medicine, psychology and the social sciences (e.g. Bowling, 1995; McDowell and Newell, 1996; Temple, 2003; Economist, 2003) tend to discuss how the measures are constructed, how to interpret them, and perhaps their statistical properties.

9.7 Conclusion

The aim of this book has been to examine the notion of measurement in its entirety, exploring the meaning of the concept, its manifestation in different areas of human endeavour, and our perceptions of measurement. Measurement is what ties our poorly formed speculations about the universe to properly constructed theoretical and predictive models. As Margenau (1959) put it, 'measurement enables both embarkation and debarkation of a theoretical traveller at the shore of empirical fact.' Indeed, one might go even further and say that measurement is what distinguishes humans from animals: the ability to construct and manipulate a formal model in place of the reality itself, and in place of ill-defined informal models. From this perspective, measurement is what defines humanity. But for the politically incorrect flavour, one might be inclined to say 'measurement maketh Man'.

Measurement is also ubiquitous, as was illustrated by the opening sentences of this book. But its ubiquity is often concealed by our very familiarity with the notion. We often simply do not notice that we are using ideas of measurement, just because we use them all the time. This familiarity can mean that the concepts and uses of measurement are subject to little critical assessment. For the physical sciences, Crosby (1997, p. 14) comments that 'unlike Plato and Aristotle, we, with few exceptions, embrace the assumption that mathematics and the material world are immediately and intimately related. We accept as self-explanatory the fact that physics, the science of palpable reality, should be intensely mathematical. But that proposition is not self-explanatory; it is a miracle about which many sages have had their doubts.' I doubt that many modern sages have doubts about the mathematical nature of the modern physical world. Eugene Wigner's 'unreasonable effectiveness' and Lord Kelvin's 'when you can measure it you know something about it' have been thoroughly absorbed. But doubt and unease still remain in some minds about the mathematical nature of the psychological, sociological and other worlds. This is despite the evidence that measurement concepts and methods are making steady advances in our ability to formulate and manipulate our ideas about such worlds. That said, it is the complexity of the psychological and sociological worlds especially which drives home the bipartite nature of measurement: the fact that it has both representational and pragmatic aspects. In some situations – the physical sciences provide many examples – it is clear what system one wishes to represent, so that a formal representational mapping to a number system, with relatively little freedom of choice, is possible. In other situations – and the psychological and sociological domains provide many examples of this – the concepts to be measured are relatively ill-defined. Measurement in such situations requires both precise definition, as well as construction of the measuring instrument. That is, in such situations, measurement involves a large pragmatic component.

Some are suspicious of pragmatic measurement precisely because it simultaneously provides a clear definition of the concept being measured, dismissing it as 'measurement by fiat'. But this disregards the merits of such measurement for decision making and control of the world about us. Furthermore, it also ignores the key merit of having made the way one views the world explicit: one is always at liberty to modify given pragmatic constraints if one feels that they fail to capture the concept in its entirety. That is, one can tinker with the definition. This is very different from the situation with unformulated subjective notions.

To conclude, measurement is the system through which we understand the world about us, and the process by which we control and manipulate it. Measurement is what distinguishes the civilized from the uncivilized.

Wherefore in all great works are Clerks so much desired? Wherefore are Auditors so well fed? What causeth Geometricians so highly to be enhaunsed? Why are Astronomers so greatly advanced? Because that by number such things they finde, which else would farre excell mans minde.

(R. Recorde, *The Declaration of the Profit of Arithmeticke*, 1540;
quoted in Swetz, 1987)

References

Abelson R.P. and Tukey J.W. (1959) Efficient conversion of non-metric into metric information. *Proceedings of the Social Statistics Section of the American Statistical Association*, 226–230. (Reprinted in E.R. Tufte (ed.) (1970) *The Quantitative Analysis of Social Problems*. Reading, MA: Addison-Wesley.)

Abelson R.P. and Tukey J.W. (1963) Efficient utilisation of non-numerical information in quantitative analysis: general theory and the case of simple order. *Annals of Mathematical Statistics*, **34**, 1347–1369.

Aczél J. (1966) *Lectures on Functional Equations and Their Applications*. New York: Academic Press.

Aczél J. and Pfanzagl J. (1966) Remarks on the measurement of subjective probability and information. *Metrika*, **2**, 91–105.

Adams E.W. (1966) On the nature and purpose of measurement. *Synthèse*, **16**, 125–169.

Adams E.W. (1979) Measurement theory. In *Current Research in Philosophy of Science*. Philosophy of Science Association, 207–227.

Adams E.W., Fagot R.F. and Robinson R.E. (1964) *Invariance, Meaningfulness, and Appropriate Statistics*. Technical Report No. 1, August 1. Measurement Theory Reports. University of Oregon.

Adams E.W., Fagot R.F. and Robinson R.E. (1965) A theory of appropriate statistics. *Psychometrika*, **30**, 99–127.

Adams N.M. and Hand D.J. (1999) Comparing classifiers when the misallocation costs are uncertain. *Pattern Recognition*, **32**, 1139–1147.

Aiken L.R. Jr (1963) The grading behaviour of a college faculty. *Educational and Psychological Measurement*, **23**, 319–322.

Aitken R.C.B. (1969) A growing edge of measurement of feelings. *Proceedings of the Royal Society of Medicine*, **62**, 989–996.

Alder K. (1995) A revolution to measure: the political economy of the metric system in France. In M.N. Wise (ed.), *The Values of Precision*. Princeton NJ: Princeton University Press, 39–71.

Alder K. (2002) *The Measure of All Things: The Seven-Year Odyssey that Transformed the World*. London: Little, Brown.

Alexander J.H. (1850) *Universal Dictionary of Weights and Measures*. Baltimore, MD: William Minifie and Co.

Allinger G.M. (1988) Do zero correlations really exist among measures of different intellectual abilities? *Educational and Psychological Measurement*, **48**, 275–280.

Allport G.W. (1937) *Personality*. New York: Henry Holt.

Allport G.W. (1942) *The Use of Personal Documents in Psychological Science* (Bull. No. 49). New York: Social Science Research Council.

Alper T.M. (1984) Groups of homeomorphisms on the real line. Unpublished Bachelor of Science thesis, Harvard University, Cambridge, MA.

Alper T.M. (1985) A note on real measurement structures of scale type $(m, m + 1)$. *Journal of Mathematical Psychology*, **29**, 73–81.

Alper T.M. (1987) A classification of all order-preserving homeomorphism groups of the reals that satisfy finite uniqueness. *Journal of Mathematical Psychology*, **31**, 135–154.

Altman D.G. (1999) *Practical Statistics for Medical Research*. Boca Raton, FL: Chapman and Hall/CRC.

American Psychological Association (1992) Call for book proposals for test instruments. *APA Monitor*, **23**, 15.

Anastasi A. (1985) Some emerging trends in psychological measurement: a fifty-year perspective. *Applied Psychological Measurement*, **9**, 121–138.

Anderson N.H. (1961) Scales and statistics: parametric and nonparametric. *Psychological Bulletin*, **58**, 305–316.

Apgar V. (1953) Proposal for a new method of evaluation of the newborn infant. *Anesthesia and Analgesia*, **32**, 260–267.

Asimov I. (1967) *Of Time, and Space and Other Things, Part II*. London: Dobson.

Audit Commission (2000a) *Aiming to Improve: The Principles of Performance Measurement*. London: Audit Commission.

Audit Commission (2000b) *On Target: the Practice of Performance Indicators*. London: Audit Commission.

Baglivi G. (1696) *De Praxi Medica*. Rome.

Baird L. and Flister W.J. (1972) Grading students: the relation of changes in average student ability to the average grades awarded. *American Educational Research Journal*, **9**, 431–442.

Baker B.O., Hardyck C.D. and Petrinovich L.F. (1966) Weak measurements vs strong statistics: an empirical critique of S.S. Stevens' proscriptions on statistics. *Educational and Psychological Measurement*, **26**, 291–309.

Baker D.J., Pynsent P.B. and Fairbank J.C.T. (1989) The Oswestry Disability Index revisited: its reliability, repeatability and validity, and a comparison with the St Thomas's Disability Index. In M.O. Roland and J.R. Jenner (eds), *Back Pain: New Approaches to Rehabilitation and Education*. Manchester: Manchester University Press, 174–186.

Balk B.M. (1995) Axiomatic price index theory: a survey. *International Statistical Review*, **63**, 69–93.

Barnes P. (1987) The analysis and use of financial ratios: a review article. *Journal of Business Finance and Accounting*, **14**, 449–461.

Barrow J.D. and Tipler F.J. (1988) *The Anthropic Cosmological Principle*. Oxford: Oxford University Press.

Bartholomew D.J. (1987) *Latent Variable Models and Factor Analysis*. London: Griffin.

Bartholomew D.J. (1996) *The Statistical Approach to Social Measurement*. San Diego, CA: Academic Press.

Bartholomew D.J. and Knott M. (1999) *Latent Variable Models and Factor Analysis*, 2nd edition London: Arnold.

Basilevsky A. (1994) *Statistical Factor Analysis and Related Models*. New York: Wiley.

Baulieu F.B. (1989) A classification of presence/absence dissimilarity coefficients. *Journal of Classification*, **6**, 233–246.

Baxter M. (1998) *The Retail Prices Index Technical Manual*. London: The Stationery Office.

Beck A.T. (1967) *Depression: Clinical, Experimental, and Theoretical Aspects*. New York: Harper & Row.

Beck A.T., Ward C.H., Mendelson M. *et al.* (1961) An inventory for measuring depression. *Archives of General Psychiatry*, **4**, 561–571.

Beck A.T. *et al.* (1979) *Cognitive Therapy of Depression*. New York: Guilford Press.

Behan F.L. and Behan R.A. (1954) Football numbers (continued). *American Psychologist*, **9**, 262–263.

Benese, Sir Rycharde de (1562) *This boke showeth the maner of measuring*. Thomas Colwell's edition.

Bennett D.J. (1998) *Randomness*. Cambridge, MA: Harvard University Press.

Bennett E.M. (1954) On the statistical measurement of index numbers. *American Psychologist*, **9**, 264.

Bennett W.J. (1999) *The Index of Leading Cultural Indicators: American Society at the End of the Twentieth Century.* New York: Broadway Books.

Bentham J. (1983) *Deonotology.* Oxford: Clarendon Press.

Benzécri J.-P. (1992) *Correspondence Analysis Handbook.* New York: Marcel Dekker.

Berriman A.E. (1953) *Historical Metrology.* London.

Berry, M.J.A. and Linoff, G. (2000) *Mastering Data Mining. The Art and Science of Customer Relationship Management,* New York: Wiley.

Binder A. (1984) Restrictions on statistics imposed by method of measurement: some reality, much mythology. *Journal of Criminal Justice*, **12**, 467–481.

Binet A. and Henri V. (1895) La Psychologie individuelle. *Année Psychologique*, **2**, 411–463.

Bird S.M., Cox D.R., Farewell V.T., Goldstein H., Holt T. and Smith P.C. (2003) *Performance Indicators: Good, Bad, and Ugly.* Royal Statistical Society Working Party on Performance Monitoring in the Public Services. London: Royal Statistical Society.

Birkhoff G. (1960) *Hydrodynamics. A Study in Logic, Fact, and Similitude.* Princeton, NJ: Princeton University Press.

Birnbaum R. (1977) Factors related to university grade inflation. *Journal of Higher Education*, **48**, 519–539.

Bland J.M. and Altman D.G. (1986) Statistical methods for assessing agreement between two methods of clinical measurement. *Lancet*, **i**, 307–310.

Boehm B.W. (1981) *Software Engineering Economics.* New York: Prentice Hall.

Bollen K.A. (1989) *Structural Equations with Latent Variables.* New York: Wiley.

Bombadier C. and Tugwell P.A. (1982) A methodological framework to develop and select indices for clinical trials: statistical and judgemental approaches. *Journal of Rheumatology*, **9**, 753–757.

Boneau C.A. (1960) The effects of violations of assumptions underlying the *t* test. *Psychological Bulletin*, **57**, 49–64.

Boneau C.A. (1961) A note on measurement scales and statistical tests. *American Psychologist*, **16**, 260–261.

Borgatta E.F. (1968) My student, the purist: a lament. *Sociological Quarterly*, **9**(Winter), 29–34.

Borgatta E.F. and Bohrnstedt G.W. (1980) Levels of measurement – once over again. *Sociological Methods and Research*, **9**, 147–160.

Boring E.G. (1929) *A History of Experimental Psychology.* New York: Century.

Boring E.G. (1945) The use of operational definitions in science. *Psychological Review*, **52**, 243–245.

Bowling A. (1995) *Measuring Disease: A Review of Disease-Specific Quality of Life Measurement Scales.* Buckingham: Open University Press.

Box G. and Hunter W. (1965) The experimental study of physical mechanisms. *Technometrics*, **7**, 57–71.

Boyce B.R., Meadow C.T. and Kraft D.H. (1994) *Measurement in Information Science.* San Diego, CA: Academic Press.

Boyle D. (2000) *The Tyranny of Numbers.* London: HarperCollins.

Bradley R.A. and Terry M.E. (1952) Rank analysis of incomplete block designs: I. The method of paired comparisons. *Biometrika*, **39**, 324–345.

Bradley W.J. and Schaefer K.C. (1998) *The Uses and Misuses of Data and Models: The Mathematization of the Human Sciences.* Thousand Oaks, CA: Sage.

Bragg G.M. (1974) *Principles of Experimentation and Measurement.* Englewood Cliffs, NJ: Prentice Hall.

Bridgman P. (1922) *Dimensional Analysis.* New Haven, CT: Yale University Press (2nd edition 1931).

Bridgman P.W. (1927) *The Logic of Modern Physics.* New York: Macmillan.

Brown G.W. and Harris T. (1978) Social origins of depression: a reply. *Psychological Medicine*, **8**, 577–588.

Brunskill A.J. (1990) Some sources of error in the coding of birth weight. *American Journal of Public Health*, **80**, 72–73.

Bryson B. (1995) *Made in America*. London: Minerva Press.

Buckingham E. (1914) On physically similar systems: illustrations of the use of dimensional equations. *Physical Review*, **IV**(4), 345–376.

Burke C.J. (1953) Additive scales and statistics. *Psychological Review*, **60**, 73–75.

Burn J.H. (1930) The errors of biological assay. *Physiological Review*, **10**, 146–169.

Burn J.H., Finney D.J. and Goodwin L.G. (1937) *Biological Standardization*. Oxford: Oxford University Press.

Burtt E.A. (1954) *The Metaphysical Foundations of Modern Physical Sciences*. Garden City, NY: Doubleday.

Cabanis P.J.G. (1798) *Du Degré de Certitude de la Médicine*. Paris.

Campbell D.T. and Fiske D.W. (1959) Convergent and discriminant validation by the multitrait-multimethod matrix. *Psychological Bulletin*, **56**, 81–105.

Campbell N.A. (1920) *Physics, the Elements*. London: Cambridge University Press.

Campbell N.R. (1938) *Symposium: Measurement and Its Importance for Philosophy*. Aristotelian Society, Suppl. Vol. 17. London: Harrison.

Cardarelli F. (2003) *Encyclopaedia of Scientific Units, Weights and Measures: Their SI Equivalences and Origins*. London: Springer.

Carroll J.B. (1993) *Human Cognitive Abilities: a Survey of Factor-Analytic Studies*. Cambridge: Cambridge University Press.

Carroll J.B., Winer R.L., Coates D., Galegher J. and Alibrio J.J. (1982) Evaluation, diagnosis, and prediction in parole decision making. *Law and Society Review*, **17**, 199–228.

Carroll J.D. and Chang J.J. (1970) Analysis of individual differences in multidimensional scaling via an *N*-way generalisation of the 'Eckart–Young' decomposition. *Psychometrika*, **35**, 283–319.

Cattell R.B. (1963) Theory of fluid and crystallized intelligence: a critical experiment. *Journal of Educational Psychology*, **54**, 1–22.

Cauchy A.L. (1821) Cours d'analyse de l'École Polytechnique. *Analyse Algébrique*, Vol. 1, V. Paris. (*Oeuvres*, Ser. 2, Vol. 3, pp. 98–113, 220. Paris, 1897.)

Chambers R. (1997) *Whose Reality Counts? Putting the First Last*. London: Intermediate Technology.

Champion D.J. (1968) Some observations on measurement and statistics: comment. *Social Forces*, **46**, 541.

Chandler G.G. and Coffman J.Y. (1979) A comparative analysis of empirical versus judgmental credit evaluation. *Journal of Retail Banking*, **1**(2), 15–26.

Chatfield C. (1995) Model uncertainty, data mining and statistical inference (with discussion). *Journal of the Royal Statistical Society, Series A*, **158**, 419–466.

Chisholm H.W. (1877) *On the Science of Weighing and Measuring and Standards of Measure and Weight*. London: Macmillan.

Chrystal K.A. and Mizen P.D. (2003) Goodhart's Law: its origins, meaning, and implications for monetary policy. In P.D. Mizen (ed.), *Monetary Theory and Practice: Essays in Honour of Charles Goodhart*, Vol. 1. Cheltenham: Edward Elgar, 221–243.

Cohen M. and Narens L. (1979) Fundamental unit structures: a theory of ratio scalability. *Journal of Mathematical Psychology*, **20**, 193–232.

Collin C., Wade D.T., Davies S. *et al.* (1988) The Barthel ADL index: a reliability study. *International Disability Studies*, **10**, 61–63.

Coombs C.H. (1950) Psychological scaling without a unit of measurement. *Psychological Review*, **57**, 148–158.

Coombs C.H. (1964) *A Theory of Data*. New York: Wiley.

Coombs C.H. and Huang L. (1970) Polynomial psychophysics of risk. *Journal of Mathematical Psychology*, **7**, 317–338.

Copas J.B. and Li H.G. (1997) Inference for non-random samples. *Journal of the Royal Statistical Society, Series B*, **59**, 55–95.

Cox D.R. (1990) Role of models in statistical analysis. *Statistical Science*, **5**, 169–174.

Cox T.F. and Cox M.A.A. (2001) *Multidimensional Scaling*, 2nd edition. Boca Raton, FL: Chapman and Hall/CRC.

Cronbach L.J. (1951) Coefficient alpha and the internal structure of tests. *Pychometrika*, **16**, 297–334.

Cronbach L.J. and Gleser G.C. (1957) *Psychological Tests and Personnel Decisions*. Urbana: University of Illinois Press.

Cronbach L.J. and Meehl P.E. (1955) Construct validity in psychological tests. *Psychological Bulletin*, **52**, 281–302.

Crosby A. (1997) *The Measure of Reality: Quantification and Western Society, 1250–1600*. Cambridge: Cambridge University Press.

Custance J. and Hillier H. (1998) Statistical issues in developing indicators of sustainable development. *Journal of the Royal Statistical Society, Series A*, **161**, 281–290.

David R.J. (1980) The quality and completeness of birthweight and gestational age data in computerized birth files. *American Journal of Public Health*, **70**, 964–973.

Davidson M.L. (1983) *Multidimensional Scaling*. New York: Wiley.

Dawes R.M. and Smith T.L. (1985) Attitude and opinion measurement. In G. Lindzey and E. Aronson (eds), *The Handbook of Social Psychology*, Volume I, 3rd edition. New York: Random House, 509–566.

Dawes R.M., Faust D. and Meehl P.E. (1989) Clinical versus actuarial judgment. *Science*, **243**, 1668–1674.

de Finetti B. (1937) La prévision: ses lois logiques, ses sources subjectives. *Annals of the Institute of Henri Poincaré*, **7**, 1–68.

de Finetti B. (1974) *Theory of Probability*. Chichester: Wiley.

Debreu G. (1960) Topological methods in cardinal utility theory. In K.J. Arrow, S. Karlin and P. Suppes (eds), *Mathematical Methods in the Social Sciences*. Stanford, CA: Stanford University Press, 16–26.

Dellaportas P., Stephens D.A., Smith A.F.M. and Guttman I. (1996) A comparative study of perinatal mortality using a two-component normal mixture. In D.A. Berry and D.K. Stangl (eds), *Bayesian Biostatistics*. New York: Marcel Dekker, 601–615.

Department of Health and Social Security (1983) *Performance Indicators: National Summary for 1981*. London: Department of Health and Social Security.

Descartes R. (1985) *The Philosophical Writings of Descartes*, trans J. Cottingham, R. Stoothoff, and D. Murdoch. Cambridge: Cambridge University Press.

Desrosières A. (1998) *The Politics of Large Numbers: A History of Statistical Reasoning*. Cambridge, MA: Harvard University Press.

Diewert W.E. and Nakamura A.O. (eds) (1993) *Essays in Index Number Construction*. Amsterdam: North-Holland.

Dingle H. (1950) A theory of measurement. *British Journal of the Philosophy of Science*, **1**, 5–26.

Divisia F. (1925) L'indice monétaire et la théorie de la monnaie. *Revue Economique Politique*, **39**, 980–1008.

Dohrenwend B.P., Shrout P.E., Egri G. *et al.* (1980) Nonspecific psychological distress and other dimensions of psychopathology. *Archives of General Psychiatry*, **37**, 1229–1236.

Downey M.T. (1965) *Ben T. Wood, Educational Reformer*. Princeton, NJ: Educational Testing Service.

Drake S. (1957) (trans.) *Discoveries and Opinions of Galileo*. Garden City, NY: Doubleday.

Dunn G. (1989) *Design and Analysis of Reliability Studies: The Statistical Evaluation of Measurement Errors*. London: Edward Arnold.

Dunn G., Sham P.C. and Hand D.J. (1993) Statistics and the nature of depression. *Journal of the Royal Statistical Society, Series A*, **156**, 63–87. (Reprinted in *Psychological Medicine*, **23**, 871–889.)

Ebbs S.R., Fallowfield L.J., Fraser S.C.A. and Baum M. (1989) Treatment outcomes and quality of life. *International Journal of Technology Assessment in Health Care*, **5**, 391–400.

Economist (2003) *Guide to Economic Indicators: Making Sense of Economics*. London: The Economist.

Eichhorn W. and Voeller J. (1976) *Theory of the Price Index*, Lecture Notes in Economics and Mathematical Systems 140. Berlin: Springer-Verlag.

Ellis B. (1966) *Basic Concepts of Measurement*. London: Cambridge University Press.

Endicott J. and Spitzer R.L. (1978) A diagnostic interview: the schedule for affective disorders and schizophrenia. *Archives of General Psychiatry*, **35**, 837–844.

Epstein S. (1979) The stability of behaviour I: On predicting most of the people much of the time. *Journal of Personality and Social Psychology*, **37**, 1097–1126.

Everitt B.S. (1984) *An Introduction to Latent Variable Models*. London: Chapman and Hall.

Everitt B.S. and Dunn G. (2001) *Applied Multivariate Data Analysis*, 2nd edition. London: Arnold.

Fairbank J.C.T., Couper J., Davies J.B. *et al.* (1980) The Oswestry Low Back Pain Disability Questionnaire. *Physiotherapy*, **66**, 271–273.

Falmagne J.C. (1979) On a class of probabilistic conjoint measurement models: some diagnostic properties. *Journal of Mathematical Psychology*, **19**, 73–88.

Falmagne J.C. (1980) A probabilistic theory of extensive measurement. *Philosophy of Science*, **47**, 277–296.

Falmagne J-C. (1985) *Elements of Psychophysical Theory*. Oxford: Clarendon Press.

Fayers P.M. and Hand D.J. (1997) Factor analysis, causal indicators, and quality of life. *Quality of Life Research*, **6**, 139–150.

Fayers P.M. and Hand D.J. (2002) Causal variables, indicator variables, and measurement scales, with discussion. *Journal of the Royal Statistical Society, Series A*, **165**, 233–261.

Fayers P.M. and Machin D. (2000) *Quality of Life: Assessment, Analysis and Interpretation*. Chichester: Wiley.

Fayers P.M., Groenvold M., Hand D.J. and Bjordal K. (1998) Clinical impact versus factor analysis for quality of life questionnaire construction. *Journal of Clinical Epidemiology*, **51**, 285–286.

FDA (1988) *Guideline for the Format and Content of the Clinical and Statistical Sections of New Drug Applications*. Food and Drug Administration, Public Health Service, US Department of Health and Human Services.

Fechner G.T. (1860) *Elemente der Psychophysik*. Leipzig: Breitkopf und Hartel.

Feinstein A.R. (1977) Clinical biostatistics XXXIX: The haze of Bayes, the aerial palaces of decision analysis, and the computerised Ouija board. *Clinical Pharmacology and Therapeutics*, **21**, 482–496.

Feinstein A.R. (1987a) Clinimetric perspectives. *Journal of Chronic Diseases*, **40**, 635–640.

Feinstein A.R. (1987b) *Clinimetrics*. New Haven, CT: Yale University Press.

Fenna D. (2002) *Oxford Dictionary of Weights, Measures and Units*. Oxford: Oxford University Press.

Fenton N. (1991) *Software Metrics: A Rigorous Approach*. London: Chapman and Hall.

Ferguson A., Myers C.S., Bartlett R.J., Banister H., Bartlett F.C., Brown W., Campbell N.R., Drever J., Guild J., Houston R.A., Irwin J.O., Kaye G.W.C., Philpott S.J.F., Richardson L.F., Shaxby J.H., Smith T., Thouless R.H. and Tucker W.S. (1938) Quantitative estimates of sensory events: interim report of the committee appointed to consider and report upon the possibility of quantitative estimates of sensory events. *British Association for the Advancement of Science*, **108**, 277–334.

Ferguson A., Myers C.S., Bartlett R.J., Banister H., Bartlett F.C., Brown W., Campbell N.R., Craik K.J.W., Drever J., Guild J., Houston R.A., Irwin J.O., Kaye G.W.C., Philpott S.J.F., Richardson L.F., Shaxby J.H., Smith T., Thouless R.H. and Tucker W.S. (1940) Quantitative estimates of sensory events: final report of the committee appointed to consider and report upon the possibility of quantitative estimates of sensory events. *Advancement of Science*, **1**, 331–349.

Finney D.J. (1977) Dimensions of statistics. *Applied Statistics*, **26**, 285–289.

Fisher I. (1922) *The Making of Index Numbers*. Boston: Houghton Mifflin.

Fisk P.R. (1977) Some approximations to an 'ideal' index number. *Journal of the Royal Statistical Society, Series A*, **140**, 217–231.

Fleiss J.L. (1981) *Statistical Methods for Rates and Proportions*, 2nd edition. New York: Wiley.

Flynn J.R. (1984) The mean IQ of Americans: Massive gains 1932 to 1978. *Psychological Bulletin*, **95**, 29–51.

Flynn J.R. (1987) Massive gains in IQ in 14 nations: what IQ tests really measure. *Psychological Bulletin*, **101**, 171–191.

Freedman D. (2002) From association to causation: some remarks on the history of statistics. Technical Report no. 521, Statistics Department, University of California, Berkeley.

Freeman P.R. (1976) A Bayesian analysis of the megalithic yard (with discussion). *Journal of the Royal Statistical Society, Series A*, **139**, 20–55.

Freyd M. (1925) The statistical viewpoint in vocational selection. *Journal of Applied Psychology*, **9**, 349–356.

Garfield E. (1979) *Citation Indexing: Its Theory and Application in Science, Technology, and Humanities*. New York: Wiley.

Garfield E. (1994) The impact factor. *Current Contents*, June 20.

Gaito J. (1960) Scale classification and statistics. *Psychological Review*, **67**, 277–278.

Gaito J. (1980) Measurement scales and statistics: resurgence of an old misconception. *Psychological Bulletin*, **87**, 564–567.

Gaito, J. (1986) Some issues in the measurement-statistics controversy. *Canadian Psychology*, **27**, 63–68.

Galton F. (1909) *Memories of My Life*. London: Methuen.

Gardner H. (1983) *Frames of Mind: The Theory of Multiple Intelligences*. New York: Basic Books.

Gardner H. (1993) *Multiple Intelligences: The Theory in Practice*. New York: Basic Books.

Gardner P.L. (1975) Scales and statistics. *Review of Educational Research*, **45**, 43–57.

Garner D.M. and Garfinkel P.E. (1979) The Eating Attitudes Test: an index of the symptoms of anorexia nervosa. *Psychological Medicine*, **9**, 273–279.

Garrison F.H. (1929) *History of Medicine*. Philadelphia.

Gifi A. (1990) *Nonlinear Multivariate Analysis*. Chichester: Wiley.

Goldberg D.P. (1972) *The Detection of Psychiatric Illness by Questionnaire*. London: Oxford University Press.

Goldstein H. (1995) *Multilevel Statistical Models*. London: Edward Arnold.

Goldstein H. and Spiegelhalter D.J. (1996) League tables and their limitations: statistical issues in comparisons of institutional performance. *Journal of the Royal Statistical Society, Series A*, **159**, 385–443.

Goleman D. (1995) *Emotional Intelligence: Why It Can Matter More Than IQ*. New York: Bantam.

Golinsky J. (1995) 'The nicety of experiment': precision of measurement and precision of reasoning in late eighteenth-century chemistry. In M.N. Wise (ed.), *The Values of Precision*, Princeton, NJ: Princeton University Press, 72–91.

Gooday G.J.N. (1995) The morals of energy metering: constructing and deconstructing the precision of the Victorian electrical engineer's ammeter and voltmeter. In M.N. Wise (ed.), *The Values of Precision*, Princeton, NJ: Princeton University Press, 239–282.

Gordon A.D. (1999) *Classification*, 2nd edition. Boca Raton, FL: Chapman and Hall/CRC.

Gould S.J. (1996) *The Mismeasure of Man*. London: Penguin.

Gower J.C. (1982) Proposal of vote of thanks to Ramsay 'Some statistical approaches to multi-dimensional scaling data'. *Journal of the Royal Statistical Society, Series A*, **145**, 303–304.

Gower J.C. (1966) Some distance properties of latent root and vector methods used in multivariate analysis. *Biometrika*, **53**, 325–338.

Gower J.C. and Hand D.J. (1996) *Biplots*. London: Chapman and Hall.

Gower J.C. and Legendre P. (1986) Metric and Euclidean properties of dissimilarity coefficients. *Journal of Classification*, **3**, 5–48.

Granger C.V., Albrecht G.L. and Hamilton B.B. (1979) Outcome of comprehensive medical rehabilitation: measurement by PULSES profile and the Barthel Index. *Archives of Physical Medicine and Rehabilitation*, **60**, 145–154.

Green J. and Wintfield N. (1995) Report cards on cardiac surgeons: assessing New York State's approach. *New England Journal of Medicine*, **332**, 1229–1232.

Greenacre M.J. (1984) *Theory and Applications of Correspondence Analysis*. London: Academic Press.

Grieve A.P. (2003) The number needed to treat: a useful clinical measure or a case of the Emperor's new clothes? *Pharmaceutical Statistics*, **2**, 87–102.

Groth-Marnat G. (ed.) (2000) *Neuropsychological Assessment in Clinical Practice: A Guide to Test Interpretation and Integration*. New York: Wiley.

Guilford J.P. (1967) *The Nature of Human Intelligence*. New York: McGraw-Hill.

Guilford J.P. and Comrey A.L. (1951) Measurement in psychology. In H. Helson (ed.), *Theoretical Foundations of Psychology*. New York: Van Nostrand, 505–556.

Gustafson A., Herrmann A. and Huber F. (eds) (2001) *Conjoint Measurement: Methods and Applications*. Berlin: Springer-Verlag.

Guttman L.A. (1941) The quantification of a class of attributes: a theory and method of scale construction. In P. Horst (ed.), *The Prediction of Personal Adjustment*. New York: Social Science Research Council, 319–348.

Guttman L.A. (1944) A basis for scaling qualitative data. *American Sociological Review*, **91**, 139–150.

Guttman L.A. (1968) A general nonparametric technique for finding the smallest coordinate space for a configuration of points. *Psychometrika*, **33**, 469–506.

Guyatt G.H., Nogradi S., Halcrow S. *et al.* (1989a) Development and testing of a new measure of health status for clinical trials in heart failure. *Journal of General Internal Medicine*, **4**, 101–107.

Guyatt G.H., Mitchell A., Irvine E.J. *et al.* (1989b) A new measure of health status for clinical trials in inflammatory bowel disease. *Gastroenterology*, **96**, 804–810.

Haken H. (1983) *Synergetics: An Introduction*. Berlin: Springer-Verlag.

Halemba, G., Hurst, H. and Gable, R. (1997) *Time and Resource Requirements for the 'Good Practice' Handling of Abuse and Neglect Cases by Juvenile and Family Courts*. Working paper. Pittsburgh, PA: National Center for Juvenile Justice.

Hamilton M. (1960) A rating scale for depression. *Journal of Neurology, Neurosurgery and Psychiatry*, **23**, 56–62.

Hand D.J. (1985) *Artificial Intelligence and Psychiatry*. Cambridge: Cambridge University Press.

Hand D.J. (1993a) Comment on 'Nominal, ordinal, interval, and ratio typologies are misleading'. *American Statistician*, **47**, 314–315.

Hand D.J. (1993b) What is synergy? revisited. *Journal of Pharmaceutical Medicine*, **3**, 97–100.

Hand D.J. (1994) Deconstructing statistical questions (with discussion). *Journal of the Royal Statistical Society, Series A*, **157**, 317–356.

Hand D.J. (1995) Discussion contribution to Chatfield (1995) *Journal of the Royal Statistical Society, Series A*, **158**, 448.

Hand D.J. (1996) Statistics and the theory of measurement (with discussion). *Journal of the Royal Statistical Society, Series A*, **159**, 445–492.

Hand D.J. (1997) *Construction and Assessment of Classification Rules*. Chichester: Wiley.

Hand D.J. (2000) Synergy in drug combinations. In W. Gaul, O. Opitz and M. Schader (eds), *Data Analysis*. Berlin: Springer-Verlag, 471–475.

Hand D.J. (2001) Modelling consumer credit risk. *IMA Journal of Management Mathematics*, **12**, 139–155.

Hand D.J. and Crowder M.J. (2003) Measuring customer quality in retail banking. Technical Report, Department of Mathematics, Imperial College, London.

Hand D.J. and Henley W.E. (1997) Statistical classification methods in consumer credit scoring: a review. *Journal of the Royal Statistical Society, Series A*, **160**, 523–541.

Hand D.J., Daly F., Lunn A.D., McConway K.J. and Ostrowski E. (eds) (1994) *A Handbook of Small Data Sets*. London: Chapman and Hall.

Hand D.J., Oliver J.J. and Lunn A.D. (1998) Discriminant analysis when the classes arise from a continuum. *Pattern Recognition*, **31**, 641–650.

Hand D.J., Blunt G., Kelly M.G. and Adams N.M. (2000) Data mining for fun and profit. *Statistical Science*, **15**, 111–131.

Hand D.J., Mannila H. and Smyth P. (2001) *Principles of Data Mining*. Cambridge, MA: MIT Press.

Handy C. (1995) *The Empty Raincoat: Making Sense of the Future*. London: Arrow.

Haney W.M., Madaus G.F. and Lyons R. (1993) *The Fractured Marketplace for Standardized Testing*. Boston: Kluwer.

Harding B.R., Thompson R. and Curtis A.R. (1973) A new mathematical model for fitting an HPL radioimmunoassay curve. *Journal of Clinical Pathology*, **26**, 273–276.

Harrington R. (1804) *The Death-Warrant of the French Theory of Chemistry*. London.

Hatcher R.A. and Brody J.G. (1910) The biological standardization of drugs. *American Journal of Pharmacy*, **82**, 360–373.

Healy P. (2001) Harvard's quiet secret: rampant grade inflation. *Boston Globe*, 10 July.

HEFCE (2001) *A Guide to the 2001 Research Assessment Exercise*. Bristol: HEFCE.

Heidbreder E. (1933) *Seven Psychologies*. New York: Appleton-Century-Crofts.

Helmholtz H. von (1887) Zählen und Messen erkenntnis-theoretisch betrachet. In *Philosophische Aufsätze Eduard Zeller gewidmet*. Leipzig. English translation by C.L. Bryan, *Counting and Measuring*, Princeton, NJ: Van Nostrand, 1930.

Hersh R. (1998) *What Is Mathematics, Really*. London: Vintage.

Hoffman P. (1998) *The Man Who Loved Only Numbers*. London: Fourth Estate.

Hogan R. and Nicholson R.A. (1988) The meaning of personality test scores. *American Psychologist*, **43**, 621–626.

Hölder O. (1901) Die Axiome der Quantität und die Lehre vom Mass, *Ver. Verh. Kgl. Sachisis. Ges. Wiss. Leipzip, Math-Phys. Classe* **53**, 1–64.

Holt R.R. (1986) Clinical and statistical prediction: a retrospective and would be integrative perspective. *Journal of Personality Assessment*, **50**, 376–386.

Hooke R. (1678) *Lectures De Potentia Restitutiva, or of Spring, Explaining the Power of Springing Bodies*. London: John Martyn Printer to the Royal Society.

Horst P. (1941) *The Prediction of Personal Adjustment* (Bull. No. 48). New York: Social Science Research Council.

Hough H.G. (1962) Clinical versus statistical prediction in psychology. In L. Postman (ed.), *Psychology in the Making*. New York: Knopf, 526–584.

Hsia D.C. (1978) Credit scoring and the Equal Credit Opportunity Act. *Hastings Law Journal*, **30**, 371–448.

Hubert L. and Arabie P. (1986) Unidimensional scaling and combinatorial optimization. In J. de Leeuw, W.J. Heiser, J. Meulman, and F. Critchley (eds), *Multidimensional Data Analysis*. Leiden: DSWO Press.

Hubert L. and Arabie P. (1988) Relying on necessary conditions for optimization: unidimensional scaling and some extensions. In H.H. Bock, (ed.) *Classification and Related Methods of Data Analysis*. Amsterdam: North-Holland.

Hubert L., Arabie P. and Meulman J. (1997) Linear and circular unidimensional scaling for symmetric proximity matrices. *British Journal of Mathematical and Statistical Psychology*, **50**, 253–284.

Humphreys L.G. (1989) Intelligence: three kinds of instability and their consequences for policy. In R. Linn (ed.), *Intelligence: Measurement, Theory, and Public Policy*. Urbana: University of Illinois Press.

Huntley H.E. (1952) *Dimensional Analysis*. London: Macdonald.

Hurst H. III (1999) Workload measurement for juvenile justice system personnel: practices and needs. *Bulletin of the Juvenile Accountability Incentive Block Grants Program*, November. US Office of Juvenile Justice and Delinquency Prevention.

Huskisson E.C. (1974) Measurement of pain. *Lancet*, **ii**, 1127–1129.

Hutton J.L. (2000) Number needed to treat: properties and problems (with discussion). *Journal of the Royal Statistical Society, Series A*, **163**, 403–419.

Hylles, Thomas (1600) *The Arte of Vulgar Arithmeticke*. G. Simpson.

Ifrah G. (1998) *The Universal History of Numbers*. London: Harvill Press.

Isaacson E. de St Q. and Isaacson M. de St Q. (1975) *Dimensional Methods in Engineering and Physics*. London: Edward Arnold.

Jardine L. (1999) *Ingenious Pursuits: Building the Scientific Revolution*. London: Little, Brown.

Jelinski Z. and Moranda P. (1972) Software reliability research. In P. Freiberger (ed.), *Statistical Computer Performance Evaluation*. New York: Academic Press, 465–484.

Johnstone W.D. (1998) *For Good Measure: The Most Complete Guide to Weights and Measures and Their Metric Equivalents*. Chicago: NTC Publishing Group.

Jones L.V. and Applebaum M.I. (1989) Psychometric methods. *Annual Review of Psychology*, **40**, 23–43.

Juola A.E. (1968) Illustrative problems in college-level grading. *Personnel and Guidance Journal*, September, 29–33.

Kaiser H. (1960) Review of Senders's book (1958). *Psychometrika*, **25**, 411–413.

Katz D. (1960) The functional approach to the study of attitudes. *Public Opinion Quarterly*, **24**, 163–204.

Katz S., Ford A.B., Moskowitz R.W. *et al.* (1963) Studies of illness in the aged. The Index of ADL: a standardized measure of biological and psychosocial function. *Journal of the American Medical Association*, **185**, 915–919.

Katz S., Downs T.D., Cash H.R. and Grotz R.C. (1970) Progress in the development of the Index of ADL. *Gerontologist*, **10**, 20–30.

Keene O.N. (1995) The log transformation is special. *Statistics in Medicine*, **14**, 811–819.

Keir J. (1789) *First Part of a Dictionary of Chemistry*. Birmingham.

Kempthorne O. (1955) The randomization theory of experimental inference. *Journal of the American Statistical Association*, **50**, 946–967.

Kendall D.G. (1971) Seriation from abundance matrices. In F.R. Hodson, D.G. Kendall and P. Tautu (eds), *Mathematics in the Archaeological and Historical Sciences*. Edinburgh: Edinburgh University Press, 215–252.

Kendall M.G. (1969) Studies in the history of probability and statistics, XXI. The early history of index numbers. *Review of the International Statistical Institute*, **37**, 1–12.

Kerrison S. and Macfarlane A. (eds) (1999) *Official Health Statistics: An Unofficial Guide*. Radical Statistics Health Group. London: Arnold.

Keyser D.J. and Sweetland R.C. (1984) *Test Critiques, Vol. 1*. Kansas City, MO: Test Corporation of America.

Kitchenham B. (1992) Never mind the metrics, what about the numbers. In T. Denvir, R. Herman and R. Whitty (eds) *Formal Aspects of Measurement*. Berlin: Springer-Verlag.

Klein H.A. (1974) *The World of Measurements: Masterpieces, Mysteries, and Muddles of Meteorology*. New York: Simon and Schuster. Reprinted (1988) as *The Science of Measurement: A Historical Survey*, 1988. New York: Dover.

Kolen M.J. and Brennan R.L. (1995) *Test Equating*. New York: Springer-Verlag.

Koopman B.O. (1940a) The axioms and algebra of intuitive probability. *Annals of Mathematics*, **41**, 269–292.

Koopman B.O. (1940b) The bases of probability. *Bulletin of the American Mathematical Society*, **46**, 763–774.

Koopman B.O. (1941) Intuitive probability and sequences. *Annals of Mathematics*, **42**, 169–187.

Krantz D.H., Luce R.D., Suppes P., and Tversky A. (1971) *Foundations of Measurement. Volume 1: Additive and Polynomial Representations*. New York: Academic Press.

Kruskal J.B. (1964a) Multidimensional scaling by optimising goodness-of-fit to a nonmetric hypothesis. *Psychometrika*, **29**, 1–27.

Kruskal J.B. (1964b) Nonmetric multidimensional scaling: a numerical method. *Psychometrika*, **29**, 115–129.

Krzanowski W.J. and Marriott F.H.C. (1994) *Multivariate Analysis, Part I: Distributions, Ordination, and Inference*. London: Edward Arnold.

Kuder G.F. and Richardson M.W. (1937) The theory of the estimation of test reliability. *Psychometrika*, **2**, 151–160.

Kyburg H. (1984) *Theory and Measurement*. Cambridge: Cambridge University Press.

Labovitz S. (1967) Some observations on measurement and statistics. *Social Forces*, **46**, 151–160.

Labovitz S. (1968) Reply to Champion and Morris. *Social Forces*, **46**, 543–544.

Labovitz S. (1970) The assignment of numbers to rank order categories. *American Sociological Review*, **35**, 515–524.

Labovitz S. (1971) In defence of assigning numbers to ranks. *American Sociological Review*, **36**, 521–522.

Labovitz S. (1972) Statistical usage in sociology: sacred cows and ritual. *Sociological Methods and Research*, **1**, 13–37.

Lader M.H. (1970) The unit of quantification of the G.S.R. *Journal of Psychosomatic Research*, **14**, 109–110.

Laming D. (2002) Review of 'Measurement in psychology: A critical history of a methodological concept'. *Quarterly Journal of Experimental Psychology A*, **55**, 689–692.

Landis J.R. and Koch G.G. (1977) The measurement of observer agreement for categorical data. *Biometrics*, **33**, 159–174.

Laspeyres E. (1871) Die Berechnung einer mittleren Warenpreissteigerung. *Jahr. Nationalökonomie und Statistik*, **16**, 296–314.

Laupacis A., Sackett D.L. and Roberts R.S. (1988) An assessment of clinically useful measures of the consequences of treatment. *New England Journal of Medicine*, **318**, 1728–1733.

Leake C.D. (1952) *The Old Egyptian Medical Papyri*. Lawrence: University of Kansas Press.

Leavitt F. (1983) Detecting psychological disturbance using verbal pain measurement: the Back Pain Classification Scale. In R. Melzack (ed.), *Pain Measurement and Assessment*. New York: Raven Press, 79–84.

Lehmann E.L. (1990) Model specification: the views of Fisher and Neyman, and later developments. *Statistical Science*, **5**, 1601–1668.

Leighton G. and McKinlay P.L. (1930) *Milk Consumption and the Growth of School Children*. London: HMSO.

Levine A. and Cureton J.S. (1998) *When Hope and Fear Collide: A Portrait of Today's College Student*. San Francisco: Jossey-Bass.

Lezak M.D. (1995) *Neuropsychological Assessment*, 3rd edition. New York: Oxford University Press.

Lindsey J.K. (1999) Some statistical heresies (with discussion). *The Statistician*, **48**, 1–40.

Little R.J.A. and Rubin D.B. (1987) *Statistical Analysis with Missing Data*. New York: Wiley.

Lloyd C.J. (1997) A survey of inference on ROC curves. Research Report No. 143, Department of Statistics and Actuarial Sciences, University of Hong Kong.

Long J.F. and Wilken P.H. (1974) A fully nonmetric unfolding technique: interval values from ordinal data. In H.M. Blalock (ed.), *Measurement in the Social Sciences*. Chicago: Aldine, 11–60.

Lord F.M. (1953) On the statistical treatment of football numbers. *American Psychologist*, **8**, 750–751.

Lord F.M. (1954) Further comment on 'Football numbers'. *American Psychologist*, **9**, 264–265.

Lord F.M. (1980) *Applications of Item Response Theory to Practical Testing Problems*. Hillsdale, NJ: Lawrence Erlbaum.

Louis P. (1835) *Recherches sur les effets de la saignée dans quelques maladies inflammatoires: et sur l'action de l'émétique et des véscatoires dans la pneumonie*. Paris: J.B. Baillière. Reprinted Birmingham, AL: Classics of Medicine Library, 1986.

Lowe J. (1922) *The Present State of England in Regard to Agriculture, Trade and Finance*. London.

Luce R.D. (1959) On the possible psychophysical laws. *Psychological Review*, **66**, 81–95.

Luce R.D. (1962) Comments on Rozeboom's criticisms of 'On the possible psychophysical laws'. *Psychological Review*, **69**, 548–551.

Luce R.D. (1978) Dimensionally invariant numerical laws correspond to meaningful qualitative relations. *Philosophy of Science*, **45**, 1–16.

Luce R.D. (1990) 'On the possible psychophysical laws' revisited: remarks on cross-modal matching. *Psychological Review*, **97**, 66–77.

Luce R.D. (1993) Let's not promulgate either Fechner's erroneous algorithm or his unidimensional approach. *Behavioural and Brain Sciences*, **16**, 155–156.

Luce R.D. and Edwards W. (1958) The derivation of subjective scales from just noticeable differences. *Psychological Review*, **65**, 222–237.

Luce R.D. and Marley A.A.J. (1969) Extensive measurement when concatenation is restricted and maximal elements may exist. In S. Morgenbesser, P. Suppes and M.G. White (eds), *Philosophy, Science, and Method: Essays in Honor of Ernest Nagel*. New York: St Martin's Press, 235–249.

Luce R.D. and Narens L. (1983) Symmetry, scale types, and generalizations of classical physical measurement. *Journal of Mathematical Psychology*, **27**, 44–85.

Luce R.D. and Narens L. (1985) Classification of concatenation measurement structures according to scale type. *Journal of Mathematical Psychology*, **29**, 1–72.

Luce R.D. and Narens L. (1986) Measurement: the theory of numerical assignments. *Psychological Bulletin*, **99**, 166–180.

Luce R.D. and Tukey J.W. (1964) Simultaneous conjoint measurement: a new type of fundamental measurement. *Journal of Mathematical Psychology*, **1**, 1–27.

Luce R.D., Krantz D.H., Suppes P. and Tversky A. (1990) *Foundations of Measurement. Volume 3: Representation, Axiomatization, and Invariance*. New York: Academic Press.

Lundberg G.A. (1926) Case work and the statistical method. *Social Forces*, **5**, 61–65.

Lundberg G.A. (1929) *Social Research*. New York: Longmans, Green.

Lynn R. and Hampson S. (1989) Secular increases in reasoning and mathematical abilities in Britain, 1972–1984. *School Psychology International*, **10**, 301–304.

MacRae A.W. (1988) Measurement scales and statistics: what can significance tests tell us about the world? *British Journal of Psychology*, **79**, 161–171.

Madaus G.F. and Raczek A.E. (1996) The extent and growth of educational testing in the United States: 1956–1994. In H. Goldstein and T. Lewis (eds), *Assessment: Problems, Developments and Statistical Issues*. Chichester: Wiley, 145–165.

Mahoney F.I., Wood O.H. and Barthel D.W. (1958) Rehabilitation of chronically ill patients: the influence of complications on the final goal. *South Medical Journal*, **51**, 605–609.

Margenau H. (1959) *Philosophical problems concerning the meaning of measurement in physics*. In C.W. Churchman and P. Ratoosh (eds), *Measurement: Definitions and Theories*. New York: Wiley, 163–176.

Mayer J.M. (1978) Assessment of depression. In P. McReynolds (ed.), *Advances in Psychological Assessment*, Vol. 4. Washington DC: Jossey-Bass, 358–425.

Mayer L.S. (1970) Comment on 'The assignment of numbers to rank order categories'. *American Sociological Review*, **35**, 916–917.

McCabe T.J. (1976) A complexity measure. *IEEE Transactions on Software Engineering*, **2**, 308–320.

McClish D.K. (1989) Analyzing a portion of the ROC curve. *Medical Decision Making*, **9**, 190–195.

McCormack H.M., Horne D.J.deL. and Sheather S. (1988) Clinical applications of visual analogue scales: a critical review. *Psychological Medicine*, **18**, 1007–1019.

McCullagh P. and Nelder J.A. (1989) *Generalized Linear Models*. London: Chapman and Hall.

McDowell I. and Newell C. (1996) *Measuring Health: A Guide to Rating Scales and Questionnaires*. New York: Oxford University Press.

McGreevy T. (1995) *The Basis of Measurement: 1. Historical Aspects*, ed. P. Cunningham. Chippenham: Picton.

McGrew K.S. and Flanaghan D.P. (1988) *The Intelligence Test Desk Reference (ITDR): Gf-Gc Cross-Battery Assessment*. Boston: Allyn and Bacon.

McGuire C.H. (1985) Medical problem solving: a critique of the literature. *Journal of Medical Education*, **60**, 587–595.

McIver J.P. and Carmines E.G. (1981) *Unidimensional Scaling*. London: Sage.

McKee M. and Hunter D. (1995) Mortality league tables: do they inform or mislead? *Quality of Health Care*, **4**, 5–12.

McKeon J.J. (1966) Canonical analysis: some relations between canonical correlation, factor analysis, discriminant function analysis and scaling theory. Monograph number 13. *Psychometrika*.

McWinnie J.R. (1981) Disability assessment in populations surveys: results of the OECD common development effort. *Revue d'Epidémiologie et de Santé Publique*, **29**, 413–419.

Meehl P.E. (1954) *Clinical versus Statistical Prediction: A Theoretical Analysis and a Review of the Evidence*. Minneapolis: University of Minnesota Press.

Meehl P.E. (1957) When shall we use our heads instead of formula? *Journal of Counseling Psychology*, **4**, 268–273.

Meehl P.E. (1965) Seer over sign: the first good example. *Journal of Experimental Research in Personality*, **1**, 27–32.

Meehl P.E. (1986) Causes and effects of my disturbing little book. *Journal of Personality Assessment*, **50**, 370–375.

Meier S.T. (1994) *The Chronic Crisis in Psychological Measurement: A Historical Survey*. San Diego, CA: Academic Press.

Melzack R. (1975) The McGill Pain Questionnaire: major properties and scoring methods. *Pain*, **1**, 277–299.

Melzack R. and Wall P. (1988) *The Challenge of Pain*, 2nd edition. New York: Penguin.

Merton R.K., Sills D.L. and Stigler S.M. (1984) The Kelvin dictum and social science: an excursion into the history of an idea. *Journal of the History of the Behavioural Sciences*, **20**, 319–331.

Michell J. (1986) Measurement scales and statistics: a clash of paradigms. *Psychological Bulletin*, **100**, 398–407.

Michell J. (1990) *An Introduction to the Logic of Psychological Measurement*. Hillsdale, NJ: Lawrence Erlbaum.

Michell J. (1999) *Measurement in Psychology: A Critical History of a Methodological Concept*. Cambridge: Cambridge University Press.

Minne H.W., Leidig G.,Wüster C., Siromachkostov L., Baldauf G., Bickel R., Sauer P., Lojen M. and Ziegler R. (1988) A newly developed spine deformity index (SDI) to quantitate vertebral crush fractures in patients with osteoporosis. *Bone and Mineral*, **3**, 335–349.

Mischel W. (1968) *Personality and Assessment*. New York: Wiley.

Mitchell W.C. (1915) *The Making and Use of Index Numbers*. In *Bulletin 173, Bureau of Labor Statistics*. Reprinted, 1938, by Augustus M. Kelley, New York.

Montesquieu, Charles de Secondat de (1875) *Esprit des Lois*. In *Oeuvres Complètes*. Paris: Garnier, **5**, 412–413.

Morris J.N. (1975) *Uses of Epidemiology*. 3rd edition. London: Churchill-Livingstone.

Morris R.N. (1968) Some observations on measurement and statistics: comment. *Social Forces*, **46**, 541–542.

Mosteller F. and Tukey J.W. (1977) *Data Analysis and Regression*. Boston: Addison-Wesley.

Mullen P.M. (1985) Performance indicators – is anything new? *Hospital and Health Service Review*, July, 165–167.

Murphy A.H. and Epstein E.S. (1967) Verification of probabilistic predictions: a brief review. *Journal of Applied Meteorology*, **6**, 748–755.

Murphy K.R. and Davidshofer C.O. (2001) *Psychological Testing*, 5th edition. Englewood Cliffs, NJ: Prentice Hall.

Murphy T. (1981) Medical knowledge and statistical methods in early nineteenth century France. *Medical History*, **25**, 301–319.

Myers J.H. and Forgy E.W. (1963) The development of numerical credit evaluation systems. *Journal of the American Statistical Association*, **58**, 799–806.

Narens L. (1974) Measurement without Archimedean axioms. *Philosophy of Science*, **41**, 374–393.

Narens L. (1976) Utility–uncertainty trade-off structures. *Journal of Mathematical Psychology*, **13**, 296–322.

Narens L. (1981a) A general theory of ratio scalability with remarks about the measurement-theoretic concept of meaningfulness. *Theory and Decision*, **13**, 296–322.

Narens L. (1981b) On the scales of measurement. *Journal of Mathematical Psychology*, **24**, 249–275.

Narens L. (2002) *Theories of Meaningfulness*. Mahwah, NJ: Lawrence Erlbaum.

Narens L. and Luce R.D. (1976) The algebra of measurement. *Journal of Pure and Applied Algebra*, **8**, 197–233.

Narens L. and Luce R.D. (1986) Measurement: the theory of numerical assignments. *Psychological Bulletin*, **99**, 166–180.

Nelder J.A. (1990) The knowledge needed to computerise the analysis and interpretation of statistical information. In *Expert systems and artificial intelligence: the need for information about data*. Library Association Report, London, March, 23–27.

Nelder J.A. (2000) Functional marginality and response surface fitting. *Journal of Applied Statistics*, **27**, 109–112.

Neligan G. (1965) A community study of the relationship between birth weight and gestational age. In M. Dawkins and W.G. MacGregor (eds), *Gestational Age, Size and Maturity*. Clinics in Developmental Medicine 19. London Spastics Society Medical Education and Information Unit and William Heinemann Medical, 28–32.

Never P.R. and Kojahn L.K. (1973) Access, attrition, test scores and grades of college entrants and persisters: 1965–1973. The American College Testing Program, April.

Neyman J. (1939) On a new class of 'contagious' distributions, applicable in entomology and bacteriology. *Annals of Mathematical Statistics*, **10**, 35–57.

Niederée R. (1994) There is more to measurement than just measurement: measurement theory, symmetry, and substantive theorizing. *Journal of Mathematical Psychology*, **38**, 527–594.

Nishisato S. (1980) *Analysis of Categorical Data: Dual Scaling and Its Applications*. Toronto: University of Toronto Press.

North J.D. (1965) *The Measure of the Universe: A History of Modern Cosmology*. Oxford: Clarendon Press.

Novick M. and Lewis G. (1967) Coefficient alpha and the reliability of composite measurements. *Psychometrika*, **32**, 1–13.

Novick S. (1970) Editorial. *Environment*, June.

OECD (1993) *The Proposed Standard Practice for Surveys of Research and Experimental Development, Frascati Manual 1993*. Paris: OECD.

Paasche A. (1874) Über die Preisentwicklung der letzten Jahre, nach den Hamburger Börsennotierungen. *Jahrbücher für Nationalökonomie und Statistik*, **23**, 168–178.

Palacios, J. (1964) *Dimensional Analysis*, trans P. Lee and D. Roth. London: Macmillan.

Paul G.L., Mariotto M.J. and Redfield J.P. (1986) Assessment purposes, domains, and utility for decision making. In G.L. Paul (ed.), *Assessment in Residential Treatment Settings*. Champaign, IL: Research Press, 1–26.

Pepe M.S. (2003) *The Statistical Evaluation of Medical Tests for Classification and Prediction*. Oxford: Oxford University Press.

Pfanzagl J. (1959) *Die axiomatischen Grundlagen einer allgemeinen Theorie des Messens*. Wurzburg: Schriftenreihe des Statistischen Instituts der Universität Wien, Neue Folge.

Pfeffer R.J., Kurosaki T.T., Harrah C.H. *et al.* (1982) Measurement of functional activities in older adults in the community. *Journal of Gerontology*, **37**, 323–329.

Porter A.W. (1958) *The Method of Dimensions*. London: Methuen.

Porter T.M. (1995) *Trust in Numbers: The Pursuit of Objectivity in Science and Public Life*. Princeton, NJ: Princeton University Press.

Priestley J. (1796) *Considerations on the Doctrine of Phlogiston and the Decomposition of Water*. Reprinted, 1929, ed. William Foster, Princeton.

Prigogine I. and Stengers I. (1985) *Order out of Chaos: Man's New Dialogue with Nature*. London: Fontana.

Ramsay J.O. (1982) Some statistical approaches to multidimensional scaling data. *Journal of the Royal Statistical Society, Series A*, **145**, 285–312.

Rasch G. (1977) On specific objectivity: an attempt at formalizing the request for generality and validity of scientific statements. *Danish Yearbook of Philosophy*, **14**, 58–94.

Rasch G. (1980) *Probabilistic Models for Some Intelligence and Attainment Tests*. Chicago: University of Chicago Press.

Reichert A.K., Cho C-C. and Wagner G.M. (1983) An examination of the conceptual issues involved in developing credit-scoring models. *Journal of Business and Economic Statistics*, **1**, 101–114.

Richardson M. and Kuder G.F. (1933) Making a rating scale that measures. *Personnel Journal*, **12**, 36–40.

Roberts F.S. (1979) *Measurement Theory*. Reading, MA: Addison-Wesley.

Roberts F.S. and Luce R.D. (1968) Axiomatic thermodynamics and extensive measurement. *Synthèse*, **18**, 311–326.

Roberts J.M. Jr and Brewer D.D. (2001) Measures and tests of heaping in discrete quantitative distributions. *Journal of Applied Statistics*, **28**, 887–896.

Roche J.J. (1998) *The Mathematics of Measurement: A Critical History*. London: Springer-Verlag.

Roethlisberger F.J., Dickson W.J. and Wright H.A. (1939) *Management and the Worker: An Account of a Research Program Conducted by the Western Electric Company, Hawthorne Works, Chicago*. Boston: Harvard University Press.

Rogerson R.J., Findlay A.M. and Morris A.S. (1989) Indicators of quality of life: some methodological issues. *Environment and Planning*, **21**, 1655–1666.

Rosenberg E. and Gleit A. (1994) Quantitative methods in credit management: a survey. *Operations Research*, **42**: 589–613.

Rosenberg R. (1992) Quality of life, ethics, and philosophy of science. *Nordic Journal of Psychiatry*, **46**, 75–77.

Rosovsky H. and Hartley M. (2002) *Evaluation and the Academy: Are We Doing the Right Thing: Grade Inflation and Letters of Recommendation*. Cambridge, MA: American Academy of Arts and Sciences.

Rozeboom W.W. (1962) The untenability of Luce's principle. *Psychological Bulletin*, **69**, 542–547.

Rozeboom W.W. (1966) Scaling and the nature of measurement. *Synthèse*, **16**, 170–233.

Rubin D.B. (1987) *Multiple Imputation for Nonresponse in Surveys*. New York: Wiley.

Rusnock A. (1995) Quantification, precision, and accuracy: determinations of population in the ancien régime. In M.N. Wise (ed.), *The Values of Precision*, Princeton, NJ: Princeton University Press, 17–38.

Russell B. (1903) *Principles of Mathematics*. Cambridge: Cambridge University Press.

Ruta D.A., Garratt A.M., Wardlaw D. and Russell I.T. (1994) A new approach to the measurement of quality of life. The patient generated index (PGI). *Medical Care*, **32**, 1109–1126.

Sarbin T.R. (1986) Prediction and clinical inference: forty years later. *Journal of Personality Assessment*, **50**, 362–369.

Savage L.J. (1954) *The Foundations of Statistics*. New York: Wiley.

Savage L.J. (1971) Elicitation of personal probabilities and expectations. *Journal of the American Statistical Association*, **66**, 781–801.

Schweitzer S. and Schweitzer D.G. (1971) Comment on the Pearson r in random number and precise functional scale transformations. *American Sociological Review*, **36**, 518–519.

Scott J. and Huskisson E.C. (1976) Graphic representation of pain. *Pain*, **2**, 175–184.

Segall D.O. and Moreno K.E. (1999) Development of the Computerized Adaptive Testing Version of the Armed Services Vocational Aptitude Battery. In F. Drasgow and J. Olson-Buchanan (eds), *Innovations in Computerized Assessment*. Mahwah, NJ: Erlbaum, 35–65.

Selvanathan E.A. and Prasada Rao D.S. (1994) *Index Numbers: A Stochastic Approach*. London: Macmillan.

Senders V.L. (1953) A comment on Burke's additive scales and statistics. *Psychological Review*, **60**, 423–424.

Senders V.L. (1958) *Measurement and Statistics*. New York: Oxford.

Shepard R.N. (1962a) The analysis of proximities: multidimensional scaling with an unknown distance function I. *Pscyhometrika*, **27**, 125–140.

Shepard R.N. (1962b) The analysis of proximities: multidimensional scaling with an unknown distance function II. *Pscyhometrika*, **27**, 219–246.

Shryock R.H. (1961) The history of quantification in medical science. In H. Woolf (ed.), *Quantification: a History of the Meaning of Measurement in the Natural and Social Sciences*. Indianapolis: Bobbs Merrill, 85–107.

Siegel S. (1956) *Nonparametric Statistics*. New York: McGraw-Hill.

Simantiraki E. (1996) Unidimensional scaling: a linear programming approach minimising absolute deviations. *Journal of Classification*, **13**, 19–25.

Simmons J. *et al.* (2002) *Crime in England and Wales 2001/2*. London: Home Office.

Skinner J.B. (1991) On combining studies. *Drug Information Journal*, **25**, 395–403.

Smith P. (1990) The use of performance indicators in the public sector. *Journal of the Royal Statistical Society, Series A*, **153**, 53–72.

Sobel D. (1995) *Longitude*. London: Fourth Estate.

Spearman C. (1904) General intelligence, objectively determined and measured. *American Journal of Psychology*, **15**, 201–293.

Spearman C. (1927) *The Abilities of Man*. New York: Macmillan.

Spielberger C.D., Gorsuch R.L. and Lushene R. (1970) *Manual for the State-Trait Anxiety Inventory: STAI*. Palo Alto, CA: Consulting Psychologists Press.

Spreen O. and Strauss E. (1991) *A Compendium of Neuropsychological Tests: Administration, Norms, and Commentary*. New York: Oxford University Press.

Spitzer W.O., Dobson A.J., Hall J. *et al.* (1981) Measuring the quality of life of cancer patients: a concise QL-index for use by physicians. *Journal of Chronic Diseases*, **34**, 585–597.

Steelman, D., Rubin, T.H. and Arnold, J.M. (1993) *Circuit Court of Cook County, Illinois: Juvenile Division Judge Workloads and Judgeship Needs: From the Cook County Circuit Court Improvement Project*. Cook County, IL: National Council for State Courts.

Sternberg R.J. (1985) *Beyond IQ: A Triarchic Theory of Human Intelligence*. New York: Cambridge University Press.

Stevens S.S. (1946) On the theory of scales of measurement. *Science*, **103**, 677–680.

Stevens S.S. (1951) Mathematics, measurement, and psychophysics. In S.S. Stevens (ed.), *Handbook of experimental psychology*. New York: Wiley.

Stevens S.S. (1957) On the psychophysical law. *Psychological Review*, **64**, 153–181.

Stevens S.S. (1959) Measurement, psychophysics, and utility. In C.W. Churchman and P. Ratoosh (eds), *Measurement: Definitions and Theories*. New York: Wiley, 18–63.

Stevens S.S. (1960) The psychophysics of sensory function. *American Scientist*, **48**, 226–253.

Stevens S.S. (1961) To honor Fechner and repeal his law. *Science*, **133**, 80–86.

Stevens S.S. (1962) The surprising simplicity of sensory metrics. *American Psychologist*, **17**, 29–39.

Stevens S.S. (1968) Measurement, statistics, and the schemapiric view. *Science*, **161**, 849–856.

Stigler S.M. (2000) *Statistics on the Table: The History of Statistical Concepts and Methods*. Cambridge, MA: Harvard University Press.

Stine W.W. (1989) Meaningful inference: the role of measurement in statistics. *Psychological Bulletin*, **105**, 147–155.

Stuvel G. (1989) *The Index-Number Problem and Its Solution*. London: Macmillan.

Suppes P. (1951) A set of independent axioms for extensive quantities. *Portugaliae Mathematica*, **10**, 163–172.

Suppes P., Krantz D.H., Luce R.D. and Tversky A. (1989) *Foundations of measurement, Volume 2: Geometrical, Threshold, and Probabilistic Representations*. San Diego, CA: Academic Press.

Suppes P. and Zinnes J.L. (1963) Basic measurement theory. In R.D. Luce, R.R. Bush and E. Galanter (eds), *Handbook of Mathematical Psychology*, Vol. 1. New York: Wiley, 1–76.

Suslow S. (1976) A report on an interinstitutional survey of undergraduate scholastic grading 1960s to 1970s. Office of Institutional Research, University of California, Berkeley, February.

Swetz F.J. (1987) *Capitalism and Arithmetic: the New Math of the 15th Century.* La Salle, IL: Open Court.

Szirtes T. and Pal R. (1997) *Applied Dimensional Analysis and Modeling.* New York: McGraw-Hill.

Takane Y., Young F.W. and De Leeuw J. (1977) Nonmetric individual differences multidimensional scaling: an alternating least squares method with optimal scaling features. *Psychometrika,* **42**, 7–67.

Taylor D. (1984) Psychoanalytic contributions to the understanding of psychiatric illness. In P. McGuffin, M.F. Shanks and R.J. Hodgson (eds), *The Scientific Principles of Psychopathology.* London: Grune and Stratton, 624–676.

Taylor E.S. (1988) *Dimensional Analysis for Engineers.* Oxford: Oxford University Press.

Temple P. (2003) *First Steps in Economic Indicators.* Boston: Prentice Hall.

Tenenhaus M. and Young F.W. (1985) An analysis and synthesis of multiple correspondence analysis, optimal scaling, dual scaling, homogeneity analysis, and other methods for quantifying categorical multivariate data. *Psychometrika,* **50**, 91–119.

Tennant C. and Bebbington P. (1978) The social causation of depression: a critique of the work of Brown and his colleagues. *Psychological Medicine,* **8**, 565–575.

Theil H. (1960) Best linear index numbers of prices and quantities. *Econometrica,* **28**, 464–480.

Thom A. (1955) A statistical examination of the megalithic sites in Britain. *Journal of the Royal Statistical Society, Series A,* **118**, 275–291.

Thom A. (1962) The megalithic units of length. *Journal of the Royal Statistical Society, Series A,* **125**, 243–251.

Thom A. (1964) The larger units of length of megalithic man. *Journal of the Royal Statistical Society, Series A,* **127**, 527–533.

Thomas L.C. (2000) A survey of credit and behavioural scoring: forecasting financial risk of lending to consumers. *International Journal of Forecasting,* **16**(2)**,** 149–172.

Thomas L.C., Edelman D.B. and Crook J.N. (2002) *Credit Scoring and Its Applications.* Philadelphia: SIAM.

Thompson M.J., Hand D.J. and Everitt B.S. (1991) Contradictory correlations between derived scales. *Statistics in Medicine,* **10**, 1315–1319.

Thomson W. (1891) *Popular Lectures and Addresses,* Vol. 1. London: Macmillan.

Thurstone L.L. (1928) Attitudes can be measured. *American Journal of Sociology,* **33**, 529–554.

Thurstone L.L. (1935) *Vectors of the Mind: Multiple-Factor Analysis for the Isolation of Primary Traits.* Chicago: University of Chicago Press.

Thurstone L.L. (1938) *Primary Mental Abilities.* Psychometric Monographs, No.1. Chicago: University of Chicago Press.

Thurstone, L.L. (1959) Attitudes can be measured. In L.L. Thurstone (ed.), *The Measurement of Values.* Chicago: University of Chicago Press, 215–233.

Torgerson W.S. (1952) Multidimensional scaling: I. Theory and method. *Psychometrika,* **17**, 401–419.

Torgerson W.S. (1958) *Theory and Methods of Scaling.* New York: Wiley.

Törnqvist L., Vartia P. and Vartia Y.O. (1985) How should relative change be measured? *American Statistician,* **39**, 43–46.

Townsend J.T. and Ashby F.G. (1984) Measurement scales and statistics: the misconception misconceived. *Psychological Bulletin,* **96**, 394–401.

Tukey J.W. (1986a) Data analysis and behavioural science or learning to bear the quantitative man's burden by shunning badmandments. In *The Collected Works of John W. Tukey,* ed. L.V. Jones. Belmont, CA: Wadsworth, 187–389. Original unpublished manuscript, 1961.

Tryon W.W. (ed.) (1991) *Activity Measurement in Psychology and Medicine.* New York: Plenum.

van den Berg G. (1991) *Choosing an Analysis Method.* Leiden: DSWO Press.

van der Ven A.H.G.S. (1980) *Introduction to Scaling.* Chichester: Wiley.

van Rijsbergen C.J. (1979) *Information Retrieval.* London: Butterworths.

Vargo L.G. (1971) Comment on 'The assignment of numbers to rank order categories'. *American Sociological Review*, **36**, 517–518.

Velleman P.F. and Wilkinson L. (1993a) Nominal, ordinal, interval, and ratio scales typologies are misleading. *American Statistician*, **47**, 65–72.

Velleman P.F. and Wilkinson L. (1993b) Reply to comments on Velleman and Wilkinson (1993a). *American Statistician*, **47**, 315–316.

Vernon P.E. (1960) *The Structure of Human Abilities*. London: Methuen.

Viteles M.S. (1925) The clinical viewpoint in vocational psychology. *Journal of Applied Psychology*, **9**, 131–138.

von Winterfeldt D. and Edwards W. (1982) Costs and payoffs in perceptual research. *Psychological Bulletin*, **91**, 609–622.

Wainer H. (2000) *Computerized Adaptive Testing: A Primer*. Mahwah, NJ: Lawrence Erlbaum.

Waters B.K. (1997) Army Alpha to CAT-ASVAB: four-score years of military personnel selection and classification testing. In R.F. Dillon (ed.), *Handbook on Testing*. Westport, CT: Greenwood Press, 187–203.

Weber E.H. (1846) *Der Tastsinn und das Gemeingefühl*, In H.E. Ross and D.J. Murray (eds) (1978) *E.H. Weber: The Sense of Touch*, trans. D.J. Murray, New York: Academic Press.

Wedding D. (1983) Clinical and statistical prediction in neuropsychology. *Clinical Neuropsychology*, **5**, 49–55.

Wedding D. and Faust D. (1989) Clinical judgment and decision making in neuropsychology. *Archives of General Neuropsychology*, **4**, 233–265.

Weitzenhoffer A.M. (1951) Mathematical structures and psychological measurements. *Psychometrika*, **16**, 387–406.

West S.G. and Graziano W.G. (1989) Long-term stability and change in personality: an introduction. *Journal of Personality*, **57**, 175–194.

WHOQOL Group (1993) *Measuring Quality of Life: The Development of the World Health Organization Quality of Life Instrument*. Geneva: WHO.

Wiesner W.H. and Cronshaw S.F. (1988) A meta-analytic investigation of the impact of interview format and degree of strucutre on the validity of the employment interview. *Journal of Occupational Psychology*, **61**, 275–290.

Wigner E.P. (1960) The unreasonable effectiveness of mathematics in the natural sciences. *Communications on Pure and Applied Mathematics*, **13**, 1–14.

Wiley D.E. and Wiley J.A. (1970) The estimation of measurement error in panel data. *American Sociological Review*, **35**, 112–117.

Wilhelm A. (2000) IASC Member Survey 1999. *Computational Statistics and Data Analysis*, **34**, 261–268.

Williams P., Hand D.J. and Tarnopolsky A. (1982) The problem of screening for uncommon disorders – a comment on the Eating Attitudes Test. *Psychological Medicine*, **12**, 431–434.

Wilson E.O. (1998) *Consilience*. London: Little, Brown.

Wilson R. (1998) New research casts doubt on value of student evaluations of professors. *Chronicle of Higher Education*, 16 January.

Wise M.N. (1995a) *The Values of Precision*. Princeton, NJ: Princeton University Press.

Wise M.N. (1995b) Precision: agent of unity and product of agreement, Part III. In M.N. Wise (ed.), *The Values of Precision*. Princeton, NJ: Princeton University Press, 352–361.

Wittenborn J.R. (1972) Reliability, validity, and objectivity of symptom-rating scales. *Journal of Nervous and Mental Diseases*, **154**, 79–87.

Wormith J.S. and Goldstone C.S. (1984) The clinical and statistical prediction of recidivism. *Criminal Justice and Behavior*, **11**, 3–34.

Wright D.E. and Bray I. (2003) A mixture model for rounded data. *The Statistician*, **52**, 3–13.

Wright G. and Ayton P. (1994) *Subjective Probability*. Chichester: Wiley.

Wright J.G. and Feinstein A.R. (1992) A comparative contrast of clinimetric and psychometric methods for constructing indexes and rating-scales. *Journal of Clinical Epidemiology*, **45**, 1201–1218.

Zhou X.H., Obuchowksi N.A. and McClish D.K. (2002) *Statistical Methods in Diagnostic Medicine*. New York: Wiley.

Ziller R.C. (1974) Self-other orientation and quality of life. *Social Indicators Research*, **1**, 301–327.

Zipf G.K. (1949) *Human Behaviour and the Principle of Least Effort*. Cambridge: Addison-Wesley.

Zuse H. (1990) *Software Complexity: Measures and Methods*. Amsterdam: de Gruyter.

Author index

Subject index

Printed and bound in the UK by
CPI Antony Rowe, Eastbourne

Printed and bound by CPI Group (UK) Ltd, Croydon, CR0 4YY

18/06/2024

14517185-0001